Proceedings of the Lebedev Physics Institute
Academy of Sciences of the USSR
Series Editor: N.G. Basov

The Proceedings of the Lebedev Physics Institute of the Academy of Sciences of the USSR

Series Editor: Academician N.G. Basov

Volume 165 **Solitons and Instantons, Operator Quantization,** Edited by V.L. Ginzburg.

Volume 166 **The Nonlinear Optics of Semiconductor Lasers,** Edited by N.G. Basov.

Volume 167 **Group Theory and the Gravitation and Physics of Elementary Particles,** Edited by A.A. Komar.

Volume 168 **Issues in Intense-Field Quantum Electrodynamics,** Edited by V.L. Ginzburg.

Volume 169 **The Physical Effects in the Gravitational Field of Black Holes,** Edited by M.A. Markov.

Volume 170 **The Theory of Target Compression by Longwave Laser Emission,** Edited by G.V. Sklizkov.

Volume 171 **Research on Laser Theory,** Edited by A.N. Orayevskiy

Volume 172 **Phase Conjugation of Laser Emission,** Edited by N.G. Basov.

Volume 173 **Quantum Mechanics and Statistical Methods,** Edited by M.M. Sushchinskiy.

Volume 174 **Nonequilibrium Superconductivity,** Edited by V.L. Ginzburg.

Volume 175 **Luminescence Centers of Rare Earth Ions in Crystal Phosphors,** Edited by M.D. Galanin.

Volume 176 **Classical and Quantum Effects in Electrodynamics,** Edited by A.A. Komar.

Proceedings of the Lebedev Physics Institute
Academy of Sciences of the USSR
Series Editor: N.G. Basov
Volume 174

NONEQUILIBRIUM SUPERCONDUCTIVITY

Edited by V.L. Ginzburg

Translated by Kevin S. Hendzel

NOVA SCIENCE PUBLISHERS
COMMACK

NOVA SCIENCE PUBLISHERS
283 Commack Road
Suite 300
Commack, New York 11725

Library of Congress Cataloging-in-Publication Data

Neravnovesnaiâ sverkhprovodimost . English.
Nonequilibrium superconductivity.

(Proceedings of the Lebedev Physics Institute of the Academy of Sciences of the USSR ; v. 174)
Translation of: Neravnovesnaiâ sverkhprovodimost .
Includes bibliographies and index.
1. Superconductivity--Congresses. 2. Superconductors--Effect of radiation on--Congresses.
I. Ginzburg, V. L. (Vitaliĭ Lazarevich), 1916- .
II. Title. II. Series: Trudy Fizicheskogo instituta. English ; v. 174.
QC1.A4114 vol. 174 530 s 87-34022
[QC612.S8] [537.6'23]
ISBN 0-941743-09-8

The original Russian-language version of this book was published by Nauka Publishing House in 1986.

Printed in the United States of America

CONTENTS

QUANTUM TUNNELING WITH DISSIPATION
A.D. Zaikin, S.V. Panyukov

THE NONEQUILIBRIUM PROPERTIES OF SUPERCONDUCTORS UNDER OPTICAL EXCITATION AND CURRENT TUNNEL INJECTION

The Kinetics of Nonequilibrium Electrons and Phonons in Superconducting Tunnel Junctions

A.M. Gulyan, G.F. Zharkov

Abstract: The behavior of nonequilibrium excitations in a tunnel junction is investigated on the basis of a derived kinetic equation for electrons and phonons. It is demonstrated that the existence of macroscopic phase coherence in the tunnel junction of superconductors produces quantum oscillations in the electron excitation density with low applied voltages. The observed effects may include the shift in the nonequilibrium chemical potential as well as the generation of satellites in radiation scattered by the junction. With high applied voltages, excess quasi particles appear in the electron system. A phonon deficit appears in the relaxational frequency range in these conditions in the phonon system. The possibility for a phonon instability causing acoustical quantum generation is investigated in detail and its achievability is demonstrated.

This chapter presents research on the kinetics of the electron-phonon system in the nonequilibrium tunnel junctions of superconductors.

The first part of this chapter provides a general examination of electron kinetics in a nonequilibrium Josephson junction. Section 1 gives the initial relations. (Here the Eliashberg kinetic equations are written in the general case, a nondiagonal inelastic scattering channel is considered, the normalization condition is analyzed and the effective collision integral is given.) Section 2 considers the tunnel nonequilibrium source. (The Hamiltonian of the problem is given here together with the intrinsic-energy parts; a quasiclassical approximation is formulated and expressions are derived for the tunnel source and the tunnel current.) Section 3 is devoted to the electron-electron collision integral. (This section also provides general expressions, the Boltzmann limit is examined together with inelastic phonon collisions; a self-consistency equation is given.)

The second half of this chapter analyzes the behavior of electron-hole excitations in the nonequilibrium tunnel junctions of superconductors. Section 4 examines the quantum oscillations of the nonequilibrium chemical potential. (The steady-state shift in the chemical potential is investigated here and the properties of the tunnel source are discussed, including the "vibrating" chemical potential.)

Section 5 is devoted to emission scattering by a nonequilibrium Josephson junction. (The equilibrium source of nonequilibrium is considered here, and the excitation distribution function is analyzed, while the satellites and the possibility of their observation are discussed). Section 6 investigates the threshold state in a symmetrical *SiS* junction. (Here we discuss the method of solving the self-consistent equations, a stationary excitation distribution function is found, the nonequilibrium order parameter is investigated and the E-I characteristic is found.)

Chapter 3 examines phonon emission in conditions of nonequilibrium. Section 8 describes the kinetics of nonequilibrium electrons and phonons by analytic continuation. (Here we examine electron-phonon collisions and phonon-electron collisions; the canonical form of the collision integrals and the relaxation of imbalance is investigated.) Section 9 examines the phonon emission spectrum. (Both subcritical and supercritical states are examined.)

Chapter 4 is devoted to the theory of an acoustic quantum-mechanical oscillator based on a nonequilibrium superconductor. Section 10 investigates population inversion in nonequilibrium superconductors. (Here we analyze the breakdown of the binding kinetics of the electrons and phonons and derive the necessary phonon instability condition.) In Section 11 self-consistent kinetic equations of the problem are derived. (This provides a simplification of the general equations, resonant electromagnetic field pumping is examined and phonon instability in symmetrical tunnel injection is investigated.) Section 12 examines an asymmetrical *SiS'*-junction. In Section 13 the role of fluctuations is discussed.

We note that studies [1-14] underlie this investigation.

Chapter 1

A GENERAL CONSIDERATION OF ELECTRON KINETICS IN A NONEQUILIBRIUM JOSEPHSON JUNCTION

This chapter is devoted to an analysis of certain general issues. First, by using the tunnel Hamiltonian, whose contribution to single-particle tunneling is accounted for in all orders of perturbation theory, we derive a term describing tunnel injection (and extraction) of excitations (electrons and holes) in the kinetic equation. The derived expression accounts for both the population imbalance of the electron-hole excitation branches and the macroscopic phase coherence in the tunnel junction. Second, an explicit form is derived (in terms of the excitation distribution functions) for the electron-electron inelastic collision integral. This expression generalizes the familiar operator obtained by Eliashberg to the case where a branch imbalance exists. Third, an expression is derived for the current through a nonequilibrium Josephson junction. These results will be needed below.

1. Initial relations

We will now use the apparatus of so-called energy-integrated Green-Gorkov-Eilenberger-Eliashberg functions.

1. The Eliashberg kinetic equations: the general case. The matrix function $\hat{g}_{\varepsilon\varepsilon\omega}$ satisfies the equation system [1]

$$\begin{pmatrix} (\omega - \mathbf{vk})\, g & (2\varepsilon - \omega - \mathbf{vk}) f \\ (2\varepsilon - \omega + \mathbf{vk})\, f^{+} & -(\omega + \mathbf{vk})\, \bar{g} \end{pmatrix} = \hat{H}_1 \hat{g} - \hat{g}\hat{H}_1 + \hat{I}, \tag{1.1}$$

$$\hat{I} = \hat{g}\hat{\Sigma}^{A} - \hat{\Sigma}^{R}\hat{g} + \hat{g}^{R}\hat{\Sigma} - \hat{\Sigma}\hat{g}^{A}, \tag{1.2}$$

where

$$\hat{g}^{(R,\,A)} = \begin{pmatrix} g & f \\ -f^{+} & \bar{g} \end{pmatrix}^{(R,\,A)}; \qquad \hat{\Sigma}^{(R,\,A)} = \begin{pmatrix} \Sigma_1 & \Sigma_2 \\ -\Sigma_2^{+} & \bar{\Sigma}_1 \end{pmatrix}^{(R,\,A)},$$

$$\hat{H}_1 = \begin{pmatrix} H_1 & 0 \\ 0 & \bar{H}_1 \end{pmatrix}, \quad H_1 = -\frac{e}{c}\,\mathbf{vA} + e\varphi, \quad \bar{H}_1 = \frac{e}{c}\,\mathbf{vA} + e\varphi, \quad \mathbf{v} = \mathbf{v}_F, \tag{1.3}$$

where the retarded (advanced) functions are determined from the diagram expansion, where all propagators and self-energies are retarded (advanced). (We will not provide the expressions for these here in order to save space.) The matrix product in (1.1), (1.2) should also be understood as convolution with respect to the internal variables: thus, in the coordinate representation ($1 \equiv \mathbf{r}_1,\ t_1$) we have

$$(\Sigma g)(1, 2) \equiv \int \Sigma(1, 3)\, g(3, 2)\, d3. \tag{1.4}$$

The self-energy matrices $\hat{\Sigma}^{(R,A)}$ in (1.2) are additive quantities

$$\Sigma = \Sigma^{(imp)} + \Sigma^{(e-ph)} + \Sigma^{(e-e)} + \Sigma^{(T)}, \tag{1.5}$$

corresponding to electron-impurity, electron-phonon, electron-electron interactions and tunneling, respectively. These self-energy parts will be modified in subsequent sections.

In (1.2) we will isolate the components corresponding to electron-phonon interaction and will then isolate the virtual processes in the latter. Dropping the renormalized terms $(\Sigma_1^R + \Sigma_2^A)^{(e-ph)}$ and introducing the superconducting ordering parameter $\Delta = {}^1/_2\,(\Sigma_2^R + \Sigma_2^A)^{(e-ph)}$, we obtain from (1.2) the following expression for the 11-component:

$$I_{\varepsilon\varepsilon-\omega} = \{-f\Delta^{*} + \Delta f^{+}\}_{\varepsilon\varepsilon-\omega} + \{-i(g\gamma + \gamma g) + i(-f\delta^{+} + \delta f^{+}) + g^{R}\Sigma_1^{(e-ph)} - \Sigma_1^{(e-ph)} g^{A} - f^{R}\Sigma_2^{+(e-ph)} + \Sigma_2^{(e-ph)} f^{+A}\}_{\varepsilon\varepsilon-\omega} + I'_{\varepsilon\varepsilon-\omega}, \tag{1.6}$$

where the quantities

$$2i\gamma_{\varepsilon\varepsilon-\omega} = (\Sigma_1^R - \Sigma_1^A)^{(e-ph)}, \qquad 2i\delta_{\varepsilon\varepsilon-\omega} = (\Sigma_2^R - \Sigma_2^A)^{(e-ph)}, \tag{1.7}$$

together with $\Sigma_{1,2}^{(e-ph)}$ describe the actual electron-phonon interaction processes that are critical for the kinetics, while $I'_{\varepsilon\varepsilon-\omega}$ no longer contains $\Sigma^{(e-ph)}$ in explicit form.

2. **Nondiagonal inelastic scattering channel.** The dissipation function γ (in 1.6) (similar to δ and $\Sigma_{1,2}^{(e-ph)}$) has a characteristic magnitude in the order of the energy damping of the electron excitations. In normal metal $\gamma \sim T^3/\omega_D^2$; in the superconducting state γ is even lower, since a significant portion of electron-phonon interaction - virtual processes - have already been accounted for in the reformulation and is an order of magnitude below the modulus of the order parameter Δ across virtually the entire temperature range of the existence of Δ. Hence, before we make the transformation to the kinetic equation in (1.6), we must account for the first expression in braces in (1.6) exactly by using equations for the nondiagonal components of the $\hat{g}$-functions following from (1.1) (see study [7] as well as study [12] for the nondiagonal channel). From equation (1.1) we have

$$\begin{aligned} 2(\varepsilon-\omega)(f-f^+)_{\varepsilon\varepsilon-\omega} &= \mathbf{vk}(f+f^+)_{\varepsilon\varepsilon-\omega} + \{i(f\overline{\gamma}+\overline{\gamma}f^+) - i(\gamma f + f^+\gamma) + \\ &+ i(\delta^+ g - g\delta) + i(\bar{g}\delta^+ - \delta\bar{g}) + (g\Delta - \Delta^* g) + (\bar{g}\Delta^* - \Delta\bar{g}) + f^R\overline{\Sigma}_1 + \\ &+ f^{+R}\Sigma_1 - \Sigma_1 f^A - \overline{\Sigma}_1 f^{+A} + g^R\Sigma_2 - \Sigma_2 g^A + \bar{g}^R\Sigma_2^+ - \Sigma_2^+ g^A\}_{\varepsilon\varepsilon-\omega} + I''_{\varepsilon\varepsilon-\omega}. \end{aligned} \tag{1.8}$$

As in deriving relation (1.6) in (1.8) we have isolated the virtual processes by representing electron-phonon interactions (in braces) in explicit form, while $I''_{\varepsilon\varepsilon-\omega}$ contains, analogous to $I'_{\varepsilon\varepsilon-\omega}$ in (1.6), all similar processes. For this quantity we have from (1.1):

$$\begin{aligned} I'' &= g\Sigma_2^A + f\overline{\Sigma}_1^A - \Sigma_1^R f - \Sigma_2^R\bar{g} + g^R\Sigma_2 + f^R\overline{\Sigma}_1 - \Sigma_1 f^A - \Sigma_2\bar{g}^A + f^+\Sigma_1^A + \\ &+ \bar{g}\Sigma_2^{+A} - \Sigma_2^{+R}g - \overline{\Sigma}_1^R f^+ + f^{+R}\Sigma_1 + \bar{g}^R\Sigma_2^+ - \Sigma_2^+ g^A - \overline{\Sigma}_1 f^{+A} \end{aligned} \tag{1.9}$$

(here we have dropped all external and internal arguments; with this notation system the order of the cofactors is important).

Below we will focus on thin-film junctions with thicknesses in the order of the correlation length of the superconductor. Such junctions, which are fabricated primarily by deposition, always contain a large quantity of electron elastic scattering centers (such as nonmagnetic impurities, lattice defects, etc.). When there are sufficient elastic scattering centers (which will be assumed below and indeed is the case if special preventive measures are not taken in the fabrication of the junctions), the superconducting films will be "dirty" and the free path length of the electrons will be less than the other lengths characterizing their kinetics in the superconductor. This fact allows significant simplifications. Specifically, we may ignore anisotropy effects, inhomogeneities associated with the junction boundaries, the nonlocality of electrodynamics, etc.

Bearing this case in mind, we will carry out averaging over the angular variable in equation (1.1) subject to (1.6), (1.8), and (1.9); from the expressions I'_ε and I''_ε that are diagonal with respect to the energy (frequency) variables, we derive the self-energy parts corresponding to electron-impurity interaction. In the approximation corresponding to the kinetic equation the Green-Gorkov-Eilenberger-Eliashberg functions are diagonal with respect to the energy variables

$$\hat{g}_{\varepsilon\varepsilon-\omega} = 2\pi\delta(\omega)\,\hat{g}_\varepsilon \tag{1.10}$$

and in the isotropic and spatially-homogeneous case have a dependence on ε only.

In proceeding directly to the derivation of expressions in terms of the distribution function of electron (n_ε) and hole ($n_{-\varepsilon}$) excitations, we will examine the relation between $n_{\pm\varepsilon}$ and $\hat{g}_\varepsilon$.

3. **Normalization condition.** When there is population symmetry of electron-hole branches, i.e., $n_\varepsilon = n_{-\varepsilon} \equiv n_\varepsilon$ (here and henceforth unless otherwise stated the parameter ε will be considered to be positively-defined), this relation takes the form [1]

$$g_\varepsilon = (2\pi i)\frac{\varepsilon}{\sqrt{\varepsilon^2-\Delta^2}}\,\theta(\varepsilon^2-\Delta^2)(1-2n_\varepsilon), \tag{1.11}$$

$$f_\varepsilon = f_\varepsilon^+, \tag{1.12}$$

$$f_\varepsilon^+ = \frac{\Delta}{\varepsilon}\,g_\varepsilon, \tag{1.13}$$

and

$$\bar{g}_\varepsilon^{(R,A)} = -g_\varepsilon^{(R,A)}, \tag{1.14}$$

$$g_\varepsilon^{R(A)} = \frac{\varepsilon}{\Delta}\,f_\varepsilon^{R(A)} = i\pi\,\frac{\varepsilon}{\xi_\varepsilon^{R(A)}}, \tag{1.15}$$

$$\xi_\varepsilon^R = -(\xi_\varepsilon^A)^* = \begin{cases}\sqrt{\varepsilon^2-\Delta^2}\,\mathrm{sign}\,\varepsilon + i\delta, & \varepsilon^2 > \Delta^2,\\ i\sqrt{\Delta^2-\varepsilon^2}, & \varepsilon^2 < \Delta^2.\end{cases} \tag{1.16}$$

Relations (1.11)-(1.14) in the case of imbalance ($n_\varepsilon \neq n_{-\varepsilon}$) requires generalization. This generalization is not reduced to the trivial substitution $2n_\varepsilon \to (n_\varepsilon + n_{-\varepsilon})$ in (1.11): such a substitution would leave the function g_ε odd with respect to ε. Moreover, in the general case, g_ε will also have an even-ε part. In order to determine this we will consider the normalization condition for the $\hat{g}$-function obtained by Larkin and Ovchinnikov [13]. With respect to this case (1.11)-(1.16) it takes the form

$$\check{g}_\varepsilon^2 = -\pi^2\check{1}, \tag{1.17}$$

where

$$\check{g} = \begin{pmatrix}\hat{g}^R & \hat{g}\\ \hat{0} & \hat{g}^A\end{pmatrix}, \quad \check{1} = \begin{pmatrix}\hat{1} & \hat{0}\\ \hat{0} & \hat{1}\end{pmatrix}, \tag{1.18}$$

while $\hat{g}^{(R,A)}$ is given by expression (1.3). Now using the fact that the normalization condition is satisfied identically by the substitution

$$\check{g} = \check{g}^R \hat{a} - \hat{a} \check{g}^A, \tag{1.19}$$

where $\hat{a}(\varepsilon)$ is a random matrix (2×2)-function that may be expanded in terms of Pauli matrices, while $\hat{g}^{R(A)}$ is determined by formulae (1.3), (1.15), and (1.16), we obtain

$$\begin{pmatrix} g & f \\ -f^+ & \bar{g} \end{pmatrix} = \check{g}(\varepsilon) = f_1(\varepsilon)(\hat{g}^R - \hat{g}^A) + f_2(\varepsilon)(\hat{g}^R \tau_z - \tau_z \hat{g}^A) =$$
$$= f_1(\varepsilon)\begin{pmatrix} g^R - g^A & f^R - f^A \\ -(f^{+R} - f^{+A}) & \bar{g}^R - \bar{g}^A \end{pmatrix} + f_2(\varepsilon)\begin{pmatrix} g^R - g^A & -(f^R + f^A) \\ -(f^{+R} + f^{+A}) & -(\bar{g}^R - \bar{g}^A) \end{pmatrix}. \tag{1.20}$$

Using the relations that follow directly from (1.1):

$$\bar{g}_\varepsilon = g_{-\varepsilon}, \qquad \bar{g}_\varepsilon^{R(A)} = -g_\varepsilon^{R(A)}, \tag{1.21}$$

we find from (1.20):

$$\begin{aligned} g_\varepsilon &= f_1(\varepsilon)(g^R - g^A)_\varepsilon + f_2(\varepsilon)(g^R - g^A)_\varepsilon, \\ \bar{g}_\varepsilon &= -f_1(\varepsilon)(g^R - g^A)_\varepsilon + f_2(\varepsilon)(g^R - g^A)_\varepsilon. \end{aligned} \tag{1.22}$$

From (1.22) subject to (1.21), we may conclude that $f_1(\varepsilon)$ is an even function with respect to ε while $f_2(\varepsilon)$ is an odd function with respect to ε. Thus, in the general case ($f_2 \neq 0$) the g-function does indeed have a part that is even with respect to ε. It is precisely this even energy part that is responsible for the shift in the electrochemical potential in the electron subsystem in nonequilibrium conditions. Indeed, as demonstrated in study [1], the electron neutrality condition produces the following expression for the shift in the electrochemical potential $\delta\mu$:

$$\delta\mu(\omega) = -\int_{-\infty}^{\infty} \frac{d\varepsilon}{4\pi i} \int \frac{dO_{\mathbf{p}}}{4\pi} g_{\varepsilon\varepsilon-\omega}(\mathbf{p}, \mathbf{k}). \tag{1.23}$$

In the steady-state spatially-homogeneous and isotropic case we have from here, subject to (1.22):

$$\delta\mu = i \int_{-\infty}^{\infty} \frac{d\varepsilon}{4\pi i} g_\varepsilon = i \int_{-\infty}^{\infty} \frac{d\varepsilon}{4\pi} f_2(\varepsilon)(g^R - g^A)_\varepsilon \tag{1.24}$$

(the integrals are understood in the sense of the principal value), and consequently, $\delta\mu \neq 0$ if $f_2 \neq 0$. As indicated from (1.20), the condition $f_\varepsilon = f_\varepsilon^+$ is satisfied in the case $f_2 \propto \theta(\varepsilon^2 - \Delta^2)$. Hence it will be convenient below to redesignate

$$f_2(\varepsilon) = \frac{\sqrt{\varepsilon^2 - \Delta^2}}{\varepsilon} f_2'(\varepsilon)\theta(\varepsilon^2 - \Delta^2). \tag{1.25}$$

Now introducing an arbitrary function n_ε (here $-\infty < \varepsilon < \infty$) we may write (dropping $'$ in (1.25))

$$f_1(\varepsilon) = a_1(n_\varepsilon + n_{-\varepsilon} - 1),$$

$$f_2(\varepsilon) = a_2 (n_\varepsilon - n_{-\varepsilon}). \tag{1.26}$$

Since the function n_ε will be subject to further determination, there is arbitrariness in the selection of the coefficients a_1 and a_2. They may easily be selected so that the expressions for $\hat{g}^{(R,A)}$ will take the form

$$g_\varepsilon = (-2\pi i)\begin{pmatrix} u_\varepsilon\beta_\varepsilon + \alpha_\varepsilon & v_\varepsilon\beta_\varepsilon \\ -v_\varepsilon\beta_\varepsilon & -u_\varepsilon\beta_\varepsilon + \alpha_\varepsilon \end{pmatrix}, \tag{1.27}$$

$$(\hat{g}^R - \hat{g}^A)_\varepsilon = 2\pi i \begin{pmatrix} u_\varepsilon & v_\varepsilon \\ -v_\varepsilon & -u_\varepsilon \end{pmatrix}, \tag{1.28}$$

$$u_\varepsilon = \frac{|\varepsilon|\,\theta(\varepsilon^2 - \Delta^2)}{\sqrt{\varepsilon^2 - \Delta^2}}, \qquad v_\varepsilon = \frac{\Delta\,\mathrm{sign}\,\varepsilon\,\theta(\varepsilon^2 - \Delta^2)}{\sqrt{\varepsilon^2 - \Delta^2}}, \tag{1.29}$$

$$\beta_\varepsilon = (n_\varepsilon + n_{-\varepsilon} - 1)\,\theta(\varepsilon^2 - \Delta^2)\,\mathrm{sign}\,\varepsilon, \qquad \alpha_\varepsilon = (n_\varepsilon - n_{-\varepsilon})\,\theta(\varepsilon^2 - \Delta^2)\,\mathrm{sign}\,\varepsilon. \tag{1.30}$$

With such a selection and without loss of generality we have, on the one hand, a limiting process to expressions (1.11)-(1.16) in the absence of imbalance, and on the other, as will be demonstrated below, the function n_ε will acquire the same simple meaning as the excitation energy distribution function in pure superconductors.

4. Effective collision integral. Based on the derived expressions (1.27)-(1.30), we may derive the following formula:

$$u_\varepsilon \dot{n}_\varepsilon = -\frac{1}{8\pi i}\{(\dot{g}_\varepsilon - \dot{g}_{-\varepsilon}) + u_\varepsilon(\dot{g}_\varepsilon + \dot{g}_{-\varepsilon})\}\,\mathrm{sign}\,\varepsilon, \tag{1.31}$$

where the dot designates time differentiation. Thus, the right half of (1.31) is expressed through the 11-component of equation (1.1). Accounting for the nondiagonal channel the effective collision integral is

$$I_{э\phi}(\varepsilon) = \omega g_\varepsilon = I_{э\phi}^{(e-ph)}(\varepsilon) + I_{э\phi}^{(e-e)}(\varepsilon) + I_{э\phi}^{(T)}, \tag{1.32}$$

where the last two terms have the structure

$$\begin{aligned} I_{э\phi}(\varepsilon) &= I'(\varepsilon) - \frac{\Delta}{2\varepsilon} I''(\varepsilon) = g\Sigma_1^A - f\Sigma_2^{+A} - \Sigma_1^R g + \Sigma_2^R f^+ + g^R\Sigma_1 - f^R\Sigma_2^+ - \\ &- \Sigma_1 g^A + \Sigma_2 f^{+A} - \frac{\Delta}{2\varepsilon}\{g\Sigma_2^A + f\bar{\Sigma}_1^A - \Sigma_1^R f - \Sigma_2^R \bar{g} + g^R\Sigma_2 + f^R\bar{\Sigma}_1 - \Sigma_1 f^A - \Sigma_2 \bar{g}^A + \\ &+ f^+\Sigma_1^A + \bar{g}\Sigma_2^{+A} - \Sigma_2^{+R} g - \bar{\Sigma}_1^R f^+ + f^{+R}\Sigma_1 + \bar{g}^R\Sigma_2^+ - \Sigma_2^+ g^A - \bar{\Sigma}_1 f^{+A}\}. \end{aligned} \tag{1.33}$$

Accounting for the nondiagonal channel produces a similar expression for $I_{э\phi}^{(e-ph)}(\varepsilon)$ as well which includes the quantities γ, δ, etc. They will not be provided here for space considerations. Formulae (1.31)-(1.33) represent the basis of subsequent calculations.

2. The tunnel source of nonequilibrium

Effective collision integral (1.33) when substituted into equation (1.31) allows us to find both the tunnel term determining the source of nonequilibrium and the collision integral describing inelastic electron-electron collisions. Here we must know the corresponding self-energy parts. In this section we will modify their form for tunnel injection from one superconductor to another.

1. The Hamiltonian and self-energy parts. In order to describe the electron tunneling process from one superconductor to another we may proceed from the tunnel Hamiltonian written in the Nambu representation:

$$H^T = \sum_{\substack{\mathbf{pq}\\ \alpha\beta}} (T_{\mathbf{pq}} a^+_{\alpha\mathbf{p}} \boldsymbol{\tau}_{\alpha\beta} a'_{\beta\mathbf{q}} + \text{c. c.}), \tag{2.1}$$

where α and β are the Nambu spin indices of the electron creation and destruction operators:

$$a_{1\mathbf{p}} = a_{\uparrow\mathbf{p}},\ a_{2\mathbf{p}} = a^+_{\downarrow-\mathbf{p}}, \tag{2.2}$$

where the operator $a'_{\beta\mathbf{q}}$ refers to the superconductor-injector (henceforth all primed quantities as well as functions with deviating arguments refer to the injector), $T_{\mathbf{pq}}$ is the tunneling matrix element relating the electron states $\mathbf{p}$ and $\mathbf{q}$. The self-energy parts induced by the Hamiltonian (2.1) have been found in the study by Volkov [14] based on the Keldysh technique. In our representation the result takes the form

$$\hat{\Sigma}^{(T)(R,A)}(t,t') = \frac{\nu}{\pi} \hat{g}'^{(R,A)}(t,t'). \tag{2.3}$$

It is clear from (2.3) that the self-energy parts in tunneling are the same as in elastic scattering by impurities with the one difference that instead of angle-averaged Green's functions, the propagators of the injector $\hat{g}'$ figure in them, while the transport time is replaced by the inverse "tunnel frequency" ν (this quantity is related to the conductivity of the SiS'-junction; its explicit expression will be derived in Section 2 of Part 4). This is related to a certain analogy between the tunneling process and scattering by impurities. Without going into great detail here we simply emphasize that the g'-function in (2.3), as in the case of impurity scattering, is assumed to be complete, i.e., it contains in the diagram expansion in addition to other energies the self-energies corresponding to tunneling, which makes it possible to account for single-particle tunneling (2.1) in all orders of perturbation theory.

2. Quasi-classical approximation. We will now assume that the test superconductor S has the potential $V = 0$, and the time-constant potential $V(e = \hbar = 1)$ is applied to the injector S'. Then by virtue of the gradient invariance the existence of potential V produces phase multipliers in the $\hat{g}$-functions of the superconductor [15] and hence

$$\Sigma_1^{(R,A)}(t_1t_2) = \frac{\nu}{\pi} g'^{(R,A)}(t_1t_2) \exp[-iV(t_1 - t_2)],$$

$$\bar{\Sigma}_1^{(R,A)}(t_1t_2) = \frac{\nu}{\pi} \bar{g}'^{(R,A)}(t_1t_2) \exp[iV(t_1 - t_2)],$$

$$\Sigma_2^{(R,A)}(t_1t_2) = \frac{\nu}{\pi} f'^{(R,A)}(t_1t_2) \exp[-iV(t_1 + t_2)],$$

$$\Sigma_2^{+(R,A)}(t_1t_2) = \frac{\nu}{\pi} f^{+'(R,A)}(t_1t_2)\exp[iV(t_1+t_2)]. \qquad (2.4)$$

These expressions must be substituted into (1.33) and a Fourier time transformation must be carried out. Here we will assume a temporal quasi-classical nature of the external action, specifically: $\Sigma(t_1 t_2)$ is a fast function of the difference variable $t_1 - t_2$ and a slow function of the sum variable $t = (t_1 + t_2)/2$. We note that when operating with formulae (1.33), (2.4) we must show some caution and bear in mind the following: the expression $\Sigma_{2\varepsilon-V}a_\varepsilon \exp(-i\varphi)$ corresponds to the $a\Sigma_2$ term in (1.33) after the Fourier transformation, at the same time that $\Sigma_2 a$ yields $\Sigma_{2\varepsilon+V}a_\varepsilon \exp(-i\varphi)$ (here $a_\varepsilon = g_\varepsilon^{(R,A)}$, $f_\varepsilon^{(R,A)}$, etc., $\varphi = 2Vt$). These rules whose validity may be directly verified also reveal that

$$\begin{aligned}
&\Sigma_2^+ a \to \Sigma_{2\,\varepsilon-V}^+ a_\varepsilon \exp(i\varphi).\\
&a\Sigma_2^+ \to \Sigma_{2\,\varepsilon+V}^+ a_\varepsilon \exp(i\varphi),\\
&\Sigma_1 a = a\Sigma_1 \to a_\varepsilon \Sigma_{1\,\varepsilon-V},\\
&\overline{\Sigma}_1 a = a\overline{\Sigma}_1 \to a_\varepsilon \overline{\Sigma}_{1\,\varepsilon-V}.
\end{aligned} \qquad (2.5)$$

3. **Tunnel source.** Accounting for rules (2.5) and relations (1.27)-(1.30) we find, by using (1.33) the following expression:

$$\begin{aligned}
&\{(I_\varepsilon - I_{-\varepsilon}) + u_\varepsilon (I_\varepsilon + I_{-\varepsilon})\}_{\text{эф}} = (2\pi i)^2 \nu\pi^{-1}\{[u_\varepsilon\beta_\varepsilon(u_{\varepsilon-V} + u_{\varepsilon+V}) +\\
&+ \alpha_\varepsilon(u_{\varepsilon-V} - u_{\varepsilon+V}) - u_\varepsilon(u_{\varepsilon-V}\beta_{\varepsilon-V} + u_{\varepsilon+V}\beta_{\varepsilon+V}) - u_\varepsilon(\alpha_{\varepsilon-V} - \alpha_{\varepsilon+V}) +\\
&+ \beta_\varepsilon(u_{\varepsilon-V} - u_{\varepsilon+V}) + (u_{\varepsilon-V} + u_{\varepsilon+V})u_\varepsilon\alpha_\varepsilon - u_{\varepsilon-V}\beta_{\varepsilon-V} + u_{\varepsilon+V}\beta_{\varepsilon+V} - \alpha_{\varepsilon-V} -\\
&- \alpha_{\varepsilon+V}] + \cos\varphi[-v_\varepsilon v_{\varepsilon+V}(\beta_\varepsilon - \beta_{\varepsilon+V}) - v_\varepsilon v_{\varepsilon-V}(\beta_\varepsilon - \beta_{\varepsilon-V}) -\\
&- \alpha_\varepsilon v_\varepsilon(v_{\varepsilon-V} + v_{\varepsilon+V})] + \sin\varphi[v_\varepsilon w_{\varepsilon+V}\beta_\varepsilon + v_{\varepsilon+V}w_\varepsilon\beta_{\varepsilon+V} - v_\varepsilon w_{\varepsilon-V}\beta_\varepsilon -\\
&- v_{\varepsilon-V}w_\varepsilon\beta_{\varepsilon-V} - \alpha_\varepsilon v_\varepsilon w_{\varepsilon-V} + v_\varepsilon\alpha_\varepsilon w_{\varepsilon+V}]\},
\end{aligned} \qquad (2.6)$$

where $w_\varepsilon = \Delta\theta(\Delta^2 - \varepsilon^2)(\Delta^2 - \varepsilon^2)^{-1/2}$. Converting to the distribution function n_ε consistent with (1.30), (1.31) we have

$$u_\varepsilon \dot{n}_\varepsilon = \frac{1}{2}\nu[Q_1(n_{\pm\varepsilon})\sin\varphi + Q_2(n_{\pm\varepsilon})\cos\varphi + Q_3(n_{\pm\varepsilon})],\ \varepsilon \geqslant \Delta. \qquad (2.7)$$

The dimensionless factors Q_1 are equal to

$$\begin{aligned}
Q_1(n_{\pm\varepsilon}) = {}& v_\varepsilon w_{\varepsilon-V}(2n_{\pm\varepsilon} - 1)\theta(\Delta' - \varepsilon + V)\theta(\Delta' + \varepsilon - V) -\\
&- v_\varepsilon w_{\varepsilon+V}(2n_{\pm\varepsilon} - 1)\theta(\Delta' + \varepsilon + V)\theta(\Delta' - \varepsilon - V),
\end{aligned} \qquad (2.8)$$

$$\begin{aligned}
Q_2(n_{\pm\varepsilon}) = {}& v_\varepsilon v_{\varepsilon-V}[(n_{\pm\varepsilon} - n_{\varepsilon-V}) + (n_{\pm\varepsilon} - n_{-\varepsilon+V})]\theta(\varepsilon - V - \Delta') -\\
&- v_\varepsilon v_{\varepsilon+V}[(n_{\varepsilon+V} - n_{\pm\varepsilon}) + (n_{-\varepsilon-V} - n_{\pm\varepsilon})]\theta(\varepsilon + V - \Delta') +\\
&+ v_\varepsilon v_{V-\varepsilon}[(1 - n_{\pm\varepsilon} - n_{V-\varepsilon}) + (1 - n_{\pm\varepsilon} - n_{-V+\varepsilon})]\theta(V - \varepsilon - \Delta'),
\end{aligned} \qquad (2.9)$$

$$\begin{aligned}
Q_3(n_{\pm\varepsilon}) = {}& [(n_{\varepsilon-V} - n_{\pm\varepsilon})(u_\varepsilon u_{\varepsilon-V} \pm u_{\varepsilon-V} + u_\varepsilon \pm 1) +\\
&+ (n_{-\varepsilon+V} - n_{\pm\varepsilon})(u_\varepsilon u_{\varepsilon-V} \pm u_{\varepsilon-V} - u_\varepsilon \mp 1)]\theta(\varepsilon - V - \Delta') -\\
&- [(n_{\pm\varepsilon} - n_{\varepsilon+V})(u_\varepsilon u_{\varepsilon+V} \mp u_{\varepsilon+V} - u_\varepsilon \pm 1) +\\
&+ (n_{\pm\varepsilon} - n_{-\varepsilon-V})(u_\varepsilon u_{\varepsilon+V} \mp u_{\varepsilon+V} + u_\varepsilon \mp 1)]\theta(\varepsilon + V - \Delta') +\\
&+ [(1 - n_{\pm\varepsilon} - n_{V-\varepsilon})(u_\varepsilon u_{V-\varepsilon} \pm u_{V-\varepsilon} - u_\varepsilon \mp 1) +\\
&+ (1 - n_{\pm\varepsilon} - n_{-V+\varepsilon})(u_\varepsilon u_{V-\varepsilon} \pm u_{V-\varepsilon} + u_\varepsilon \pm 1)]\theta(V - \varepsilon - \Delta').
\end{aligned} \qquad (2.10)$$

We emphasize that in these final expressions that represent the tunnel source of nonequilibrium in the Josephson junction, the quantity ε is positively-defined. Here n_ε appears as the electron excitation distribution function, while $n_{-\varepsilon}$ functions as the hole excitation distribution function.

4. Tunnel current. In formula (2.7) the tunnel frequency must be expressed through the measured parameters. For this purpose it is convenient to use the expression for the current through the nonequilibrium Josephson junction. The current expression is also of interest in that by expanding this expression we may calculate the E-I characteristic of the nonequilibrium junction. We will derive this expression.

For this purpose we take advantage of the fact that the 11-component of equation (1.1) after integration with respect to ε and averaging over the angular variables is transformed into a continuity equation (here the electroneutrality div $\mathbf{J}$ - 0 is provided by the shift in the electrochemical potential that occurs in accordance with (1.23) [1]). After these procedures the terms remaining in (1.1) that contain the self-energy parts represent the divergence of the total tunnel current with multiplier accuracy. Integrating this divergence with respect to the volume of the nonequilibrium film (assuming spatial homogeneity of the pattern), we find the integral tunnel current $J = \int \mathbf{J} d\mathbf{S}$ as

$$J/J_0 = J_1 \sin\varphi + J_2 \cos\varphi + J_3, \tag{2.11}$$

where $J_0 = \nu\mathcal{V}mp_F/\pi^2$, $\mathcal{V}$ is the volume of the superconducting film. For J_1 - J_3 we obtain the following functionals

$$J_1 = \Delta\Delta' \int_{-\infty}^{\infty} d\varepsilon \left\{ \frac{\theta(\varepsilon^2 - \Delta^2)\,\theta[\Delta'^2 - (\varepsilon - V)^2]}{\sqrt{\varepsilon^2 - \Delta^2}\sqrt{\Delta'^2 - (\varepsilon - V)^2}} (n_\varepsilon + n_{-\varepsilon} - 1) + \right.$$
$$\left. + \frac{\theta[(\varepsilon - V)^2 - \Delta'^2]\,\theta(\Delta^2 - \varepsilon^2)}{\sqrt{(\varepsilon - V)^2 - \Delta'^2}\sqrt{\Delta^2 - \varepsilon^2}} (n_{\varepsilon-V} + n_{-\varepsilon+V} - 1)\right\}, \tag{2.12}$$

$$J_2 = \Delta\Delta' \int_{-\infty}^{\infty} d\varepsilon \frac{\theta(\varepsilon^2 - \Delta^2)\,\theta[(\varepsilon + V)^2 - \Delta'^2]}{\sqrt{\varepsilon^2 - \Delta^2}\sqrt{(\varepsilon + V)^2 - \Delta'^2}} \{(1 - n_\varepsilon - n_{-\varepsilon})\,\mathrm{sign}(\varepsilon + V) -$$
$$- (1 - n_{\varepsilon+V} - n_{-\varepsilon-V})\,\mathrm{sign}\,\varepsilon\}, \tag{2.13}$$

$$J_3 = \int_{-\infty}^{\infty} d\varepsilon \left\{ \frac{|\varepsilon|\,|\varepsilon - V|\,\theta(\varepsilon^2 - \Delta^2)\,\theta[(\varepsilon - V)^2 - \Delta'^2]}{\sqrt{\varepsilon^2 - \Delta^2}\sqrt{(\varepsilon - V)^2 - \Delta'^2}} \times \right.$$
$$\times [(1 - n_\varepsilon - n_{-\varepsilon})\,\mathrm{sign}\,\varepsilon - (1 - n_{\varepsilon-V} - n_{-\varepsilon+V})\,\mathrm{sign}(\varepsilon - V)] -$$
$$- \frac{|\varepsilon - V|\,\theta(\varepsilon^2 - \Delta^2)\,\theta[(\varepsilon - V)^2 - \Delta'^2]}{\sqrt{(\varepsilon - V)^2 - \Delta'^2}} (n_\varepsilon - n_{-\varepsilon})\,\mathrm{sign}\,\varepsilon +$$
$$\left. + \frac{|\varepsilon|\,\theta(\varepsilon^2 - \Delta^2)\,\theta[(\varepsilon - V)^2 - \Delta'^2]}{\sqrt{\varepsilon^2 - \Delta^2}} (n_{\varepsilon-V} - n_{-\varepsilon+V})\,\mathrm{sign}(\varepsilon - V)\right\}. \tag{2.14}$$

These expressions include the nonequilibrium distribution functions n_ε and n'_ε (as well as Δ and Δ') to be determined. We emphasize that the derived expressions for current are not a trivial generalization of the formulae figuring into equilibrium theory. It is the last two components in (2.14) that are a result of the branch imbalance, and hence they are characteristic of the nonequilibrium situation only. In the equilibrium approximation $n_\varepsilon \to \overset{0}{n}_\varepsilon(T)$ where $\overset{0}{n}_\varepsilon(T)$ is the Fermi function at temperature T, and expressions (2.11)-(2.14) become the familiar expressions for the current in a Josephson junction obtained by Larkin and Ovchinnikov [6]. In the case $\Delta' = 0$ (*NiS*-junction) (2.14) coincides with the expression obtained by Bulyzhenkov and Ivlev [7].

We will now determine the explicit form of ν. Converting to the equilibrium case $n_\varepsilon \to \overset{0}{n}_\varepsilon(T)$ and then assuming $t \gg \Delta$, Δ' and V, we obtain Ohm's law from (2.11)-(2.14), where $J_0 = 1/2eR$, and from here:

$$\nu = 1/4e^2 N(0)\, SdR. \tag{2.15}$$

Here $N(0) = mp_F/2\pi^2$ is the density of electron levels in a normal metal calculated for a single spin, S is the cross-section of the tunnel junction (the area of the dielectric interlayer), d is its thickness (for clarity the electron charge e is restored in expression (2.15)).

3. Electron-electron collision integral

Here we will derive the canonical form of the electron-electron collision integral. These collisions and electron-phonon collisions are responsible for the existence of stationary dissipative states. In addition, inelastic electron-electron collisions may produce a significant increase in the number of nonequilibrium electron-hole excitations due to "impact ionization" processes [16, 2].

1. General expressions. We will use general expressions (1.31)-(1.33) derived above and will employ the self-energy parts found by Eliashberg in them. They take the form [1]

$$\Sigma_1^R - \Sigma_1^A = \hat{L}\,[A\,\{g_1 g_2 \bar{g}_3\}^{(R-A)} - B\,\{f_1 f_2^+ g_3\}^{(R-A)}], \tag{3.2}$$

$$\Sigma_1 = \hat{L}\,[A\,\{g_1 g_2 \bar{g}_3\} - B\,\{f_1 f_2^+ g_3\}], \tag{3.3}$$

$$\Sigma_2^R - \Sigma_2^A = \hat{L}\,[B\,\{g_1 \bar{g}_2 f_3\}^{(R-A)} - A\,\{f_1 f_2 f_3^+\}^{(R-A)}], \tag{3.4}$$

$$\Sigma_2 = \hat{L}\,[B\,\{g_1 \bar{g}_2 f_3\} - A\,\{f_1 f_2 f_3^+\}], \tag{3.5}$$

where the operator $\hat{L}$ is determined as

$$\hat{L} = \left(\frac{mp_F}{2\pi^2}\right)^2 \frac{1}{2\varepsilon_F} \iint\limits_{-\infty}^{\infty} \frac{d\varepsilon_1\, d\varepsilon_2}{(4\pi i)^2} \iint \frac{dO\mathbf{p}_1\, dO\mathbf{p}_2}{(4\pi)^2}\, \delta\left(\frac{p_3}{p_F} - 1\right), \tag{3.6}$$

while the quantities A and B in (3.2)-(3.5) are related to the scattering amplitudes of two normal excitations on the Fermi surface; in Born's approximation they take the form

$$A = -2|V_{\mathbf{p}-\mathbf{p}_2}|^2 + V_{\mathbf{p}-\mathbf{p}_2}V_{\mathbf{p}-\mathbf{p}_1}, \tag{3.7}$$

$$B = -2|V_{\mathbf{p}_1+\mathbf{p}_2}|^2 + V_{\mathbf{p}-\mathbf{p}_1}V_{\mathbf{p}-\mathbf{p}_2} + V_{\mathbf{p}+\mathbf{p}_1}V_{\mathbf{p}-\mathbf{p}_2} + V_{\mathbf{p}-\mathbf{p}_1}V_{\mathbf{p}+\mathbf{p}_2}, \tag{3.8}$$

where V_q is the interaction potential. The forms $\{\ldots\}$ and $\{\ldots\}^{(R-A)}$ appear as

$$\{g_1g_2g_3\}^{(R-A)} = g_1g_2(g_3^R - g_3^A) + g_1(g_2^R - g_2^A)g_3 + (g_1^R - g_1^A)g_2g_3 + \\ + (g_1^R - g_1^A)(g_2^R - g_2^A)(g_3^R - g_3^A), \tag{3.9}$$

$$\{g_1g_2g_3\} = g_1g_2g_3 + g_1(g_2^R - g_2^A)(g_3^R - g_3^A) + (g_1^R - g_1^A)g_2(g_3^R - g_3^A) + \\ + (g_1^R - g_1^A)(g_2^R - g_2^A)g_3. \tag{3.10}$$

2. The Boltzmann limit. Substituting expressions (3.2)-(3.5) subject to (3.9), (3.10), and (1.27)-(1.29) into formula (1.33) we have

$$\begin{aligned} I_{\text{эф}}(\varepsilon) = 2\pi^4\hat{L}\,\{&-A\,(-uu_1u_2u_3\beta\beta_1\beta_2 - uu_2u_3\beta\beta_2\alpha_1 - uu_1u_3\beta\alpha_2\beta_1 - uu_3\beta\alpha_1\alpha_2 - \\ &- uu_1u_2u_3\beta\beta_1\beta_3 - uu_2u_3\beta\beta_3\alpha_1 + uu_1u_2\beta\beta_1\alpha_3 + uu_2\beta\alpha_1\alpha_3 - \\ &- uu_1u_2u_3\beta\beta_2\beta_3 - uu_1u_3\beta\beta_3\alpha_2 + uu_1u_2\beta\beta_2\alpha_3 + uu_1\beta\alpha_2\alpha_3 - \\ &- uu_1u_2u_3\beta - u_1u_2u_3\beta_1\beta_2\alpha - u_2u_3\beta_2\alpha\alpha_1 - u_1u_3\beta_1\alpha\alpha_1 - \\ &- u_3\alpha\alpha_1\alpha_2 - u_1u_2u_3\beta_1\beta_3\alpha - u_2u_3\beta_3\alpha\alpha_1 + u_1u_2\beta_1\alpha\alpha_3 + u_2\alpha\alpha_1\alpha_3 - \\ &- u_1u_2u_3\beta_2\beta_3\alpha - u_1u_3\beta_3\alpha\alpha_2 + u_1u_2\beta_2\alpha\alpha_3 + u_1\alpha\alpha_2\alpha_3 - u_1u_2u_3\alpha) + \\ &+ B\,(uu_3v_1v_2\beta\beta_1\beta_2 + uu_3v_1v_2\beta\beta_1\beta_3 + uv_1v_2\beta\beta_1\alpha_3 + uv_1v_2u_3\beta\beta_2\beta_3 + \\ &+ uv_1v_2\beta\beta_2\alpha_3 + uu_3v_1v_2\beta + u_3v_1v_2\beta_1\beta_2\alpha + u_3v_1v_2\beta_1\beta_3\alpha + \\ &+ v_1v_2\beta_1\alpha\alpha_3 + u_3v_1v_2\beta_2\beta_3\alpha + v_1v_2\beta_2\alpha\alpha_3 + u_3v_1v_2\alpha) - \\ &- B\,(u_1u_2vv_3\beta\beta_1\beta_2 + u_2vv_3\beta\beta_2\alpha_1 - u_1vv_3\beta\beta_1\alpha_2 - vv_3\beta\alpha_1\alpha_2 + \\ &+ u_1u_2vv_3\beta\beta_1\beta_3 + u_2vv_3\beta\beta_3\alpha_1 + u_1u_2vv_3\beta\beta_2\beta_3 - u_1vv_3\beta\beta_3\alpha_2 + \\ &+ u_1u_2vv_3\beta) - A\,(vv_1v_2v_3\beta\beta_1\beta_2 + vv_1v_2v_3\beta\beta_1\beta_3 + vv_1v_2v_3\beta\beta_2\beta_3 + \\ &+ vv_1v_2v_3\beta) - A\,(uu_1u_2u_3\beta_1\beta_2\beta_3 - uu_1u_2\beta_1\beta_2\alpha_3 + uu_2u_3\beta_2\beta_3\alpha_1 - \\ &- uu_2\beta_2\alpha_1\alpha_3 + uu_1u_3\beta_1\beta_3\alpha_2 - uu_1\beta_1\alpha_2\alpha_3 + uu_3\beta_3\alpha_1\alpha_2 - \\ &- u\alpha_1\alpha_2\alpha_3 + uu_1u_2u_3\beta_1 + uu_2u_3\alpha_1 + uu_1u_2u_3\beta_2 + uu_1u_3\alpha_2 + \\ &+ uu_1u_2u_3\beta_3 - uu_1u_2\alpha_3) - B\,(uu_3v_1v_2\beta_1\beta_2\beta_3 + uv_1v_2\beta_1\beta_2\alpha_3 + \\ &+ uu_3v_1v_2\beta_1 + uu_3v_1v_2\beta_2 + uu_3v_1v_2\beta_3 + uv_1v_2\alpha_3) + \\ &+ B\,(u_1u_2vv_3\beta_1\beta_2\beta_3 + u_2vv_3\beta_2\beta_3\alpha_1 - u_1v_3v\beta_1\beta_3\alpha_2 - vv_3\beta_3\alpha_1\alpha_2 + \\ &+ u_1u_2vv_3\beta_1 + u_2vv_3\alpha_1 + u_1u_2vv_3\beta_2 - u_1vv_3\alpha_2 + u_1u_2vv_3\beta_3) + \\ &+ A\,(vv_1v_2v_3\beta_1\beta_2\beta_3 + vv_1v_2v_3\beta_1 + vv_1v_2v_3\beta_2 + vv_1v_2v_3\beta_3) - \\ &- (\Delta/\varepsilon)[-A\,(-u_2u_3v\beta\beta_2\alpha_1 - u_1u_3v\beta\beta_1\alpha_2 - u_2u_3v\beta\beta_3\alpha_1 + \\ &+ u_1u_2v\beta\beta_1\alpha_3 + u_1u_2v\beta\beta_2\alpha_3 - u_1u_3v\beta\beta_3\alpha_2) + B\,(vv_1v_2\beta\beta_1\alpha_3 + \\ &+ vv_1v_2\beta\beta_2\alpha_3) - B\,(-u_1u_2v_3\beta_1\beta_2\alpha - u_2v_3\beta_2\alpha\alpha_1 + u_1v_3\beta_1\alpha\alpha_2 + \\ &+ v_3\alpha\alpha_1\alpha_2 - u_1u_2v_3\beta_3\alpha - u_2v_3\beta_3\alpha\alpha_1 - u_1u_2v_3\beta_2\beta_3\alpha + \\ &+ u_1v_3\beta_3\alpha\alpha_2 - u_1u_2v_3\alpha) + A\,(v_1v_2v_3\beta_1\beta_2\alpha + v_1v_2v_3\beta_1\beta_3\alpha + \\ &+ v_1v_2v_3\beta_2\beta_3\alpha + v_1v_2v_3\alpha) - A\,(u_2u_3v\beta_2\beta_3\alpha_1 + u_1u_3v\beta_1\beta_3\alpha_2 - \\ &- v\alpha_1\alpha_2\alpha_3 + vu_2u_3\alpha_1 + u_1u_3v\alpha_2 - u_1u_2v\alpha_3) - B\,(vv_1v_2\beta_1\beta_2\alpha_3 + \\ &+ vv_1v_2\alpha_3)]\} \end{aligned} \tag{3.11}$$

(here $u = u(\varepsilon)$, $u_1 = u(\varepsilon_1)$, etc.). Reversing the sign of ε in this expression and substituting the values of $I(\varepsilon)$ and $I(-\varepsilon)$ into (1.31), we obtain subject to relations (1.30) the following canonical form for the inelastic electron-electron collision integral:

$$J^{(e-e)}(n_{\pm\varepsilon}) = \frac{1}{16\,\varepsilon_F}\iiint_{\Delta}^{\infty} \{E_1\delta(\varepsilon-\varepsilon_1-\varepsilon_2-\varepsilon_3) + 3E_2\delta(\varepsilon+\varepsilon_1-\varepsilon_2-\varepsilon_3) + 3E_3\delta(\varepsilon+\varepsilon_2+\varepsilon_3-\varepsilon_1)\}\, d\varepsilon_1 d\varepsilon_2 d\varepsilon_3, \quad (3.12)$$

in which the factors E_1, E_2 and E_3 take the form

$$\begin{aligned} E_1 = {} & M_1^1\{[(1-n_{\pm\varepsilon})\,n_{\varepsilon_1}n_{\varepsilon_2}n_{\varepsilon_3} - n_{\pm\varepsilon}(1-n_{\varepsilon_1})(1-n_{\varepsilon_2})(1-n_{\varepsilon_3})] + \\ & + [(1-n_{\pm\varepsilon})\,n_{\varepsilon_1}n_{-\varepsilon_2}n_{-\varepsilon_3} - n_{\pm\varepsilon}(1-n_{\varepsilon_1})(1-n_{-\varepsilon_2})(1-n_{-\varepsilon_3})]\} + \\ & + M_1^2\{[(1-n_{\pm\varepsilon})\,n_{-\varepsilon_1}n_{\varepsilon_2}n_{\varepsilon_3} - n_{\pm\varepsilon}(1-n_{-\varepsilon_1})(1-n_{\varepsilon_2})(1-n_{\varepsilon_3})] + \\ & + [(1-n_{\pm\varepsilon})\,n_{-\varepsilon_1}n_{-\varepsilon_2}n_{-\varepsilon_3} - n_{\pm\varepsilon}(1-n_{-\varepsilon_1})(1-n_{-\varepsilon_2})(1-n_{-\varepsilon_3})]\} + \\ & + 2M_1^3\{(1-n_{\pm\varepsilon})\,n_{\varepsilon_1}n_{-\varepsilon_2}n_{\varepsilon_3} - n_{\pm\varepsilon}(1-n_{\varepsilon_1})(1-n_{-\varepsilon_2})(1-n_{\varepsilon_3})\} + \\ & + 2M_1^4\{(1-n_{\pm\varepsilon})\,n_{-\varepsilon_1}n_{-\varepsilon_2}n_{\varepsilon_3} - n_{\pm\varepsilon}(1-n_{-\varepsilon_1})(1-n_{-\varepsilon_2})(1-n_{\varepsilon_3})\}, \end{aligned} \quad (3.13)$$

$$\begin{aligned} E_2 = {} & M_2^1\{[n_{\varepsilon_2}n_{\varepsilon_3}(1-n_{\pm\varepsilon})(1-n_{-\varepsilon_1}) - (1-n_{\varepsilon_2})(1-n_{\varepsilon_3})\,n_{\pm\varepsilon}n_{\varepsilon_1}] + \\ & + [n_{-\varepsilon_2}n_{-\varepsilon}(1-n_{\pm\varepsilon})(1-n_{\varepsilon_1}) - (1-n_{-\varepsilon_2})(1-n_{-\varepsilon_3})\,n_{\pm\varepsilon}n_{\varepsilon_1}]\} + \\ & + M_2^2\{[n_{\varepsilon_2}n_{\varepsilon_3}(1-n_{\pm\varepsilon})(1-n_{-\varepsilon_1}) - (1-n_{\varepsilon_2})(1-n_{\varepsilon_3})\,n_{\pm\varepsilon}n_{-\varepsilon_1}] + \\ & + [n_{-\varepsilon_2}n_{-\varepsilon_3}(1-n_{\pm\varepsilon})(1-n_{-\varepsilon_1}) - (1-n_{-\varepsilon_2})(1-n_{-\varepsilon_3})\,n_{\pm\varepsilon}n_{-\varepsilon_1}]\} + \\ & + 2M_2^3\{n_{-\varepsilon_2}n_{\varepsilon_3}(1-n_{\pm\varepsilon})(1-n_{\varepsilon_1}) - (1-n_{-\varepsilon_2})(1-n_{\varepsilon_3})\,n_{\pm\varepsilon}n_{\varepsilon_1}\} + \\ & + 2M_2^4\{n_{-\varepsilon_2}n_{\varepsilon_3}(1-n_{\pm\varepsilon})(1-n_{-\varepsilon_1}) - (1-n_{-\varepsilon_2})(1-n_{\varepsilon_3})\,n_{\pm\varepsilon}n_{-\varepsilon_1}\}, \end{aligned} \quad (3.14)$$

$$\begin{aligned} E_3 = {} & M_3^1\{[n_{\varepsilon_1}(1-n_{\pm\varepsilon})(1-n_{\varepsilon_2})(1-n_{\varepsilon_3}) - (1-n_{\varepsilon_1})\,n_{\pm\varepsilon}n_{\varepsilon_2}n_{\varepsilon_3}] + \\ & + [n_{\varepsilon_1}(1-n_{\pm\varepsilon})(1-n_{-\varepsilon_2})(1-n_{-\varepsilon_3}) - (1-n_{\varepsilon_1})\,n_{\pm\varepsilon}n_{-\varepsilon_2}n_{-\varepsilon_3}]\} + \\ & + M_3^2\{[n_{-\varepsilon_1}(1-n_{\pm\varepsilon})(1-n_{\varepsilon_2})(1-n_{\varepsilon_3}) - (1-n_{-\varepsilon_1})\,n_{\pm\varepsilon}n_{\varepsilon_2}n_{\varepsilon_3}] + \\ & + [n_{-\varepsilon_1}(1-n_{\pm\varepsilon})(1-n_{-\varepsilon_2})(1-n_{-\varepsilon_3}) - (1-n_{-\varepsilon_1})\,n_{\pm\varepsilon}n_{-\varepsilon_2}n_{-\varepsilon_3}]\} + \\ & + 2M_3^3\{n_{\varepsilon_1}(1-n_{\pm\varepsilon})(1-n_{-\varepsilon_2})(1-n_{\varepsilon_3}) - (1-n_{\varepsilon_1})\,n_{\pm\varepsilon}n_{-\varepsilon_2}n_{\varepsilon_3}\} + \\ & + 2M_3^4\{n_{-\varepsilon_1}(1-n_{\pm\varepsilon})(1-n_{-\varepsilon_2})(1-n_{\varepsilon_3}) - (1-n_{-\varepsilon_1})\,n_{\pm\varepsilon}n_{-\varepsilon_2}n_{\varepsilon_3}\}. \end{aligned} \quad (3.15)$$

The coefficients M_j^i figuring in (3.13)-(3.15) are given by the following relations:

$$\begin{aligned} M_1^1 &= a\,(uu_1u_2u_3 - vv_1v_2v_3 - uu_1 \pm u_2u_3 \mp 1) + \\ &\quad + b\,(uv_1v_2u_3 - vu_1u_2v_3 + vv_1 \pm v_2v_3), \\ M_1^2 &= a\,(uu_1u_2u_3 - vv_1v_2v_3 - uu_1 \mp u_2u_3 \pm 1) + \\ &\quad + b\,(uv_1v_2u_3 - vu_1u_2v_3 + vv_1 \mp v_2v_3), \\ M_1^3 &= a\,(uu_1u_2u_3 - vv_1v_2v_3 + uu_1 \pm u_2u_3 \pm 1) + \\ &\quad + b\,(uv_1v_2u_3 - vu_1u_2v_3 - vv_1 \pm v_2v_3), \\ M_1^4 &= a\,(uu_1u_2u_3 - vv_1v_2v_3 + uu_1 \mp u_2u_3 \mp 1) + \\ &\quad + b\,(uv_1v_2u_3 - vu_1u_2v_3 - vv_1 \mp v_2v_3). \end{aligned} \quad (3.16)$$

The quantities M_2^i and M_3^i are equal to

$$M_2^i = -M_1^i(-\varepsilon_1); \quad M_3^i = M_1^i(-\varepsilon_2, -\varepsilon_3). \tag{3.17}$$

The factors a and b entering into (3.16), (3.17) are numbers (in the order of unity) and are related to A (3.7) and B (3.8) by relations of the type

$$a = -2\pi\left(\frac{mp_F}{2\pi^2}\right)^2 \iiint \frac{dO_{\mathbf{p}}\, dO_{\mathbf{p}_1}\, dO_{\mathbf{p}_2}}{(4\pi)^2}\, \delta\left(\frac{|\mathbf{p}-\mathbf{p}_1-\mathbf{p}_2|}{p_F} - 1\right) A. \tag{3.18}$$

The meaning of the elementary acts described by the collision operator (3.12), (3.13)-(3.18) is quite transparent. We will consider, for example, the component in E_1 (3.13) proportional to M_1^1. With a positive sign of ε the first term in this component describes the merger of the three electron excitations into a single electron excitation. With a negative sign of ε the three electron excitations in merging form an excitation on the hole branch. As a result in the first case the difference in the number of electrons and holes changes by 2, while in the second case it changes by 4. Analogous processes are described by other terms in this component. This is why it vanishes in the case of a normal metal ($M_1^1 = 0$ when $u_i = 1$, $v_i = 0$).

We note that effective collision integral (1.31)-(1.33) has such a simple meaning only by virtue of the selection of the form of the functions $n_{\pm\varepsilon}$ in the expressions for $g^{(R,A)}$ (1.27)-(1.30). This selection also canonizes the effective collision integral in the case of inelastic electron-phonon collisions.

3. Inelastic phonon collisions. In this section we will focus on situations where the nonequilibrium phonons emitted by the electron subsystem have no reverse influence on the electrons that have escaped the superconductor (related to the external thermostat). The phonon subsystem thereby plays the role of thermostat and may be characterized by an equilibrium (Bose) distribution function N_ω with temperature T. With respect to this case the self-energies were obtained in the study by Eliashberg [1] and take the form

$$\begin{aligned}
2i\gamma_{\varepsilon\varepsilon-\omega} &= \int_{-\infty}^{\infty} \frac{d\varepsilon'}{4\pi i} \int \frac{dO_{\mathbf{p}'}}{4\pi}\left[\operatorname{cth}\frac{\varepsilon'-\varepsilon}{2T}(g^R - g^A) - g\right](D^R - D^A)_{\varepsilon'-\varepsilon},\\
2i\delta_{\varepsilon\varepsilon-\omega} &= \int_{-\infty}^{\infty} \frac{d\varepsilon'}{4\pi i} \int \frac{dO_{\mathbf{p}'}}{4\pi}\left[\operatorname{cth}\frac{\varepsilon'-\varepsilon}{2T}(f^R - f^A) - f\right](D^R - D^A)_{\varepsilon'-\varepsilon},\\
\Sigma_{1\varepsilon\varepsilon-\omega} &= \int_{-\infty}^{\infty} \frac{d\varepsilon'}{4\pi i} \int \frac{dO_{\mathbf{p}'}}{4\pi}\left[\operatorname{cth}\frac{\varepsilon'-\varepsilon}{2T} g - (g^R - g^A)\right](D^R - D^A)_{\varepsilon'-\varepsilon},\\
\Sigma_{2\varepsilon\varepsilon-\omega} &= \int_{-\infty}^{\infty} \frac{d\varepsilon'}{4\pi i} \int \frac{dO_{\mathbf{p}'}}{4\pi}\left[\operatorname{cth}\frac{\varepsilon'-\varepsilon}{2T} f - (f^R - f^A)\right](D^R - D^A)_{\varepsilon'-\varepsilon},
\end{aligned} \tag{3.19}$$

where $D^{R(A)}$ is the retarded (advanced) Green's function of the equilibrium phonons ($P \equiv \{\varepsilon, \mathbf{p}\}$):

$$D^{R(A)}(P-P')=\lambda \frac{2\omega^2_{\mathbf{p}'-\mathbf{p}}}{\omega^2_{\mathbf{p}'-\mathbf{p}}-(\varepsilon'-\varepsilon \underset{(-)}{+} i\delta)^2} \tag{3.20}$$

(here λ is the dimensionless electron-phonon interaction constant). Now using relations (1.31), (1.32), (3.19), (3.20), (1.15), (1.16), (1.27)-(1.30), we arrive at the following form of the electron-phonon collision integral:

$$J^{(e-ph)}(n_{\pm\varepsilon})=\frac{\pi\lambda}{4(up_F)^2}\int_0^\infty \omega^2 d\omega \int_\Delta^\infty d\varepsilon' [p_1\delta(\varepsilon'-\varepsilon-\omega)+p_2\delta(\varepsilon-\varepsilon'-\omega)+ + p_3\delta(\varepsilon+\varepsilon'-\omega)], \tag{3.21}$$

where the multipliers $p_{1,2,3}$ are determined as:

$$p_1=(u_\varepsilon u_{\varepsilon'}-v_\varepsilon v_{\varepsilon'}\pm 1)[n_{\varepsilon'}(1-n_{\pm\varepsilon})(1+N_\omega)-n_{\pm\varepsilon}(1-n_{\varepsilon'})N_\omega]+ + (u_\varepsilon u_{\varepsilon'}-v_\varepsilon v_{\varepsilon'}\mp 1)[n_{-\varepsilon'}(1-n_{\pm\varepsilon})(1+N_\omega)-n_{\pm\varepsilon}(1-n_{-\varepsilon'})N_\omega], \tag{3.22}$$

$$p_2=(u_\varepsilon u_{\varepsilon'}-v_\varepsilon v_{\varepsilon'}\pm 1)[n_{\varepsilon'}(1-n_{\pm\varepsilon})N_\omega-n_{\pm\varepsilon}(1-n_{\varepsilon'})(1+N_\omega)]+ + (u_\varepsilon u_{\varepsilon'}-v_\varepsilon v_{\varepsilon'}\mp 1)[n_{-\varepsilon'}(1-n_{\pm\varepsilon})N_\omega-n_{\pm\varepsilon}(1-n_{-\varepsilon'})(1+N_\omega)], \tag{3.23}$$

$$p_3=(u_\varepsilon u_{\varepsilon'}+v_\varepsilon v_{\varepsilon'}\mp 1)[(1-n_{\pm\varepsilon})(1-n_{\varepsilon'})N_\omega-n_{\pm\varepsilon}n_{\varepsilon'}(1+N_\omega)]+ + (u_\varepsilon u_{\varepsilon'}+v_\varepsilon v_{\varepsilon'}\pm 1)[(1-n_{\pm\varepsilon})(1-n_{-\varepsilon'})N_\omega-n_{\pm\varepsilon}n_{-\varepsilon'}(1+N_\omega)]. \tag{3.24}$$

As a result of the relaxation processes, as indicated by these expressions, the excitations may jump from one state to another not only on the same branch but between branches: this process (and its opposite) is described by the second component in expression (3.22) for p_1 with a positive sign of the argument n_ε. As a result of this elementary act the difference in the number of excitations changes by 2.

In some sense the reverse equation situation is described by the term p_3. Thus when two excitations on the same branch are merged (the first component with a positive sign of the argument n_ε) the number of such excitations drops by 2, at the same time that in the recombination of two excitations from different branches there will be no difference in the number of excitations.

We note that in a normal metal (u_ε = 1, v_ε = 0) these terms vanish: there is no homogeneous imbalance relaxation mechanism. As first noted by Tinkham [4] this causes long imbalance relaxation times in superconductors near T_c.

We note that the expression for electron-phonon collision integral (3.21) with respect to the situation with an imbalance was derived previously in the study by Bulyzhenkov and Ivlev [7] in the particle representation.

4. Self-consistency equation. It remains to be noted that in the kinetic equation

$$u_\varepsilon \dot{n}_{\pm\varepsilon} = Q(n_{\pm\varepsilon}) + J^{(e-h)}(n_{\pm\varepsilon}) + J^{(e-e)}(n_{\pm\varepsilon}) \tag{3.25}$$

when the electron subsystem is far from equilibrium the order parameter Δ (nonequilibrium gap) will be self-consistent. The self-consistent value of Δ is determined from the equation

$$1 = \lambda \int_{\Delta}^{\infty} \frac{1 - n_\varepsilon - n_{-\varepsilon}}{\sqrt{\varepsilon^2 - \Delta^2}} d\varepsilon, \tag{3.26}$$

which will be examined in addition to kinetic equation (3.25).

Generally speaking this system of equations should also be supplemented by Maxwell's equations which then dictate the selection of the various solutions of kinetic equations (3.25), (3.26). In this regard we remember that the current through the system is determined by relation (2.11) which is also dependent on the form of the function $n_{\Pi\varepsilon}$ and the value of Δ.

We therefore now have the necessary expressions that allow us to consider nonequilibrium states in the tunnel junctions of superconductors. Certain such issues will be examined in subsequent chapters.

Chapter 2

ELECTRON-HOLE EXCITATIONS IN NONEQUILIBRIUM SUPERCONDUCTOR TUNNEL JUNCTIONS

We will now investigate the kinetics of the electron subsystem of a nonequilibrium tunnel Josephson junction. We will assume that the Josephson junction consists of two sufficiently thin films with a constant difference in potentials V between these films. We will investigate the various possible cases of an applied voltage: both the case of supercritical values ($V > \Delta + \Delta'$) where the applied electrical field generates excess quasi particles from the condensate, and the case of subcritical voltages ($V < \Delta + \Delta'$). In the latter case in addition to the Josephson effect there is also another inherent effect: in the nonequilibrium system of electron excitations, quantum oscillations arise. We will discuss certain manifestations of this effect in detail.

4. Quantum oscillations in the nonequilibrium chemical potential

Probably the most interesting property of nonequilibrium source (2.7) is that it oscillates in time. In this section and subsequent sections we will consider the consequences of this behavior of the tunnel source.

1. The steady-state shift in the chemical potential. As first demonstrated by Clarke's experiments [3] when electrons are injected from a normal metal into a superconducting film (a so-called *NiS*-junction) nonequilibrium appears in the latter and a difference in the potentials occurs between the Cooper condensate and the excitations. Tinkham [4] has discovered that this property of the *NiS*-junction was due to the appearance of an imbalance in the populations of the electron-hole branches of the excitations in the superconducting film.

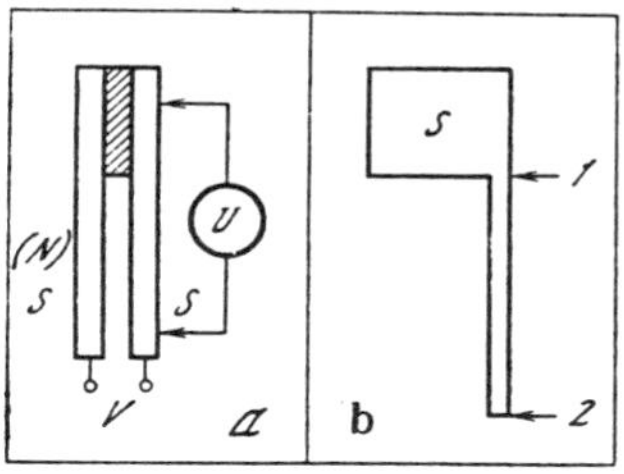

Fig. 1. The test Josephson junction
a - the connection configuration for the external circuit (*V*) and the voltmeter (*U*) for measuring the nonequilibrium shift in the electrochemical potential; b - it is assumed that in the projection the junction package has the shape shown in the figure, and the distance between 1 and 2 exceeds the decay length l_E of the electrical field, while the volume of the region 1-2 is much less than the volume of region *S*, so diffusion of quasi particles in region 1-2 only insignificantly distorts the homogeneous pattern in the primary region *S*.

With regard to the junction phenomena, such an imbalance also occurs when single-particle excitations are injected to an *NiN'*-junction, although in the normal metal the excess charge relaxes over lengths in the order of the Thomas-Fermi lengths, so the direct observation of this effect is quite difficult in a normal metal. In superconductors the electroneutrality condition does not cause a rapid decay in the chemical potential shift, since there is an additional degree of freedom related to the Cooper pairs which removes most of the excess charge. As a result the shift in the chemical potential relaxes over macroscopic lengths of l_E (in pure superconductors $l_E \sim$ 0.1 cm), and the effect may be registered experimentally (Fig. 1). Moreover, it is possible to formulate spatially-homogeneous problems for films whose thickness is small compared to l_E.

2. The properties of the tunnel source. We will now consider the tunnel junction *Sis'* consisting of sufficiently thin films so that the phonon subsystem may be considered an equilibrium subsystem while the pattern in each of the films is homogeneous in thickness (compared to study [7]). We will assume that the junction is connected to a thermostat having a temperature *T*. We will consider the solutions of equation (3.25) assuming a slight deviation in the distribution function $n_{\pm\varepsilon}$ from the equilibrium function (Fermi function) $n^0_\varepsilon(T)$. First, however, we will examine certain properties of the expression $Q(n_{\pm\varepsilon})$ in (3.25).

a) Normally typical values of ν (2.15) that determine the injection intensity are significantly less than the intensity of relaxation processes (which may be roughly characterized by the damping $\gamma \sim T^3_c/\omega^2_d$, where T_c is the transition temperature and ω_D is the Debye frequency). In this limiting case $\nu/\gamma \ll 1$ in (2.7) we may assume that the distribution functions are in equilibrium: $n_{\pm\varepsilon} \to n_\varepsilon^0 = \frac{1}{2}\left(1 - \text{th}\frac{|\varepsilon|}{2T}\right)$. Also assuming that $\Delta' = 0$ in (2.8)-(2.10) we obtain the familiar expression [4, 14] for the tunnel source of an NiS-junction. A remarkable property of this expression is that $Q(\varepsilon) \neq Q(-\varepsilon)$: this also, in the final analysis, causes a branch imbalance in excitation injection.

b) We may easily see that the property $Q(\varepsilon) \neq Q(-\varepsilon)$ of the tunnel source is also conserved in the SiS'-case (even for a symmetrical SiS-junction). Moreover this property does not belong to the source obtained in study [5] and figuring in a number of studies (see survey [8]). We may arrive at expression [5] if we drop in (2.10) all terms aside from $u_\varepsilon u_{\varepsilon\pm V}$ in the corresponding parentheses as well as the terms $Q_{1,2}$ (2.8), (2.9) and then set $n_\varepsilon = n_{-\varepsilon}$. Such a procedure may be justified only in injection to the range of values of ε directly over the gap when imbalance is insignificant. Here, however, the interesting properties of the tunnel source resulting from macroscopic phase coherence are lost.

c) These features are related to the existence of the time-oscillating components $Q_1 \sin 2Vt$, $q_2 \cos 2Vt$ in (2.7) that have a quantum nature (these vanish when $\hbar \to 0$). Since as a result of the solution of nonlinear problem (3.25) the distribution functions $n_{\pm\varepsilon}$ will have an explicit time dependence, in the general case the entire source (including the "stationary" term Q_3) "vibrates", and the Fourier-spectrum of these oscillations when significantly out of equilibrium is not, generally speaking, exhausted by the first (Josephson) harmonic.

Thus, if electron injection in the case of an NiS-junction results in the appearance of the measured shift in the chemical potential, then in the SiS' case there will not only be a shift in the chemical potential but this shift will oscillate in time [39]. With small values of ν/γ this will appear as oscillations with a small amplitude and the Josephson frequency near the stationary value.

3. The "vibrating" chemical potential. We will now establish certain quantitative characteristics of the predicted effects. We will consider a symmetrical SiS-junction. We will use collision integral (3.21) for the calculations and will assume that the electron-phonon collisions in (3.25) are the dominant relaxation mechanism. Assuming that we are only slightly out of equilibrium, we will represent the difference $\delta n_\varepsilon = n_\varepsilon - n_{-\varepsilon}$ in the form

$$\delta n_\varepsilon = \delta n_\varepsilon^0 + \delta n_\varepsilon^1 \sin\varphi + \delta n_\varepsilon^2 \cos\varphi. \tag{4.1}$$

Carrying out the substitution of (4.1) into equation (3.25) subject to (3.21) and (2.7) we obtain a related system of integral equations for determining $\delta n^{0,1,2}$:

$$0=\nu\frac{V}{2T}\,\mathrm{ch}^{-2}\frac{\varepsilon}{2T}\left[u_{\varepsilon-V}\theta(\varepsilon-V-\Delta)-u_{\varepsilon+V}\theta(\varepsilon+V-\Delta)\right]-$$
$$-\gamma\left[\delta n^0_\varepsilon M(\varepsilon)-\hat{M}(\varepsilon\varepsilon')\,\delta n^0_{\varepsilon}\right],$$
$$-Vu_\varepsilon\delta n^2_\varepsilon=\nu\delta n^0_\varepsilon\left[v_\varepsilon v_{\varepsilon-V}\theta(\Delta-|\varepsilon-V|)\right]-\gamma\left[\delta n^1_\varepsilon M(\varepsilon)-\hat{M}(\varepsilon\varepsilon')\,\delta n^1_{\varepsilon'}\right],$$
$$Vu_\varepsilon\delta n^1_\varepsilon=\nu\delta n^0_\varepsilon\left[v_\varepsilon v_{\varepsilon+V}\theta(\varepsilon+V-\Delta)+v_\varepsilon v_{\varepsilon-V}\theta(\varepsilon-V-\Delta)\right]-$$
$$-\gamma\left[\delta n^2_\varepsilon M(\varepsilon)-\hat{M}(\varepsilon\varepsilon')\,\delta n^2_{\varepsilon'}\right], \qquad (4.2)$$

where the function $M(\varepsilon)$ is

$$M(\varepsilon)=\frac{1}{T_c^3}\left\{\int_\Delta^\infty d\varepsilon'(\varepsilon'-\varepsilon)^2(u_\varepsilon u_{\varepsilon'}-v_\varepsilon v_{\varepsilon'})\left[N_{|\varepsilon'-\varepsilon|}+n^0_\varepsilon\theta(\varepsilon'-\varepsilon)-\right.\right.$$
$$\left.\left.-(1-n^0_\varepsilon)\theta(\varepsilon-\varepsilon')\right]+\int_\Delta^\infty d\varepsilon'(\varepsilon+\varepsilon')^2(u_\varepsilon u_{\varepsilon'}+v_\varepsilon v_{\varepsilon'}\left[N_{\varepsilon+\varepsilon'}+n^0_\varepsilon\right]\right\}, \qquad (4.3)$$

while the operator $\hat{M}(\varepsilon\varepsilon')$ is determined as

$$\hat{M}(\varepsilon\varepsilon')X(\varepsilon')=\frac{1}{T_c^3}\left\{\int_\Delta^\infty d\varepsilon'(\varepsilon'-\varepsilon)^2\left[N_{|\varepsilon'-\varepsilon|}+(1-n^0_\varepsilon)\theta(\varepsilon'-\varepsilon)+\right.\right.$$
$$\left.\left.+n^0_\varepsilon\theta(\varepsilon-\varepsilon')\right]X(\varepsilon')+\int_\Delta^\infty d\varepsilon'(\varepsilon+\varepsilon')^2\left[N_{\varepsilon+\varepsilon'}+n^0_\varepsilon\right]X(\varepsilon')\right\}. \qquad (4.4)$$

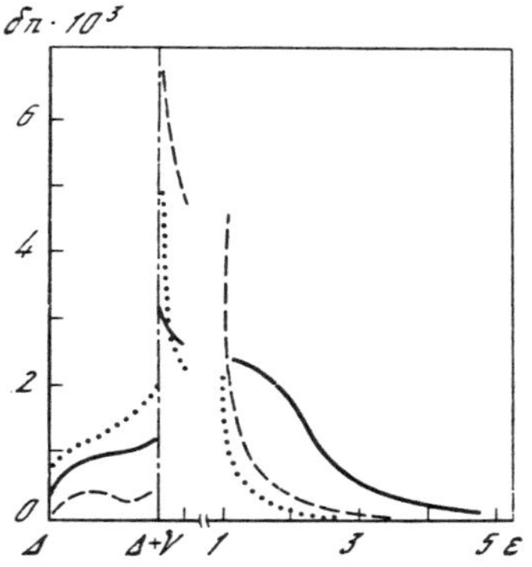

Fig. 2. The steady-state function δn^0 (4.1) (solid line curve in units of ν/γ) and the amplitudes (dotted line curve in units of $(\nu/\gamma)^2$, δn^1, δn^2 (dots)) plotted as a function of ε with injection parameters γ = 0.01; T = 0.9, V = 0.016, Δ = 0.98 (all parameters in units of T_c)

Moreover,

$$\gamma = T_c^3\pi\lambda/2\,(up_F)^2, \qquad (4.5)$$

where $\lambda \sim 1$ is the dimensionless electron-phonon interaction constant; u is the speed of sound.

If we have the results from the numerical solution of equations (4.2) shown in Fig. 2, we may easily calculate the shift of the electrochemical potential that, consistent with (1.23) is equal to

$$\delta\mu=\int_\Delta^\infty \delta n_\varepsilon\,d\varepsilon=\delta\mu^0+\delta\mu^1\sin\varphi+\delta\mu^2\cos\varphi. \qquad (4.6)$$

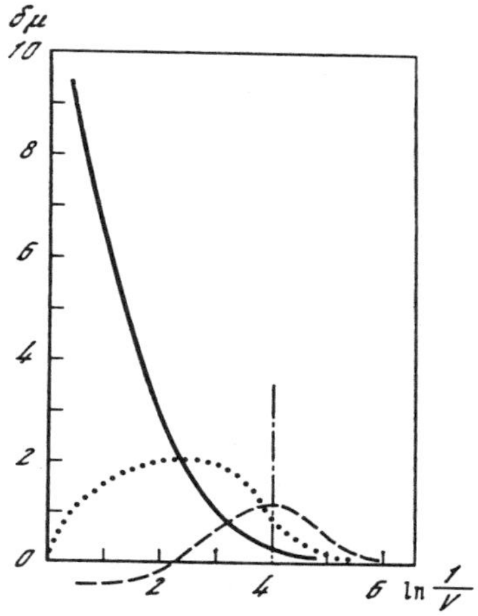

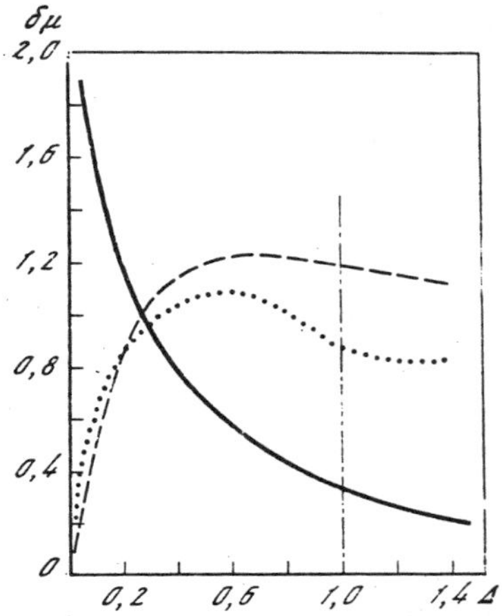

Fig. 3. The steady-state shift $\delta\mu^0$ and amplitudes $\delta\mu^1$ and $\delta\mu^2$ plotted as a function of the voltage V applied to the junction
The solid line curve represents amplitude $\delta\mu^0$ (in units of $T_c\nu$), while the dotted line curve represents amplitude $\delta\mu^1$ (in units of 10 $T_c\nu^2$) and the dots represent amplitude $\delta\mu^2$ (in units of 10 $T_c\nu^2$) with the dot and dash curve corresponding to the situation shown in Fig. 1. Temperature T = 0.9, γ = 0.01 (in units of T_c).

Fig. 4. The steady-state shift and amplitudes plotted as a function of gap size $\delta(T)$
The conventions and scales are the same as in Fig. 3, where V/T_c = 0.016.

Figures 3 and 4 show the behavior of $\delta\mu^1$ as a function of V and T, respectively. The steady-state shift $\delta\mu^0$ in the case of an *SiS*-junction, as indicated from this examination, behaves completely analogously to the case of an *NiS*-junction. With sufficiently small values of V (Fig. 3) the progression of $\delta\mu^{1,2}$ is analogous to $\delta\mu^0$, although with an increase in V oscillations of $\delta\mu$ reach a maximum and then diminish, at the same time that the stationary part of ($\delta\mu^0$) increases without limit (the limitation is the condition $v < 2\Delta$). Another feature of the oscillating contributions to the chemical potential is related to their dependence on the values of $\Delta(T)$ (i.e., on temperature, Fig. 4). If when $T \to T_c$ $\delta\mu^0 \to \infty$ ($\delta\mu^0 \propto \Delta^{-1}$, compare to study [4]), then $\varepsilon\mu^{1,2} \to 0$ when $T \to T_c$, which is related to the difference in the structures $Q^{1,2}$ and Q^3 in (2.7).

We should note that this examination, strictly speaking, is valid when $\nu/\gamma \ll 1$ and, for example, when $\nu/\gamma \sim$ (0.1-0.3) $\delta\mu^{1,2}/\delta\mu^0$ may, as is clear from Figs. 3 and 4, amount to 1-10%. In the most favorable cases the characteristic values of $\delta\mu^0$ may be in the order of a microvolt, then $\delta\mu^{1,2}$ may be fractions of this quantity (tens and hundreds of nanovolts, for, say, low-resistance aluminum junctions).

Then the oscillations in the nonequilibrium electrochemical potential may be easily identified experimentally (here in Fig. 1 the voltmeter must be replaced by an appropriate oscilloscope).

5. Emission scattering by a nonequilibrium Josephson junction

The oscillatory behavior of the shift in the electromechanical potential examined above is an effect of the second order of smallness in the parameter ν/γ at the same time that the shift itself (its average value over time) is a first order effect in ν/γ. The question arises: is it possible for oscillations of the tunnel source to appear in the first approximation in ν/γ? We will now describe such an effect.

1. The equilibrium source of nonequilibrium. We will now focus on kinetic equation (3.25). In the first approximation in the small parameter ν/γ the distribution functions $n_{\pm\varepsilon}$ in (2.8)-(2.10) may be equilibrium functions. As a result we obtain the following expressions for $Q_i(\varepsilon \geqslant \Delta)$:

$$Q_3(\pm\varepsilon)=2\Big\{(u_\varepsilon u_{\varepsilon-V}\pm u_{\varepsilon-V})\Big(\operatorname{th}\frac{\varepsilon}{2T}-\operatorname{th}\frac{\varepsilon-V}{2T}\Big)\theta(\varepsilon-V-\Delta')-$$
$$-(u_\varepsilon u_{\varepsilon+V}\mp u_{\varepsilon+V})\Big(\operatorname{th}\frac{\varepsilon+V}{2T}-\operatorname{th}\frac{\varepsilon}{2T}\Big)\theta(\varepsilon+V-\Delta')\Big\}, \quad (5.1)$$

$$Q_1(\pm\varepsilon)=\Big\{v_\varepsilon w_{\varepsilon+V}\operatorname{th}\frac{\varepsilon}{2T}\,\theta[\Delta'^2-(\varepsilon+V)^2]-$$
$$-v_\varepsilon w_{\varepsilon-V}\operatorname{th}\frac{\varepsilon}{2T}\,\theta[\Delta'^2-(\varepsilon-V)^2]\Big\}, \quad (5.2)$$

$$Q_2(\pm\varepsilon)=2\Big\{v_\varepsilon v_{\varepsilon-V}\Big(\operatorname{th}\frac{\varepsilon-V}{2T}-\operatorname{th}\frac{\varepsilon}{2T}\Big)\theta(\varepsilon-V-\Delta')-$$
$$-v_\varepsilon v_{\varepsilon+V}\Big(\operatorname{th}\frac{\varepsilon}{2T}-\operatorname{th}\frac{\varepsilon+V}{2T}\Big)\theta(\varepsilon+V-\Delta')\Big\}. \quad (5.3)$$

In drafting these expressions we have limited our examination to the case $V < \Delta + \Delta'$ (we remember that the primed quantities and functions with deviating arguments relate to the injector). We may easily see that the functions (5.1)-(5.3) have the properties

$$Q_3(\varepsilon)\neq Q_3(-\varepsilon), \quad (5.4)$$
$$Q_1(\varepsilon)=Q_1(-\varepsilon), \qquad Q_2(\varepsilon)=Q_2(-\varepsilon). \quad (5.5)$$

(It is precisely property (5.5) that causes (see Section 4) the oscillating contribution to the shift in the chemical potential to be of the second order of smallness in ν/γ.) It follows from (5.5) that the oscillating nonequilibrium additions to the distribution function of the electron-hole excitations in the first order in n/γ are symmetrical with respect to ε. We will find these additions.

2. The excitation distribution function. We will be interested in the range of temperatures $T \geqslant \Delta(T)$ when there are sufficient equilibrium electron excitations. The collision integrals $J(n_\varepsilon)$ in (3.25) may then be linearized over the minor deviations of the distribution function. The stationary correction to n^0_ε in this case is determined by the term Q_3 in (5.1) and yields insignificant renormalization of n^0_ε. The asymmetry of Q_3 with respect to ε (property (5.4)) causes an imbalance which we will also ignore in view of its additional smallness (as follows from (5.1)) (and further we assume $\delta\nu_\varepsilon = \delta\nu_{-\varepsilon}$). This means that now we may use an approximation of the relaxation time in (3.25) for the collision integrals

$$J(n_\varepsilon) = -\gamma \frac{\varepsilon}{\sqrt{\varepsilon^2 - \Delta^2}} \delta n_\varepsilon, \tag{5.6}$$

where γ is a certain characteristic energy damping of the nonequilibrium excitations. Substituting (5.1)-(5.3), and (5.6) into (3.25) and (3.27) and carrying out a Fourier-expansion of the nonequilibrium addition

$$\delta n_\varepsilon = \delta n^0_\varepsilon + \delta n^1_\varepsilon \sin 2Vt + \delta n^2_\varepsilon \cos 2Vt + \ldots, \tag{5.7}$$

we obtain in the first order in ν/γ the following values of the amplitudes $\delta\nu^{1,2}_\varepsilon$

$$\delta n^1_\varepsilon = \frac{\nu}{2}\left\{\frac{Q_2(\varepsilon)\,\omega_J + \gamma Q_1(\varepsilon)}{u_\varepsilon(\omega_J^2 + \gamma^2)}\right\}, \tag{5.8}$$

$$\delta n^2_\varepsilon = \frac{\nu}{2}\left\{\frac{Q_2(\varepsilon)\,\gamma - Q_1(\varepsilon)\,\omega_J}{u_\varepsilon(\omega_J^2 + \gamma^2)}\right\}, \tag{5.9}$$

where $\omega_J = 2V$. Using these expressions we may now consider scattering of external electromagnetic emission in a nonequilibrium tunnel junction [10].

3. Satellites and the possibility of their observation. For a strict description of the scattering process we must know the generalized susceptibility [17] of the nonequilibrium junction with respect to the electromagnetic emission action. For this purpose we will use the photon field-electron collisional operator. We may show that this operator for "dirty" superconductors takes the form

$$\begin{aligned}
\hat{\Phi}(N_{\omega_0}) &= N \iint_\Delta^\infty d\varepsilon\, d\varepsilon' \,[2\delta(\varepsilon - \varepsilon' - \omega_0)\, T_1(\varepsilon, \varepsilon') + \delta(\varepsilon + \varepsilon' - \omega_0)\, T_2(\varepsilon, \varepsilon')],\\
T_1(\varepsilon, \varepsilon') &= \{[n_\varepsilon(1 - n_{-\varepsilon'})(1 + N_{\omega_0}) - (1 - n_\varepsilon)\, n_{-\varepsilon'} N_{\omega_0}] +\\
&+ [n_{-\varepsilon}(1 - n_{\varepsilon'})(1 + N_{\omega_0}) - (1 - n_{-\varepsilon})\, n_{\varepsilon'} N_{\omega_0}]\}(u_\varepsilon u_{\varepsilon'} + v_\varepsilon v_{\varepsilon'} - 1) +\\
&+ \{[n_\varepsilon(1 - n_{\varepsilon'})(1 + N_{\omega_0}) - (1 - n_\varepsilon)\, n_{\varepsilon'} N_{\omega_0}] +\\
&+ [n_{-\varepsilon}(1 - n_{-\varepsilon'})(1 + N_{\omega_0}) - (1 - n_{-\varepsilon})\, n_{-\varepsilon'} N_{\omega_0}]\}(u_\varepsilon u_{\varepsilon'} + v_\varepsilon v_{\varepsilon'} + 1),\\
T_2(\varepsilon, \varepsilon') &= \{[n_\varepsilon n_{-\varepsilon'}(1 + N_{\omega_0}) - (1 - n_\varepsilon)(1 - n_{-\varepsilon'})\, N_{\omega_0}] +\\
&+ [n_{-\varepsilon} n_{\varepsilon'}(1 + N_{\omega_0}) - (1 - n_{-\varepsilon})(1 - n_{\varepsilon'})\, N_{\omega_0}]\}(u_\varepsilon u_{\varepsilon'} - v_\varepsilon v_{\varepsilon'} + 1) +\\
&+ \{[n_\varepsilon n_{\varepsilon'}(1 + N_{\omega_0}) - (1 - n_\varepsilon)(1 - n_{\varepsilon'})\, N_{\omega_0}] +\\
&+ [n_{-\varepsilon} n_{-\varepsilon'}(1 + N_{\omega_0}) - (1 - n_{-\varepsilon})(1 - n_{-\varepsilon'})\, N_{\omega_0}]\}(u_\varepsilon u_{\varepsilon'} - v_\varepsilon v_{\varepsilon'} - 1).
\end{aligned} \tag{5.10}$$

We have written operator (5.10) in general form that also accounts for the branch imbalance. We will now be interested in the classical limit of (5.10) when $N_{\omega 0} \gg 1$. In this case factoring $N_{\omega 0}$ out of the integral signs (henceforth we will assume that $N_{\omega 0}$ is included in N and we will not need the explicit form of this factor and it will be dropped), and conserving only terms that correspond to the action of a field with frequency $\omega_0 < 2\Delta$ we obtain

$$\hat{\Phi}(N_{\omega_0}) = \Phi_0 + \Phi_1 \sin \omega_J t + \Phi_2 \cos \omega_J t, \tag{5.11}$$

where the functions Φ_1 determined subject to (5.2), (5.3), (5.7)-(5.9) take the form

$$\Phi_0(\omega_0, T) = \frac{1}{2}\int_\Delta^\infty d\varepsilon \, \frac{\Delta^2 + \varepsilon(\varepsilon + \omega_0)}{\sqrt{\varepsilon^2 - \Delta^2}\sqrt{(\varepsilon + \omega_0)^2 - \Delta^2}}\left(\operatorname{th}\frac{\varepsilon}{2T} - \operatorname{th}\frac{\varepsilon + \omega_0}{2T}\right), \tag{5.12}$$

$$\begin{aligned}
\Phi_{1(2)}(\omega_0, T, V, \gamma, \Delta', \nu) &= \Delta\Delta' \int_\Delta^\infty d\varepsilon \, \frac{\varepsilon(\varepsilon + \omega_0) + \Delta^2}{\sqrt{\varepsilon^2 - \Delta^2}\sqrt{(\varepsilon + \omega_0)^2 - \Delta^2}} \times \\
&\times \Bigg\{ a_{1(2)} \Bigg[\Bigg(\frac{\theta(\varepsilon + \omega_0 - V - \Delta')}{(\varepsilon + \omega_0)\sqrt{(\varepsilon - V)^2 - \Delta'^2}} \left(\operatorname{th}\frac{\varepsilon + \omega_0 - V}{2T} - \operatorname{th}\frac{\varepsilon + \omega_0}{2T}\right) - \\
&- \frac{\theta(\varepsilon + \omega_0 - V - \Delta')}{(\varepsilon + \omega_0)\sqrt{(\varepsilon + \omega_0 + V)^2 - \Delta'^2}} \left(\operatorname{th}\frac{\varepsilon + \omega_0}{2T} - \operatorname{th}\frac{\varepsilon + \omega_0 + V}{2T}\right)\Bigg) - \\
&- \left(\frac{\theta(\varepsilon - V - \Delta')}{\varepsilon\sqrt{(\varepsilon - V)^2 - \Delta'^2}}\left(\operatorname{th}\frac{\varepsilon - V}{2T} - \operatorname{th}\frac{\varepsilon}{2T}\right) - \right. \\
&\left. - \frac{\theta(\varepsilon + V - \Delta')}{\varepsilon\sqrt{(\varepsilon + V)^2 - \Delta'^2}}\left(\operatorname{th}\frac{\varepsilon}{2T} - \operatorname{th}\frac{\varepsilon + V}{2T}\right)\right)\Bigg] \overset{+}{(-)} a_{2(1)} \Bigg[\frac{1}{2(\varepsilon + \omega_0)} \operatorname{th}\frac{\varepsilon + \omega_0}{2T} \times \\
&\times \left(\frac{\theta[\Delta'^2 - (\varepsilon + \omega_0 + V)^2]}{\sqrt{\Delta'^2 - (\varepsilon + \omega_0 + V)^2}} - \frac{\theta[\Delta'^2 - (\varepsilon + \omega_0 - V)^2]}{\sqrt{\Delta'^2 - (\varepsilon + \omega_0 - V)^2}}\right) - \\
&- \frac{1}{2\varepsilon}\operatorname{th}\frac{\varepsilon}{2T}\left(\frac{\theta(\Delta'^2 - (\varepsilon + V)^2]}{\sqrt{\Delta'^2 - (\varepsilon + V)^2}} - \frac{\theta[\Delta'^2 - (\varepsilon - V)^2]}{\sqrt{\Delta'^2 - (\varepsilon - V)^2}}\right)\Bigg]\Bigg\}.
\end{aligned} \tag{5.13}$$

The constants a_1 and a_2 are equal to:

$$a_1 = \frac{\nu\omega_J}{\omega_J^2 + \gamma^2}, \qquad a_2 = \frac{\nu\gamma}{\omega_J^2 + \gamma^2}. \tag{5.14}$$

Before providing results from the numerical analysis of these integrals, we will briefly comment on the physical interpretation.

According to (5.7)-(5.9) the density of electron excitations in the thin superconductor film oscillates with a small amplitude near its steady-state value by virtue of the existence of macroscopic phase coherence in the system of superconductor excitations. The interaction between the external electromagnetic emission and the oscillating density of electron states causes scattering with formation of satellites at frequencies $\omega_0 \pm \omega_J$.

Collisional operator (5.10) allows us to calculate the response of the nonequilibrium system, such as the relative intensity of satellites in the reflected emission. From the viewpoint of spectroscopic experiments, however, it is easier to employ the absorption coefficient of the system (actually involving the imaginary part of the dielectric constant) or, in other words, the characteristics of a wave that has passed through the film.

We will assume that a test wave passes through one of the films of an equilibrium superconducting junction. Then the absorption coefficient with multiplier accuracy is given by formula (5.12). We will assume that oscillations of the type noted above occur in the film. Then the absorption coefficient also oscillates in accordance with (5.11). As a result the amplitude of the transit wave is modeled (we assume that $2\Delta > \omega_0 > \omega_J$)and satellites arise whose relative intensity may be estimated as

$$P \sim \sqrt{\Phi_1^2 + \Phi_2^2}/2\Phi_0. \qquad (5.15)$$

The existence of a second film in the junction in which oscillations also occur somewhat distorts this simple pattern. We will return to this issue somewhat later.

We will now examine results from a numerical analysis of the behavior of the function $P = P(\omega_0, T, V, \Delta, \Delta', \gamma, \nu)$. Calculations were carried out for three metals: lead, niobium, and tantalum, using the parameters outlined in the table.

Select superconductor parameters
(see, specifically, studies [19, 20])

Metal	T_c, K	$N(0)\cdot 10^{-33}$, erg^{-1} cm^{-3}	ω_D, K	ε_F, eV	λ
Pb	7,2	5,4	105	10,9	1,55
Nb	9,2	19,9	275	6,2	0,80
Ta	4,5	25,4	240	18,0	0,70
Al	1,2	8,1	428	13,5	0,38

The growth of P with a drop in temperature (Fig. 5) is related to the reduction in the number of equilibrium excitations and the effective "bleaching" of the film (for convenience Fig. 5 shows the $\Delta(T)$ relation in the BCS model). We may determine this by direct analysis of the behavior of the Φ_i amplitudes (the dotted line curves in Fig. 5).

The existence of a maximum (logarithmic) near $V \sim \gamma$ (Fig. 6) is characteristic of the dependence of P on the voltage V across the junction. These voltages are optimum for detection of satellites.

The transformation coefficient P is independent of the frequency ω_0 of the test emission (Fig. 7) at sufficiently high frequencies ($\omega_0 \gg \Delta_0$), although at lower frequencies it increases sharply, which in the final analysis is due to the feature in the density of the electron levels characteristic of superconductors (in the calculations this feature was lost at energies $\varepsilon \sim \gamma$). As indicated from this analysis the intensity of the satellites increases with a drop in γ (Fig. 8).

The dependence of P on the gap of the injector Δ' is curious (Fig. 9). When $\Delta' < \Delta$ by 1%, P drops by an order of magnitude. At the same time if $\Delta' > \Delta$ by 1%, P increases noticeably and diminishes with a further increase in Δ'. This fact may be useful in the recording of satellites. Indeed, as noted above, estimate (5.15), generally speaking, is not valid for the real Josephson junction shown in Fig. 8, but rather is obtained for the response of a single film only. However, in asymmetrical junctions oscillations in the electron excitation density are manifest only in a film with a small gap. (The difference between Δ and Δ' must be quite small, ~1%; such a differential ($\sim\gamma/\Delta$) may exist for films of identical metals due to the difference in deposition conditions, thicknesses, etc.) In these cases estimate (5.15) may be used directly.

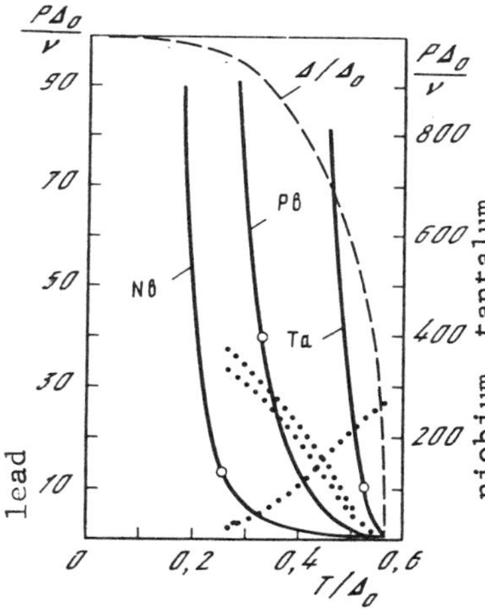

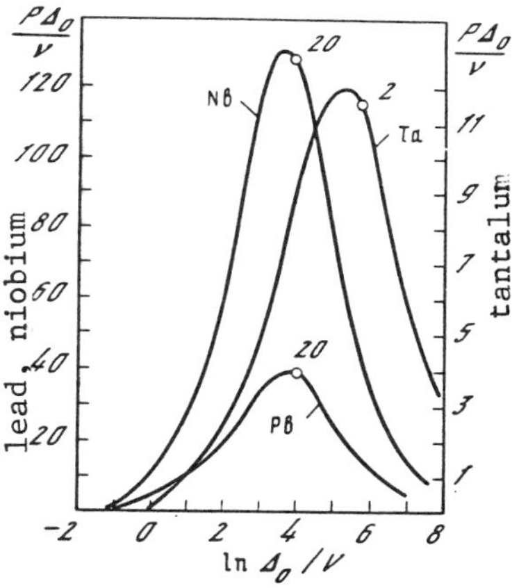

Fig. 5. The transformation coefficients P for Pb, Nb, and Ta plotted as a function of temperature (with $\omega_0 \sim 300$ GHz for Pb and Nb and $\omega_0 \sim 50$ GHz for Ta, $T = 4.2$ K, $\Delta = \Delta'$)
The dotted line curve shows the $\Delta(T)$ relation in the BCS model; $\Delta_0 = \Delta(T = 0)$. The behavior of the amplitudes $|\Phi_i|$ (in arbitrary units for the case of Pb) is given by the dotted line curves while the diminishing curves (upper and lower, respectively) represent Φ_1 and Φ_2. The circles indicate the temperature of the helium bath. (4.2 K)

ig. 6. P plotted as a function of the voltage V across the junction ($\omega_0 \sim 300$ GHz for Pb and Nb and $\omega_0 \sim 500$ GHz for Ta; $T = 4.2$ K, $\Delta = \Delta'$)
he circles indicate values $V \sim \gamma$ in μv

For a lead film with $d \sim 10^3$ Å and $RS \sim 10^{-4}$ ohm·cm^2 we have $/\Delta_0 \sim 2.6\cdot10^{-5}$ and $P \sim 10^{-3}$ with $T = 4.2$ K. Quantities of the same

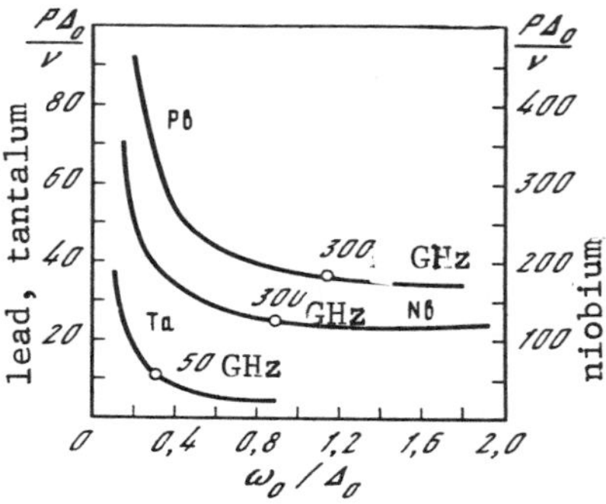

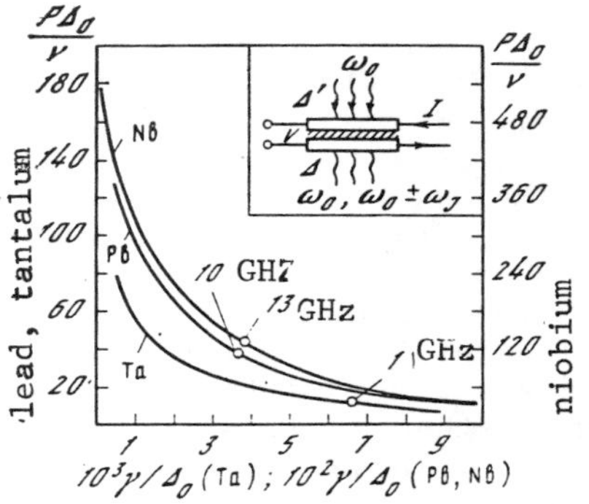

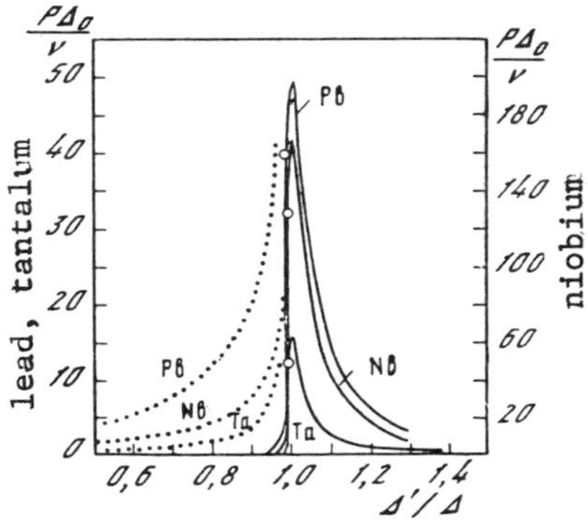

Fig. 7. P plotted as a function ω of the test emission (T = 4.2 K, = Δ', $V \sim \gamma$)

Fig. 8. P plotted as a function o the energy damping of electron ex citations γ
The circles give the tabular value of γ. The insert shows the Joseph son junction in the test situation

Fig. 9. The transformation coeffi cient as a function of the para meter Δ'/Δ
For Pb and Nb ω_0 ~ 300 GHz, $V \sim 20$ μv; for Ta ω_0 ~ 50 GHz, $V \sim 1$ μv; = 4.2 K. The circles show the case $\Delta' = \Delta$. When $\Delta' < \Delta$ the dots re present the values of $P\Delta_0/\nu$ on a 100% greater scale

order are obtained for niobium and tantalum. These quantities ma easily be registered spectroscopically. For metals with low value of T_c (and large values of ω_D such as aluminum) ν/γ_0 is large (~10^{-4}) while γ is small ($\nu/\gamma \sim 0.1$) and hence $P \sim 0.1$. In such junction the nonequilibrium level is quite high and other interesting effect may also be observed (for example, satellites with even-multiple fre quencies, etc.). We will not discuss this issue here.

We will also note the following. In principle the manifestatio of satellites in an electromagnetic wave scattered by the Josephso junction may be explained within the scope of the electrodynamic ap proach by accounting for the interference of the a.c. induced by th external field with the Josephson current. Clearly in this case th specific functional relations (similar to those shown in Figs. 5-9 will differ. It is also important to emphasize the fundamental dif ference between these effects (kinetic and electrodynamic). Th "kinetic" effect is conserved if we consider scattering at the junc tion of acoustic emission rather than electromagnetic emission (ω_0 ω_q). Here, scattering of a high frequency acoustic wave will als generate satellites with frequencies $\omega_q \pm \omega_D$. The "electrodynamic" effect, however, does not always exist in the scattering of acoustical waves. Hence it is clear that the satellite formation effect we have described here resulting from oscillations in excitation density may in principle be identified and unambiguously interpreted. Many studies in the field of Josephson junctions [see, for example, [18] makes it possible to hope that the issue of the microscopic behavior of tunnel junctions will be explained experimentally in the foreseeable future.

6. The critical state in a symmetrical SiS-junction

We will now proceed to the case of the voltages

$$V \sim \Delta + \Delta' \tag{6.1}$$

Across virtually the entire superconductivity temperature range with the exception of a very small vicinity near the transition point, the scale of the inverse frequencies corresponding to (6.1) is small compared to the times characterizing the kinetics of the single-frequency excitations. Hence with the voltages in (6.1) the effects associated with the coherent phase difference of the superconductors in the excitation system are negligible (the results given in the preceding two sections indicate this fact to some degree).

In spite of the reduction in the role of quantum oscillations, the range of voltages in (6.1) is interesting by virtue of the fact that the applied electrical field is capable of pair breaking in tunneling, and this breaking process manifests a resonant nature. As a result the degree of nonequilibrium may be quite high.

1. The solution method for the self-consistent equations. In the general case we must consider (3.25), (3.26) in addition to analogous equations for the injector and hence the number of coupled equations doubles, since the quantity from (2.10) entering into (2.7) (only its contribution is accounted for as discussed above) contains the distribution functions of both superconductors. In these two limiting cases - symmetrical and highly asymmetrical junctions - the situation is simplified. In this chapter we will consider the case of a symmetrical *SiS* junction. Then by virtue of symmetry the following identities hold:

$$n_\varepsilon = n'_{-\varepsilon}, \quad n_{-\varepsilon} = n'_\varepsilon. \tag{6.2}$$

Accounting for (6.2), we represent (3.25) as

$$n_\varepsilon a_{11}(\varepsilon) + n_{V-\varepsilon} a_{13}(\varepsilon) + n_{-V+\varepsilon} a_{14}(\varepsilon) = c_1(\varepsilon), \tag{6.3}$$

$$n_{-\varepsilon} a_{22}(\varepsilon) + n_{V-\varepsilon} a_{23}(\varepsilon) + n_{-V+\varepsilon} a_{24}(\varepsilon) = c_2(\varepsilon), \tag{6.4}$$

where all distribution functions refer to the test superconductor. We will not provide explicit expressions for the quantities a_{ik} and j for space considerations. These expressions are obvious from a comparison of (6.3), (6.4) to (3.25).

In representation (6.3), (6.4) we may explicitly identify terms related to resonant pair breaking. This representation is convenient in that it allows us to exactly account for these processes. For this purpose we will carry out a conversion to new arguments in (6.3), (6.4) by transformation as $V - \varepsilon \to \varepsilon$. We then have two additional equations:

$$n_\varepsilon a_{31}(\varepsilon) + n_{-\varepsilon} a_{32}(\varepsilon) + n_{V-\varepsilon} a_{33}(\varepsilon) = c_3(\varepsilon), \tag{6.5}$$

$$n_\varepsilon a_{41}(\varepsilon) + n_{-\varepsilon} a_{42}(\varepsilon) + n_{-V+\varepsilon} a_{44}(\varepsilon) = c_4(\varepsilon) \tag{6.6}$$

with the determinant

$$\det a_{ik} = \begin{vmatrix} a_{11} & 0 & a_{13} & a_{14} \\ 0 & a_{22} & a_{23} & a_{24} \\ a_{31} & a_{32} & a_{33} & 0 \\ a_{41} & a_{42} & 0 & a_{44} \end{vmatrix}, \tag{6.7}$$

that is nonzero in the range

$$\Delta \leqslant \varepsilon \leqslant V - \Delta, \tag{6.8}$$

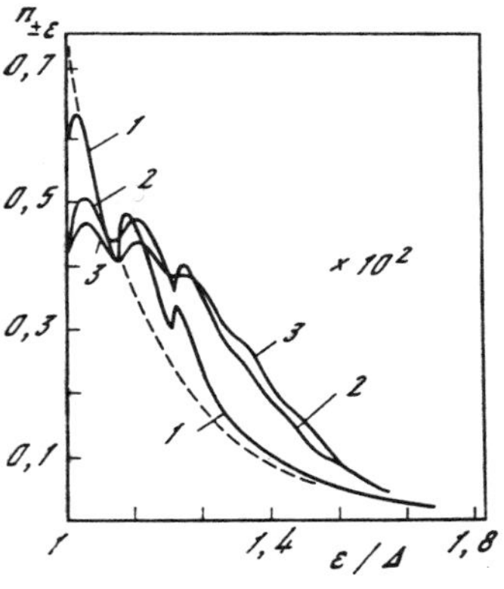

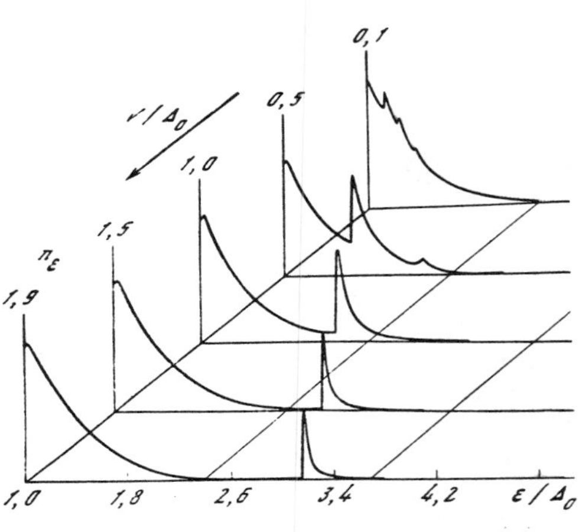

Fig. 10. The distribution function of electron excitations n_ε for various values of the parameter ν
1 - $\nu = 0.01\gamma$; 2 - $\nu = 0.1\gamma$; 3 - $\nu = 0.2\gamma$ (temperature $T = 0.2\Delta_0$; the voltage across the junction $V = 0.1\ \Delta_0$; the dotted line represents the Fermi distribution)

Fig. 11. The evolution of the distribution function of electron excitations n_ε with an increase in the voltage applied across the junction (when $T = 0.3\ \Delta_0$, $\nu = 0.1\gamma$)

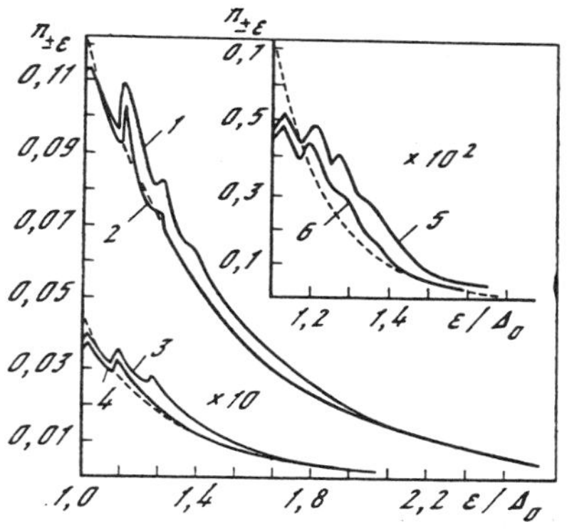

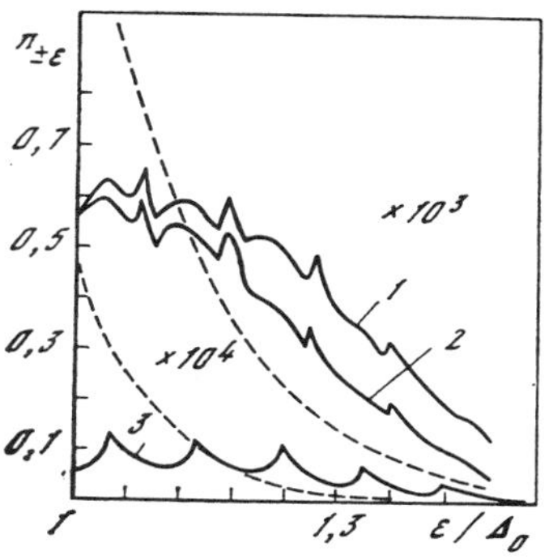

Fig. 12. The distribution functions of electron n_ε (upper curves) and hole $n_{-\varepsilon}$ (lower curves) excitations for various temperatures
1, 2 - at $T = 0.4\ \Delta_0$; 3, 4 - at $T = 0.3\ \Delta_0$; 5, 6 - at $T = 0.2\ \Delta_0$. In all cases $\nu/\gamma = 0.1$; $V/\Delta_0 = 0.1$. The dotted line curve represents the Fermi distribution

Fig. 13. The distribution functions of electron n_ε (upper curves) and hole $n_{-\varepsilon}$ (lower curve) excitations for various temperatures
1, 2 - at $T = 0.15\ \Delta_0$; 3 - at $T = 0.1\ \Delta_0$ (in the last case the behavior of the electron and hole excitation branches is virtually identical)

where subject to (6.2):

$$\begin{aligned} a_{33}(\varepsilon) &= a_{11}(V-\varepsilon), & a_{44}(\varepsilon) &= a_{22}(V-\varepsilon), \\ a_{31}(\varepsilon) &= a_{13}(V-\varepsilon), & a_{41}(\varepsilon) &= a_{23}(V-\varepsilon), \\ a_{32}(\varepsilon) &= a_{14}(V-\varepsilon), & a_{42}(\varepsilon) &= a_{24}(V-\varepsilon), \\ c_3(\varepsilon) &= c_1(V-\varepsilon), & c_4(\varepsilon) &= c_2(V-\varepsilon). \end{aligned} \tag{6.9}$$

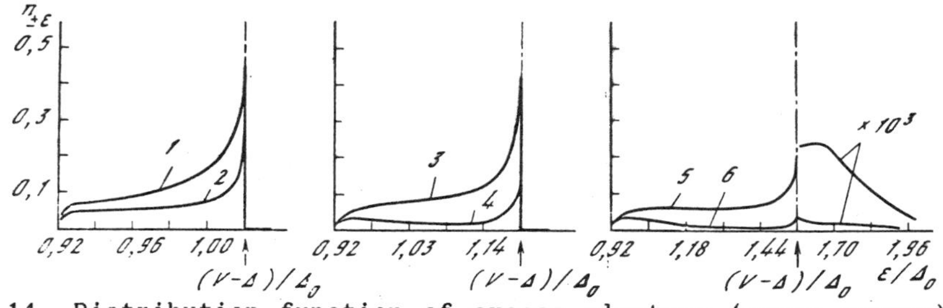

Fig. 14. Distribution function of excess electron (upper curves) and hole (lower curves) excitations in a supercritical state ($T = 0.1\ \Delta_0$, $\nu = 0.01\gamma$)
1, 2 - $V = 9\ \Delta_{1,0}.1$; 3, 4 - $V = 2.01\ \Delta_0$; 5, 6 - $V = 2.5\ \Delta_0$ (in the last case the small "tail" of the distribution is also shown). As the voltage is increased from the critical value ($V = 2\ \Delta \approx 1.84\ \Delta_0$) the degree of imbalance increases. The distribution of the equilibrium excitations is not shown: in these conditions the Fermi function is exponentially small

We will now draft the determinants det 1 and det 2 for a system of four equations in the regular manner, thus writing

$$n_\varepsilon = \frac{\det 1}{\det a_{ik}}, \qquad n_{-\varepsilon} = \frac{\det 2}{\det a_{ik}} \tag{6.10}$$

in the region (6.8). At low temperatures $T \ll \Delta$ the "tail" of the distribution function when $\varepsilon \geqslant V - \Delta$ is small. If we drop this tail as well as the tunnel redistribution of excitations and the relaxation processes, we may arrive at a solution that generalizes the expression derived by Aronov and Spivak [21] (for the case of microwave pumping) to the case of an imbalance. Iterations of the solution reveal that accounting for the "tail" in fact produces small changes in the entire pattern.

The procedure for finding solutions was not entirely exhausted by simple iterations. We remember that the nonequilibrium gap Δ enters into the kinetic equation for $n_{\pm\varepsilon}$; this gap may itself depend on the form of $n_{\pm\varepsilon}$. We may show that when $\Delta \ll \omega_D$ equation (3.25) is equivalent to equation system

$$\bar{n}_\pm = \int_\Delta^\infty \frac{n_{\pm\varepsilon}}{\sqrt{\varepsilon^2 - \Delta^2}}\, d\varepsilon, \qquad \Delta = \Delta_0 \exp(-\bar{n}_+ - \bar{n}_-). \tag{6.11}$$

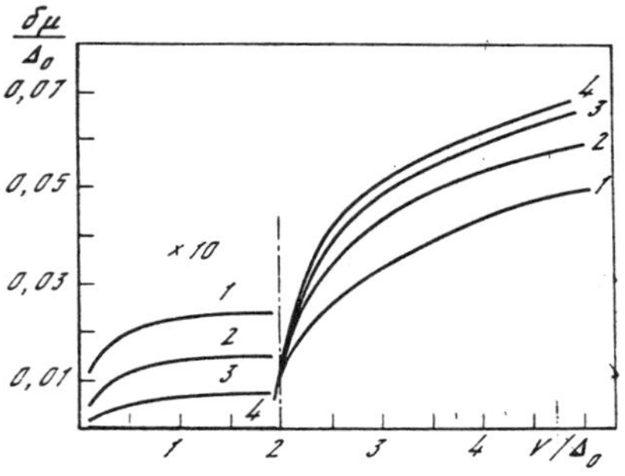

Fig. 15. The nonequilibrium shift in the chemical potential $\delta\mu$ plotted as a function of applied voltage V with $\nu = 0.01\gamma$
1 - $T = 0.4\ \Delta_0$; 3 - $T = 0.2\ \Delta_0$
2 - $T = 0.3\ \Delta_0$; 4 - $T = 0.1\ \Delta_0$

This system was included in the general iteration scheme that also utilizes (6.10) when $V > 2\ \Delta$.

2. The steady-state excitation distribution function. The behavior of the excitation distribution function is illustrated by Figs. 10-14. We will provide brief commentary for these. First we note the imbalance between the electron-like (n_ε) and the hole-like ($n_{-\varepsilon}$) excitations which occurs in both the subcritical ($V < 2\Delta$) and supercritical ($V > 2\Delta$) states.

In the subcritical case, in addition to the generation of "spikes", the distribution function manifests a tendency towards a global shift in the direction of higher energies (Fig. 10) which in the final analysis stimulates superconductivity [41] and creates a phonon deficit (Chapter 3). As the temperature drops the relative degree of nonequilibrium increases (with approximately equal conditions): a large number of peaks separated by distance V in the energy space (compare Figs. 12 and 13) appears. At very low temperatures the distribution function assumes a "sawtooth" shape and is nonzero in the range where the equilibrium Fermi "tail" is negligible. We note that the number of observed spikes increases with a drop in voltage (Fig. 11.)

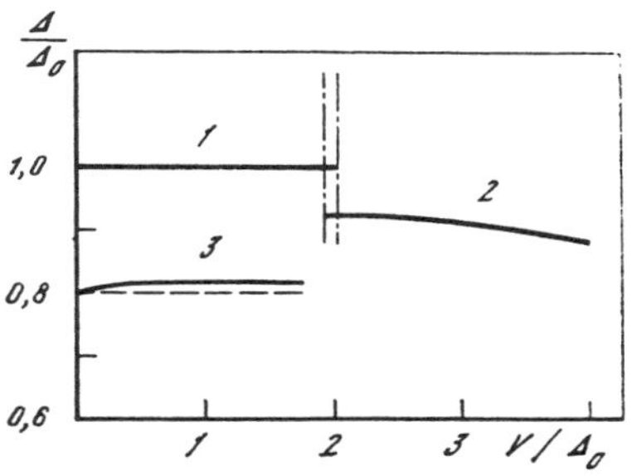

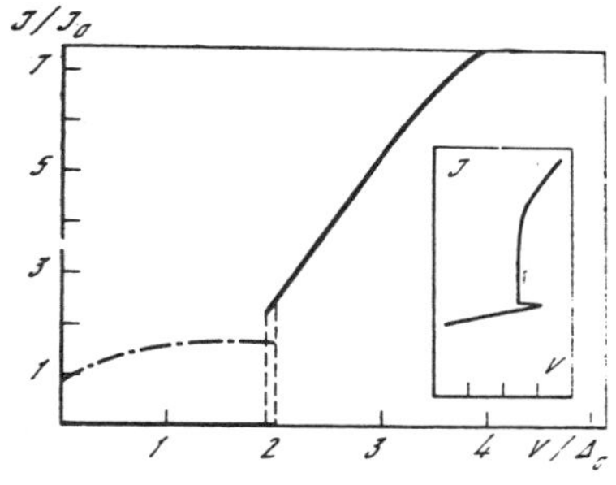

Fig. 16. The nonequilibrium gap Δ plotted as a function of voltage V
1, 2 - at $T/\Delta_0 = 0.1$ (the same is achieved at $T/\Delta_0 = 0.2$), $\nu = 0.01\gamma$; 3 - at $T/\Delta_0 = 0.4$, $\nu = 0.1\gamma$. The dotted lines represent the equilibrium value.

Fig. 17. The E-I characteristic with hysteresis ($T = 0.2\ \Delta_0$; $\nu = 0.01\gamma$)
The insert shows the experimental curve.

In the supercritical state virtually the entire mass of excess excitations is concentrated in the region over the gap and, as demonstrated in Fig. 14, the "tail" adjacent to this region is negligible. Hence we may speak of a "quasi-local" distribution of nonequilibrium quasi particles. Localization results in an interesting manifestation of the phonon deficit effect in a system with excess quasi particles (Chapter 3). The manifestation of nonequilibrium chemical potential $\delta\mu$ is illustrated in Fig. 15.

3. Nonequilibrium order parameter. We will identify two characteristic features of the behavior of nonequilibrium gap Δ. First, there exists a range where subcritical and supercritical values coexist (compare curves 1 and 2 in Fig. 16). This produces hysteresis in the E-I characteristics. Second, when $T \sim T_c$ with an increase in voltage V the $\Delta(T, V)$ curve jumps insignificantly (~1% (curve 3 in Fig. 16) with an increase in V by an order of magnitude), while an increase in Δ rapidly results in saturation. Here we have the familiar phenomenon of tunnel current stimulation of superconductivity.

4. Current-voltage characteristic. The ambiguous dependence of Δ on V produces hysteresis [8]. For comparison Fig. 17 shows the experimentally observed curve [37].

Chapter 3

PHONON EMISSION IN NONEQUILIBRIUM CONDITIONS

We will now focus on an investigation of the kinetics of the phonon subsystem of a nonequilibrium superconductor. We will be interested in phonon emission from nonequilibrium superconductors. Its spectral dependence yields very detailed information on the nonequilibrium picture and hence we will provide results from the derived calculations for the case of the tunnel junction considered in Chapter 2. For this purpose we will use the general analysis scheme proposed in study [22] that employs a kinetic equation for the phonons. The polarization operators are the nonequilibrium source; expressions for these were derived in study [22] both on the basis of the Keldysh technique and using the analytic continuation method proposed by Gorkov and Eliashberg [23]. Here the issue of the possibility of obtaining the entire kinetic equation for phonons (without assuming the existence of a thermostat) within the scope of the analytic continuation method remains open here. We will begin our examination with the technique for deriving kinetic equations in the case where the phonons cease to be in equilibrium.

8. A description of the kinetics of nonequilibrium electrons and phonons by means of the analytic continuation technique

1. Electron-phonon collisions. As indicated by the results from Chapter 1, the primary aspect in finding the electron-phonon collision operator is the derivation of formulae (3.19). Here we will derive the expressions generalizing (3.19) without assuming equilibrium of the phonon subsystem. This assumption [1] has led to the contention that the Green's function not containing Σ (Fig. 18a) has the same analytic structure (as a function of ε with fixed ($\omega_l = i\cdot 2m_l\pi T$)), as the propagator shown in Fig. 18b. In this case we cannot make this statement although it is possible to formulate an expression for the self-energy parts without its use.

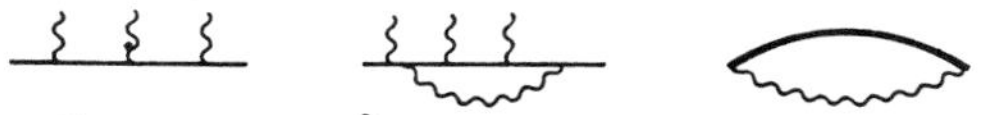

a b c

Fig. 18

We will write an expression for Σ in the initial representation of discrete imaginary frequencies ($P = \{\varepsilon, \mathbf{p}\}$, $K = \{\omega, \mathbf{k}\}$, $\varepsilon = (2n + 1)i\pi T$, $\omega = \omega_l$, l and n are integers):

$$\hat{\Sigma}(P, P-K) = T \sum_{\varepsilon'} \int \frac{d^3\mathbf{p}'}{(2\pi)^3} D(P'-P, P'-P-K'+K) \hat{G}(P', P'-K'), \tag{8.1}$$

corresponding to the diagram in Fig. 18c. Both functions $\hat{G}$ and D are assumed to be complete in (8.1) including, in addition to the external field, the self-energy parts and the polarization operators. We will represent G and D as sums of the diagrams of all field orders (assuming that in the absence of fields the initial state is an equilibrium state) and will consider the analytic structure of the Nth-order diagrams as a function of the complex variable ε with fixed imaginary frequencies ω_l $\left(\omega_l = 2m_l\pi Ti, \sum_l^N \omega_l = \omega\right)$. Since the complete phonon propagator with external action is not assumed to be in equilibrium, the set of branch lines of the sample consists of the branch lines of both the internal electron G-line and the branch lines of the D-function. We will label the entire set of branch lines Ω_n. Then these branch lines may be considered to be distributed in the complex plane ε along the lines $\mathrm{Im}(\varepsilon - \Omega_n) = 0$ between the far upper line $\mathrm{Im}(\varepsilon - \omega) = 0$ and the X axis (since in the final analysis due to the causality principle we are interested in the analytic continuation for each of ω_l onto the real axis). The set of combinations ω_l representing Ω_n is determined by the distribution of field nodes among the internal lines of the diagram $\hat{\Sigma}^{(N)}$. We note that the G- and D-functions correspond to the sets of branch lines $\mathrm{Im}(\varepsilon' - \omega_1) = 0$ and $\mathrm{Im}(\varepsilon' - \varepsilon - \omega_{2k}) = 0$, respectively. We replace summation with respect to ε' in (8.1) by contour integration and then deform the integration contour to the borders of the branch lines in the regular manner [1]. As a result we obtain the formula

$$\hat{\Sigma}^{(N)} \backsim \sum_{i,k} \int_{-\infty}^{\infty} \frac{dz}{4\pi i} \left\{ \mathrm{th}\,\frac{z}{2T}\,\delta_i(\hat{G}_{z+\omega_{1i}}) D_{z-\varepsilon+\omega_{1i}} + \hat{G}_{z+\omega_{2k}+\varepsilon}\,\mathrm{cth}\,\frac{z}{2T}\,\delta_k(D_{z+\omega_{2k}}) \right\}, \quad (8.2)$$

where $\delta_i(\hat{G}_{z+\omega_{1i}})$ and $\delta_k(D_{z+\omega_{2k}})$ are the jumps of the Green's functions on the corresponding branch lines (for simplicity here and henceforth we will drop the second indices on the Green's functions). Now continuing (8.2) analytically with respect to ε from the upper border of the upper branch line (the lower border of the lower branch line) we obtain after returning to real $\omega_{i,k}$ the complete shift of the integration variable, summation with respect to all orders of perturbation theory and integration with respect to energy ξ:

$$\hat{\Sigma}^{R(A)} \backsim \int_{-\infty}^{\infty} \frac{d\varepsilon'}{4\pi i} \{\hat{g}_{\varepsilon'} D^{A(R)}_{\varepsilon'-\varepsilon} + D_{\varepsilon'-\varepsilon} \hat{g}^{R(A)}_{\varepsilon'}\}, \quad (8.3)$$

in (8.3) the D-function is determined as

$$D_\omega = \sum_{N=0}^{\infty} \sum_{k=0}^{\infty} \mathrm{cth}\,\frac{\omega - \omega_k}{2T}\,\delta_k(D^{(N)}). \quad (8.4)$$

determining $\Sigma_{\varepsilon\varepsilon-\omega}$ as

$$\Sigma_{\varepsilon\varepsilon-\omega} = \sum_{N=0}^{\infty} \sum_{n=0}^{\infty} \delta_n(\Sigma^{(N)}_{\varepsilon\varepsilon-\omega})\,\mathrm{th}\,\frac{\varepsilon - \Omega_n}{2T}, \quad (8.5)$$

we find, based on (8.2) expressions for the matrix $\hat{\Sigma}_{\varepsilon\varepsilon-\omega}$ which may be represented as

$$\Sigma_1 = \int_{-\infty}^{\infty} \frac{d\varepsilon'}{4\pi i} \int \frac{dO_{\mathbf{p}'}}{4\pi} [D_{\varepsilon'-\varepsilon} g_{\varepsilon'} - (g^R - g^A)_\varepsilon (D^R - D^A)_{\varepsilon'-\varepsilon}],$$

$$2i\gamma = \int_{-\infty}^{\infty} \frac{d\varepsilon'}{4\pi i} \int \frac{dO_{\mathbf{p}'}}{4\pi} [D_{\varepsilon'-\varepsilon} (g^R - g^A)_{\varepsilon'} - g_\varepsilon (D^R - D^A)_{\varepsilon'-\varepsilon}], \tag{8.6}$$

$$\Sigma_2 = \Sigma_1 (g^{(R,A)} \to f^{(R,A)}), \qquad \delta = \gamma (g^{(R,A)} \to f^{(R,A)}).$$

In the diagonal approximation the expressions for $\hat{g}^{(R,A)}$ are obtained in Section 3 of Chapter 1, while the phonon propagator may be expressed through the function N_ω in accordance with

$$D_{\varepsilon'-\varepsilon} = (1 + 2N_{\omega_{\mathbf{p}'-\mathbf{p}}}) \operatorname{sign}(\varepsilon' - \varepsilon) (D^R - D^A)_{\varepsilon'-\varepsilon}, \tag{8.7}$$

$$D^{R(A)}_{\varepsilon'-\varepsilon} = \lambda \frac{2\omega^2_{\mathbf{p}'-\mathbf{p}}}{\omega^2_{\mathbf{p}'-\mathbf{p}} - (\varepsilon' - \varepsilon \underset{(-)}{+} i\delta)^2}. \tag{8.8}$$

Using these expressions we may obtain the electron-phonon collision operator in the form of (3.21)-(3.24) where the function N_ω is still a random function. We will obtain the equation for this function in the following section.

2. Phonon-electron collisions. We will see that the function N_ω introduced by relation (8.7) plays the role of nonequilibrium phonon distribution function by obtaining the kinetic equation for this quantity.

We will proceed from the expressions for $d_{\omega\omega-\omega'}$ (8.4) and the determination of the polarization operator $\Pi_{\omega\omega-\omega'}$ (see [2]). After isolating the anomalous parts $D^{(a)}$ and $\Pi^{(a)}$ associated with the nonequilibrium we obtain the following relations:

$$D_{\omega\omega-\omega'} = D^R_{\omega\omega-\omega'} \operatorname{cth} \frac{\omega - \omega'}{2T} - D^A_{\omega\omega-\omega'} \operatorname{cth} \frac{\omega}{2T} + D^{(a)}_{\omega\omega-\omega'}, \tag{8.9}$$

$$\Pi_{\omega\omega-\omega'} = \Pi^R_{\omega\omega-\omega'} \operatorname{cth} \frac{\omega - \omega'}{2T} - \Pi^A_{\omega\omega-\omega'} \operatorname{cth} \frac{\omega}{2T} + \Pi^{(a)}_{\omega\omega-\omega'}. \tag{8.10}$$

The regular functions in (8.9), (8.10) such as the advanced function D^A are determined by the diagram expansion in which all functions (propagators and polarization operators) are advanced. In the diagram expansion by factoring out the left free line D^R_ω for the function $D^{(a)}_{\omega\omega-\omega'}$, we have the equation

$$(D^0_\omega)^{-1} D^{(a)}_{\omega\omega-\omega'} = \{\Pi^R D^{(a)} + \Pi^{(a)} D^A\}_{\omega\omega-\omega'}, \tag{8.11}$$

utilizing the abbreviated notation system

$$\{AB\}_{\omega\omega-\omega'} = \int \frac{d\omega_1}{2\pi} A_{\omega\omega-\omega_1} B_{\omega-\omega_1\omega-\omega'} \tag{8.12}$$

(we also assume integration with respect to the internal momentum or coordinate (position) variable). Likewise factoring out the right free line $D^A_{\omega-\omega'}$, and subtracting the derived expression from (8.11), and utilizing formulae (8.9) and (8.10) together with the expressions for the regular $D^{R,A}$-functions we obtain the relation

$$[(D^0_\omega)^{-1} - (D^0_{\omega-\omega'})^{-1}]\, D_{\omega\omega-\omega'} = \{\Pi^R D - \Pi D^A - D\Pi^A - D^R\Pi\}_{\omega\omega-\omega'}, \quad (8.13)$$

which is the kinetic equation for phonons in the general form (compare to study [2]). The set of polarization operators in terms of the $\hat{g}$-functions figuring here were obtained previously [2] by analytic continuation.

We note that at this stage the "bare" temperature T entering into the imaginary frequency variables of the initial equations drops completely out of both (8.13) and (8.16). At the same time the situation is entirely equivalent to that obtained on the basis of the Keldysh technique [9]. Here the "bare" temperature in the analytic continuation technique plays the role of equilibrium density matrix in the Keldysh method: the latter also vanishes from the final expressions.

3. The canonical form and imbalance relaxation. Carrying out a transformation to the energy-diagonal functions $D^{(R,A)}$ in (8.13) (the quasi-classical case) and using formulae (8.7), (8.8) in this approximation, we obtain after integration of (8.13) with respect to the positive semiaxis ω the kinetic equation for the phonons

$$\dot{N}(\omega_q) = I(\omega_q) = i\left\{(\Pi^R - \Pi^A)_{\omega_q} N(\omega_q) - \frac{1}{2}[\Pi - (\Pi^R - \Pi^A]_{\omega_q}\right\}. \quad (8.14)$$

Using the formulae for the energy-diagonal $\hat{g}$-functions found in Section 3 of Chapter 1 and the expressions for the polarization operators [2],[1] we obtain the canonical form of the phonon-electron collision integral

$$I(N_{\omega_q}) = \frac{\pi\lambda}{8}\frac{\omega_D}{\varepsilon_F}\iint\limits_{\Delta}^{\infty} d\varepsilon\, d\varepsilon'\, \{\delta(\varepsilon+\varepsilon'-\omega_q)S_1 + 2\delta(\varepsilon-\varepsilon'-\omega_q)S_2\}, \quad (8.15)$$

$$\begin{aligned} S_1 = {}& (u_\varepsilon u_{\varepsilon'} + v_\varepsilon v_{\varepsilon'} + 1)\{[(N_{\omega_q}+1)n_\varepsilon n_{-\varepsilon'} - N_{\omega_q}(1-n_\varepsilon)(1-n_{-\varepsilon'})] + \\ & + [(N_{\omega_q}+1)n_{-\varepsilon}n_{\varepsilon'} - N_{\omega_q}(1-n_{-\varepsilon})(1-n_{\varepsilon'})]\} + \\ & + (u_\varepsilon u_{\varepsilon'} + v_\varepsilon v_{\varepsilon'} - 1)\{[(N_{\omega_q}+1)n_\varepsilon n_{\varepsilon'} - N_{\omega_q}(1-n_\varepsilon)(1-n_{\varepsilon'})] + \\ & + [(N_{\omega_q}+1)n_{-\varepsilon}n_{-\varepsilon'} - N_{\omega_q}(1-n_{-\varepsilon})(1-n_{-\varepsilon'})]\}, \end{aligned} \quad (8.16)$$

$$\begin{aligned} S_2 = {}& (u_\varepsilon u_{\varepsilon'} - v_\varepsilon v_{\varepsilon'} - 1)\{[(N_{\omega_q}+1)n_\varepsilon(1-n_{-\varepsilon'}) - N_{\omega_q}(1-n_\varepsilon)n_{-\varepsilon'}] + \\ & + [(N_{\omega_q}+1)n_{-\varepsilon}(1-n_{\varepsilon'}) - N_{\omega_q}(1-n_{-\varepsilon})n_{\varepsilon'}]\} + \\ & + (u_\varepsilon u_{\varepsilon'} - v_\varepsilon v_{\varepsilon'} + 1)\{[(N_{\omega_q}+1)n_\varepsilon(1-n_{\varepsilon'}) - N_{\omega_q}(1-n_\varepsilon)n_{\varepsilon'}] + \\ & + [(N_{\omega_q}+1)n_{-\varepsilon}(1-n_{-\varepsilon'}) - N_{\omega_q}(1-n_{-\varepsilon})n_{-\varepsilon'}]\}. \end{aligned} \quad (8.17)$$

Expressions (8.15)-(8.16) generalize the canonical form derived in study [2] of the phonon-electron collision integral to the case of

---

[1] Here we must drop the multiplier λ in them which is already accounted for in the left half of (8.13).

imbalance existing in the electron system. We will enumerate certain properties of this operator.

a) It is clear that the quantity N_{ω_q} (8.7) indeed plays the role of the phonon distribution function. Substituting equilibrium values of $N^0_{\omega_q}$ and n^0_ε (with coincident values of T) causes the collision integral to vanish.

b) If the coupled system of kinetic equations is decomposed such as in a model with a phonon thermostat [1], the solutions $n_{\pm\varepsilon}$ of equation (3.25) with $N_{\omega_q} = N^0_{\omega_q}$ substituted into (8.15) (where $N_{\omega_q} = N^0_{\omega_q}$ as well), determine the number of phonons emitted from the superconductor per unit of time. If $d\omega_q$ is the spectral interval of the phonon frequencies, this number is equal to

$$d\dot{N}_{\omega_q} = I(N_{\omega_q})\,\rho(\omega_q)\,d\omega_q, \tag{8.18}$$

where $\rho(\omega_q) = \mathscr{V}\,\omega_q^2/2\pi^2u^3$; $\mathscr{V}$ is the volume of the emission region (for a detailed discussion see study [2]).

In a situation with imbalance, the phonon kinetics have a specific nature which may easily be traced by using (8.15)-(8.17). For example, it is clear from (8.16) that the recombination phonons are emitted not only as a result of the merging of two excitations but also from the recombination of two excitations of the same branch. In the latter case the recombination process reduces the difference in the number of electron- and hole-like excitations. It is clear that this process does not occur for normal metal when $v_\varepsilon \equiv 0$, $u_\varepsilon \equiv 1$.

9. Phonon emission spectrum

1. Subcritical state. The phonon emission spectrum that arises in this case from the nonequilibrium film is shown in Fig. 19. The valley in the spectral relation when $\omega_q \geqslant 2\Delta$ designates the existence of the phonon deficit effect in the system. We predicted this effect from the perturbing action of a microwave field in a previous study [42] and examined it in detail in studies [2, 22]. The essence of this case remains the same in the present discussion and hence we will not provide detailed commentary. We note only that the effect is associated with the shift in the "center of gravity" of the excitation distribution function (see Figs. 10-13; this also results in the superconductivity stimulation effect [43]). One small difference in this case[2] is the fact that there are two (rather than 1) relaxation peaks (see Fig. 19, curve 1); the peaks are not differentiable on the scale used (see Fig. 19, curve 2) and their origin is related to the

---

[2] Evidently the same peaks would occur under microwave pumping. Their absence in studies [2, 22] is due to the use of approximations

"sawtooth" nature of the excitation distribution function (see the insert to Fig. 19).

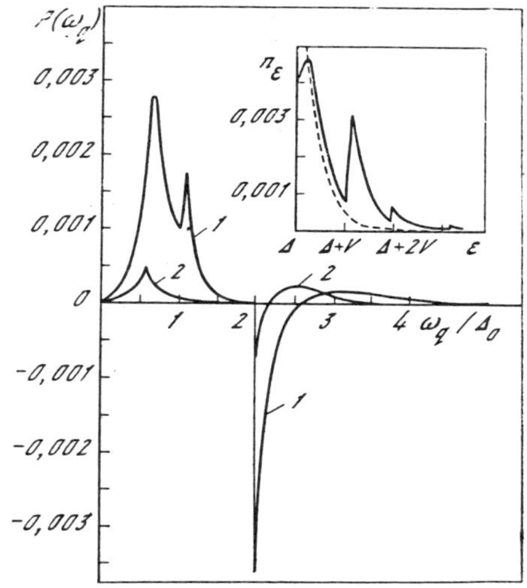

Fig. 19. The spectral dependence of phonon emission in the subcritical state
1 - at $T = 0.2\Delta_0$, $V = 0.5\Delta_0$, $\nu = 0.2\gamma$;
2 - same as above although $V = 0.1\Delta_0$. The insert shows the electron excitation distribution function corresponding to case (1):
$P(\omega_q) = \omega_q^2 I(N_{\omega_q})/(\pi\lambda\omega_D^2/5\varepsilon_F\Delta_0^3)$

2. **Supercritical state.** The behavior of the phonon emission spectrum in the supercritical state (Fig. 20) is curious. We note that curves similar to those in Fig. 20a were obtained previously by Chang and Scalapino [19] (in a simplified model ignoring imbalance). However this study did not discuss the fact that with small values of ω_q the phonon fluxes became negative: this fact is illustrated by Fig. 21. The phonon deficit effect in conditions where excess quasi particles created by the field from the condensate in the system is not trivial.

Evidently this is due to the strong localized nature of excess excitations in the energy space (see Fig. 14) near the gap edge and as a result when photons are scattered by the electrons scattering acts that involve the emission of phonons with a frequency above the boundary frequency are not possible. At the same time scattering with phonon absorption is possible: at high energies ε there is no bandgap and hence the scattering mechanism causes a phonon deficit in a certain spectral range. A question arises in this regard: isn't the deficit related to the phonon instability (i.e., the sign reversal of the absorption coefficient of the phonons) at the relaxation frequencies? As indicated by calculations (curve 3 in Fig. 21), the sign of the absorption coefficient does not change. Evidently the reason is the fact that the small (equilibrium) "tail" of the excitation distribution function makes no contribution to the nonequilibrium emission of phonons and nonetheless makes a contribution to the absorption coefficient that compensates the small valley that results from nonequilibrium.

We note that effects similar to those in Figs. 19-21 are also valid over a broad range of junction parameters. Here we will not carry out a quantitative analysis of such parameters.

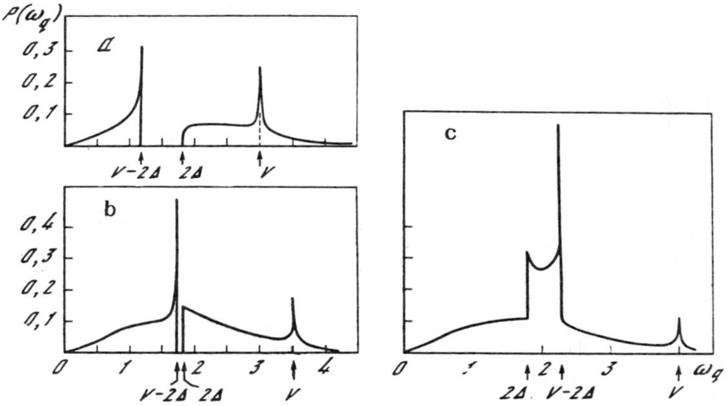

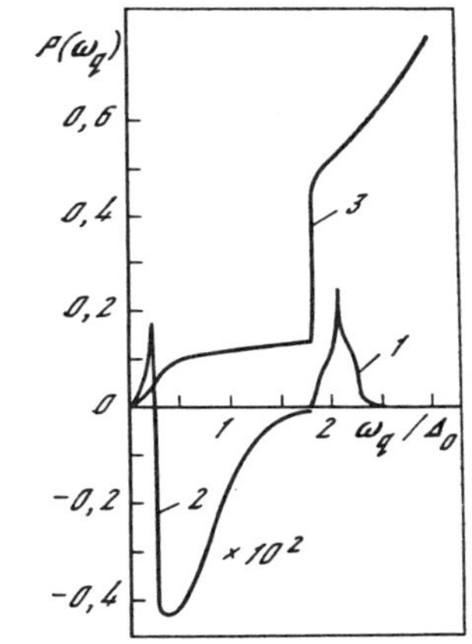

Fig. 20. The phonon emission spectrum in the supercritical state ($T = 0.1\Delta_0$; $\nu = 0.01\gamma$)
a - $V = 3\Delta_0$;
b - $V = 3.5\Delta_0$;
- $V = 4\Delta_0$. The X-axis is plotted in units of Δ_0

Fig. 21. The phonon deficit effect in the subcritical state ($V = 2.05\Delta_0$, $T = 0.2\Delta_0$, $\nu = 0.01\gamma$)
1 - recombination peak;
2 - relaxation valley (increased by 100 fold;)
3 - spectral dependence of the absorption coefficient (in arbitrary units)

Chapter 4

THE THEORY OF AN ACOUSTICAL QUANTUM-MECHANICAL OSCILLATOR BASED ON A NONEQUILIBRIUM SUPERCONDUCTOR

We will now discuss the possibility of the phonon instability[3] in a nonequilibrium superconductor. Such an instability may cause a "quasi-laser" phonon generation state in which phonon emission from a nonequilibrium superconducting film is coherent and monochromatic.

---

[3] Phonon instability refers to the sign reversal of the sound absorption coefficient of the nonequilibrium electron system of the superconductor.

The possibility for transforming to a "quasi-laser" generation state is related to the fact common to Bose-fields that the phonon source contains terms proportional to the occupation numbers of the phonons [22]. This possibility was discussed independently in studies [20, 2] for the case of tunnel injection and microwave pumping, respectively (we also note the study by Bulyzhenkov [26]). As demonstrated in these studies a population inversion must be present for phonon instability ($n_\varepsilon > 1/2$ in a certain region of energy values ε over the gap), which, however, is not a sufficient condition for the development of the instability. The population inversion may occur both in the case of microwave pumping and in tunnel injection although in the case of microwave pumping the phonon instability is, evidently, not realized at the same time that in tunnel injection from a bulk superconductor to a thin film (an SiS'-junction) the situation is significantly more favorable [20].

In this chapter we will analyze the problem of achieving the critical state of "quasi-laser" generation from the viewpoint of kinetic equations for the phonons and electrons. We note that as early as 1974 Chang and Scalapino, in assuming that the electron excitation distribution function is a quasi-Fermi function (with a certain effective chemical potential), calculated the sound absorption coefficient which was negative at a high nonequilibrium level which made it possible to conclude that it is possible to amplify sound with a nonequilibrium superconductor [29]. However in a subsequent study devoted to the stability of this situation these same authors concluded that with such an instability level the system ceases to be in a spatially-homogeneous state and hence the "phonon instability becomes meaningless" [28]. Subsequent studies, particularly [29-33], have provided some basis for concluding that the problem of stability is nontrivial and hence the conclusion from study [28] is, in our view, categorically unwarranted.

From the experimental viewpoint the subject of primary interest is the selection of external parameters such that would allow us to achieve subcritical operation of an acoustical quantum-mechanical oscillator in the simplest manner. It is clear that the simplest case is when the phonon instability (subcritical generation) is achieved in the spatially-homogeneous and steady-state situation and the role of fluctuations may be ignored. As indicated by the results given in the preceding chapter, the critical state of an acoustical quantum-mechanical oscillator in this case evidently may be achieved by the proper selection of parameters.

10. Population inversion in nonequilibrium superconductors

We will modify the formal formulation of the problem.

1. The decomposition of the binding kinetics of electrons and phonons. In order to solve the problem of the dynamics of the phonon system of a nonequilibrium superconductor we must, generally speaking, go beyond the model containing the phonon thermostat. Here we have a coupled system of kinetic equations. Its general form is

$$\dot{n}_\varepsilon = Q\{n_\varepsilon\} + I^{(e-ph)}\{n_\varepsilon, N_{\omega_q}\}, \tag{10.1}$$

$$\dot{N}_{\omega_q} = I^{(ph-e)}\{n_\varepsilon, N_{\omega_q}\} + I^{(ph-t)}\{N_{\omega_q}\}, \tag{10.2}$$

where Q is the external nonequilibrium source, while $I^{(ph-t)}$ is the operator describing phonon interaction with the external thermostat (here we will avoid the dynamics of the pair condensate). In the presence of feedback the portion of nonequilibrium phonons emitted by the electron system is absorbed in the electron excitation system and has a reverse influence on the electron system. If the absorption coefficient at any phonon mode is negative in this case, the number of such phonons in the system grows exponentially, resulting in an instability and the development of a dynamical state.

The solution of dynamical problem (10.1), (10.2) is an independent problem and is not required in this context. The problem is that here we are interested in the possibility of achieving critical instability itself. In this case we may consider the following kinetic scheme:

$$\dot{n}_\varepsilon = Q\{n_\varepsilon\} + I^{(e-ph)}\{n_\varepsilon, N^{(i)}_{\omega_q}\}, \tag{10.3}$$

$$\dot{N}^{(i)}_{\omega_q} = I^{(ph-e)}\{N^{(i)}_{\omega_q}, n_\varepsilon\} + I^{(ph-t)}\{N^{(i)}_{\omega_q}\}, \tag{10.4}$$

$$\dot{N}^{(e)}_{\omega_q} = I^{(ph-e)}\{N^{(e)}_{\omega_q}, n_\varepsilon\}. \tag{10.5}$$

where $N^{(i)}_{\omega q}$ is the phonon eigenfield; $N^{(e)}_{\omega q}$ corresponds to the external phonon flux whose linear response is of interest to us (the operators $I^{(ph-e)}$ in (10.4) and (10.5) have an identical structure). In this case if we use the model with the phonon thermostat (it is applicable, in any case, in the absence of phonon instability) in the subcritical state a dissipative steady state $\dot{n}_\varepsilon = 0$, $\dot{N}_{\omega q} = 0$ holds and

$$I^{(ph-e)} = -I^{(ph-t)}, \tag{10.6}$$

hence equations (10.3)-(10.5) are completely decoupled and in (10.3) the function $N^{(i)}_{\omega q}$ is equal to $N^{0}_{\omega q}$ (i.e,. equilibrium), equation (10.4) is an identity while (10.5) is used to calculate the linear response (i.e., analyze the phonon instability).

2. The necessary condition of phonon instability. We will assume that the action of external factors that cause the electron subsystem to deviate from equilibrium in a superconductor coupled to an external thermostat has a dissipative steady state in which the nonequilibrium electron excitations are characterized by the distribution function n_ε (henceforth it will be assumed that n_ε is an even function and is also spatially-homogeneous and isotropic, like the order parameter Δ).

We will consider the influence of the external phonon flux on the electron subsystem; this flux is characterized by the occupation numbers N_{ω_q} (we will assume $N_{\omega_q} \gg 1$). In the nonequilibrium superconductors the phonons may be absorbed (and emitted) both in decoupling (and recombination) processes as well as relaxational processes. When the electron subsystem is only slightly out of equilibrium, relaxation processes dominate [12, 13]. As demonstrated by Oronov and Spivak [21] the phonon instability in this case is not valid (in any case with microwave-pumping). When significantly out of equilibrium the recombination processes in the superconductors occur significantly more rapidly than the relaxation processes [34] (we emphasize that unlike semiconductors in superconductors both these processes are single-phonon processes), and in this case it is necessary to consider the stability with respect to recombination emission of phonons.

The number of photons absorbed per unit of time at frequency ω_q in the interval $d\omega_q$ is given by the expression (see Chapter 3):

$$d\dot{N}_{\omega_q} = \rho(\omega_q) I(\omega_q)\, d\omega_q, \tag{10.7}$$

where $I(\omega_q)$ is the phonon-electron collision operator (8.15). In the case of interest to us it takes the form

$$I(\omega_q) = \iint\limits_{\Delta}^{\infty} d\varepsilon_1\, d\varepsilon_2 L(\varepsilon_1, \varepsilon_2)\Big\{\big[(1-n_{\varepsilon_1})(1-n_{\varepsilon_2}) - n_{\varepsilon_1}n_{\varepsilon_2}\big]\Big(1+\frac{\Delta^2}{\varepsilon_1\varepsilon_2}\Big) \times$$
$$\times\,\delta(\varepsilon_1+\varepsilon_2-\omega_q) - 2\big[n_{\varepsilon_1}(1-n_{\varepsilon_2}) - (1-n_{\varepsilon_1})n_{\varepsilon_2}\big]\Big(1-\frac{\Delta^2}{\varepsilon_1\varepsilon_2}\Big)\delta(\varepsilon_2-\varepsilon_1+\omega_q)\Big\},$$
$$L(\varepsilon_1, \varepsilon_2) = N_{\omega_q}\frac{\pi\lambda}{2}\frac{\omega_D}{\varepsilon_F}\frac{\varepsilon_1}{\sqrt{\varepsilon_1^2-\Delta^2}}\frac{\varepsilon_2}{\sqrt{\varepsilon_2^2-\Delta^2}}\,\theta(\varepsilon_1^2-\Delta^2)\,\theta(\varepsilon_2^2-\Delta^2) \tag{10.8}$$

(we remember that for $N_{\omega_q} \gg 1$ the external field may be assumed to be classical). The recombination part of collision integral (10.8) may be represented as

$$I(\omega_q)^{\rm rec} = \int\limits_{\Delta}^{\omega_q-\Delta} L(\varepsilon, \omega_q-\varepsilon)\Big(1+\frac{\Delta^2}{\varepsilon(\omega_q-\varepsilon)}\Big)(1-2n_\varepsilon)\, d\varepsilon, \tag{10.9}$$

while the relaxation part may be represented as

$$I(\omega_q)^{\rm rel} = 2\int\limits_{\Delta}^{\infty} L(\varepsilon, \varepsilon+\omega_q)\Big(1-\frac{\Delta^2}{\varepsilon(\omega_q+\varepsilon)}\Big)(n_\varepsilon - n_{\varepsilon+\omega_q})\, d\varepsilon. \tag{10.10}$$

As we see from expressions (10.9), (10.10) the phonon instability ($I(\omega_q) < 0$) due to recombination processes (at frequencies $\omega_q > 2\Delta$) may exist only when the following condition is satisfied

$$n_\varepsilon > 1/2 \tag{10.11}$$

in a certain region of values of $\tilde{\varepsilon}$[4] over the gap. According to the familiar Elesin theorem [34, 8] such a situation cannot exist in the case of pumping by a "broad" source, when the external action produces quasi particles over a broad energy range ($\tilde{\varepsilon}_{max} - \Delta \gg \Delta$). The picture is somewhat different with "narrow" quasi-particle sources when the quasi particles are produced in a narrow energy range ($\tilde{\varepsilon}_{max} - \Delta \ll \Delta$). In this case, as demonstrated in the studies by Aronov and Spivak [21] as well as Genkin and Protogenov [31] (see also [20, 2]), n_{ε} may exceed the value 1/2 and hence the necessary condition of "phonon instability" in principle may be satisfied.

Assuming "narrowness" of the quasi-particle distribution, we will simplify expressions (10.9), (10.10) and will represent the "sufficient" condition of the phonon instability as

$$I(\omega_q) = \frac{\pi}{2}\frac{\omega_D}{\varepsilon_F}\Delta N_{\omega_q}\left\{\int\limits_{\Delta}^{\omega_q-\Delta}\frac{(1-2n_{\varepsilon})\,d\varepsilon}{\sqrt{\varepsilon^2-\Delta^2}\sqrt{\omega_q-\varepsilon-\Delta}}+\frac{1}{\sqrt{\Delta}}\int\limits_{\Delta}^{\tilde{\varepsilon}_{max}}\frac{n_{\varepsilon}\,d\varepsilon}{\sqrt{\varepsilon-\Delta}}\right\}<0 \tag{10.12}$$

for any frequency $\omega_q \geqslant 2\Delta$. An analysis of expression (10.12) demonstrates that this condition may be satisfied if the deviation from equilibrium is sufficiently large: $n_{\tilde{\varepsilon}} \sim 1$, $\tilde{\varepsilon} < \tilde{\varepsilon}_{max}$. Indeed, in this case for frequencies $\omega_q \approx \tilde{\varepsilon}_{max} + \Delta$ the first integral in the braces is equal to $(-\pi)$ at the same time that the second integral does not exceed the value $2\sqrt{\tilde{\varepsilon}_{max}/\Delta - 1}$ and is small due to the "narrowness" of the electron distribution. Thus, in this case instead of absorption of the acoustic wave we have amplification of the acoustic wave.

We will analyze to what degree condition (10.12) may be satisfied in tunnel injection.

11. Self-consistent kinetic equations

1. Simplification of general equations. We will be interested in possible steady-state electron energy distributions in a superconducting film connected to a thermostat under the action of external perturbations of various types driving the electron system from equilibrium. Assume the film thickness is sufficiently narrow so that the nonequilibrium phonons emitted by the electron subsystem will have no reverse influence on the electrons. With ideal acoustical matching between the film and the thermostat (see studies [1, 22] for more detail) such a situation exists for films of thickness $d \sim \xi_0$ (ξ_0 is the correlation radius of the superconductor). As stated above in this

---

[4] We note that the n_{ε} distribution that occurs in the superconductor need not be a monotonic function, and hence the mode $\omega_q = 2\Delta$ need not be the least stable and, moreover, the condition $n_{\Delta} > 1/2$ indicated in study [20] is not, generally speaking, a necessary condition. The necessary condition will be the weaker condition (10.11).

case the binding kinetics of the electron and phonon subsystem of the superconductor are decoupled and a model with a phonon thermostat is used for investigating electron kinetics. Moreover, the existence of a sufficient number of electron elastic scattering centers is assumed in the film; these make the electron distribution function and order parameter isotropic. Then kinetic equation (3.25) is written as

$$0 = Q(\varepsilon) + J(\varepsilon), \tag{11.1}$$

where $Q(\varepsilon)$ is the external source of nonequilibrium electron excitations, while collision integral $J(\varepsilon)$ contains the electron-phonon and electron-electron parts. In the preceding chapters we focused attention on the existence of asymmetry in the population of the electron-hole branches in the electron system. In this chapter this asymmetry is insignificant for the following reason. We are primarily interested in tunnel injection at critical voltage levels of $V \approx \Delta + \Delta'$, although as indicated from the analysis given in Chapter 2 in this case the imbalance is quite small. It is also small in the case of microwave pumping. In this regard general equations (3.25) and (3.26) may be significantly simplified which allows us to progress rather far in their analytic investigation. We will perform these simplifications assuming $n_{\varepsilon} = n_{-\varepsilon}$.

The electron-phonon part of (3.21) when $T = 0$ (we are limited to this case) takes the form

$$J^{(e-ph)} = \frac{\pi\lambda}{2(up_F)^2} \int_{\Delta}^{\infty} d\varepsilon' \int_{0}^{\infty} d\omega \frac{\omega^2 \varepsilon\varepsilon'}{\sqrt{\varepsilon^2-\Delta^2}\sqrt{\varepsilon'^2-\Delta^2}} \Big[n'(1-n)\Big(1-\frac{\Delta^2}{\varepsilon\varepsilon'}\Big) \times$$
$$\times \delta(\varepsilon'-\varepsilon-\omega) - n(1-n')\delta(\varepsilon-\varepsilon'-\omega)\Big(1-\frac{\Delta^2}{\varepsilon\varepsilon'}\Big) -$$
$$- nn'\Big(1+\frac{\Delta^2}{\varepsilon\varepsilon'}\Big)\delta(\varepsilon+\varepsilon'-\omega)\Big]. \tag{11.2}$$

The electron-electron part of (3.12) is written as

$$J^{(e-e)}(\varepsilon) = \frac{1}{2\varepsilon_F} \iiint_{\Delta}^{\infty} \frac{d\varepsilon_1\, d\varepsilon_2\, d\varepsilon_3}{\sqrt{\varepsilon^2-\Delta^2}\sqrt{\varepsilon_1^2-\Delta^2}\sqrt{\varepsilon_2^2-\Delta^2}\sqrt{\varepsilon_3^2-\Delta^2}} \times$$
$$\times \{M_1[(1-n)n_1n_2n_3 - n(1-n_1)(1-n_2)(1-n_3)]\delta(\varepsilon-\varepsilon_1-\varepsilon_2-\varepsilon_3) +$$
$$+ 3M_2[n_1n_2(1-n)(1-n_3) - nn_3(1-n_1)(1-n_2)]\delta(\varepsilon+\varepsilon_3-\varepsilon_1-\varepsilon_2) +$$
$$+ 3M_3[n_1(1-n)(1-n_2)(1-n_3) - (1-n_1)nn_2n_3]\delta(\varepsilon+\varepsilon_2+\varepsilon_3-\varepsilon_1)\}, \tag{11.3}$$

$$M_1 = a(\varepsilon_1\varepsilon_2\varepsilon_3\varepsilon - \Delta^4) - \frac{1}{3} b(-\varepsilon^2 + \varepsilon_1\varepsilon_2 + \varepsilon_1\varepsilon_3 + \varepsilon_2\varepsilon_3)\Delta^2,$$
$$M_2 = -M_1(-\varepsilon_3), \quad M_3 = M_1(-\varepsilon_2, -\varepsilon_3).$$

The quantity Δ entering into expressions (11.1)-(11.3) will be determined from the self-consistency equation

$$1 = \lambda \int_{\Delta}^{\infty} \frac{1-2n_{\varepsilon}}{\sqrt{\varepsilon^2-\Delta^2}} d\varepsilon. \tag{11.4}$$

Self-consistent equation system (11.1)-(11.4) is highly nonlinear. However, the "narrowness" (presumably!) of the energy distribution of the electron excitations makes it possible to effectively find the solution even accounting for (11.3).

We will introduce the variables $\delta = \tilde{\varepsilon}_{max} - \Delta$ and $z = (\varepsilon - \Delta)/\delta$, $\delta > 0$. We will set

$$\bar{n} = \int_{\Delta}^{\infty} \frac{n_\varepsilon d\varepsilon}{\sqrt{\varepsilon^2 - \Delta^2}} = \left(\frac{\delta}{2\Delta}\right)^{1/2} \int_0^1 \frac{n_z dz}{\sqrt{z}} \tag{11.5}$$

and will represent equation (11.4) as

$$\ln(\Delta/\Delta_0) = -2\bar{n}, \tag{11.6}$$

where Δ_0 is the gap at $T = 0$ in the absence of an external perturbation. The narrowness of the distribution means, as we see from (11.5), smallness of $\bar{n}$ and hence

$$\Delta \approx \Delta_0 (1 - 2\bar{n}). \tag{11.7}$$

We will now examine the collision integrals. If $n_\varepsilon \sim 1$ and is concentrated in the region directly over the gap, the relaxation terms in (11.2) are insignificant compared to the recombination terms and $J^{(e-ph)}$ may be simplified:

$$J^{(e-ph)}(\varepsilon) \approx -\frac{4\pi\lambda\Delta^4}{(up_F)^2} \bar{n} \frac{n_\varepsilon}{\sqrt{\varepsilon^2 - \Delta^2}}. \tag{11.8}$$

The electron-electron collisions may cause a significant change in the distribution function if they are sufficiently intensive. In this case when the quasi particles are concentrated in a narrow layer near the Fermi surface and $n_\varepsilon \sim 1$ in this layer, we must first account for the "collisional pairing" processes (the second term in the component proportional to the factor M_3 in collision integral $J^{(e-e)}(\varepsilon)$) when the three electron excitations with energies $\varepsilon \geq 0$ in colliding form a bound state (a Cooper pair) and a free quasi particle with energy $\geq 3\Delta$. Their opposite "collisional multiplication" processes in this case are not efficient and hence $J^{(e-e)}(\varepsilon)$ may be reduced to the form

$$J^{(e-e)}(\varepsilon) \approx -\frac{3}{2\sqrt{2}}(a + b)\frac{\Delta^3}{\varepsilon_F}\bar{n}^2 \frac{n_\varepsilon}{\sqrt{\varepsilon^2 - \Delta^2}}. \tag{11.9}$$

Comparing (10.20) and (10.21) we may determine that the electron-electron collisions are not significant if the following parameter is small

$$c = c_0 \frac{\bar{n}\Delta_0}{\Delta}, \qquad c_0 = \frac{3(a+b)}{8\sqrt{2}\pi\lambda} \frac{\omega_D^2}{\varepsilon_F \Delta_0} \tag{11.10}$$

(we set $\omega_D \approx up_F$). For metals with relatively large Debye frequencies (such as Al, see table) the value of c_0 may be only moderately small and it becomes necessary to account for c. Unfortunately, the factors

a and b in (11.10) are not very well known experimentally and hence below we are limited to considering values of c_0 = 0, 1, and 10.

2. Resonant electromagnetic field pumping [irradiation]. This case is analyzed in detail in study [2]. In the limiting case of intense fields the distribution function n_z takes the form

$$n_z = \frac{\sqrt{z}}{\sqrt{z}+\sqrt{1-z}}\,\theta(1-z), \qquad z \geqslant 0. \tag{11.11}$$

Substituting (11.11) into (10.12) causes the first of the integrals at frequency $\omega_q = \omega_0$ to vanish while the second integral, although it is small, is not positive, and hence instability condition (10.12) is not satisfied (at other frequencies ω_q the first integral is also positive and hence the frequency $\omega_q = \omega_0$ is the least stable).

Thus, in a microwave field the phonon instability evidently[5] does not exist even in intense fields.

3. The phonon instability in symmetrical tunnel injection. In the case of an SiS-junction quasi-particle source (2.10) takes the form (ignoring imbalance):

$$Q(\varepsilon) = \dot{\nu}\,[U_-(n_{\varepsilon-V} - n_\varepsilon) + U_+(n_\varepsilon - n_{\varepsilon+V}) + V_0(1 - n_\varepsilon - n_{V-\varepsilon})], \tag{11.12}$$

$$U_\pm = \frac{\varepsilon(\varepsilon \pm V)\,\theta(\varepsilon \pm V - \Delta)}{\sqrt{\varepsilon^2-\Delta^2}\sqrt{(\varepsilon \pm V)^2-\Delta^2}}, \qquad V_0 = \frac{\varepsilon(V-\varepsilon)\,\theta(V-\varepsilon-\Delta)}{\sqrt{\varepsilon^2-\Delta^2}\,\sqrt{(\Delta-\varepsilon)^2-\Delta^2}}\,. \tag{11.13}$$

The analysis of kinetic equations is carried out here in the same manner as for the case of a microwave field [2]. The nonequilibrium electron distribution function is given by the expression

$$n_z = \frac{A\sqrt{z}\,\theta(1-z)}{A(\sqrt{z}+\sqrt{1-z})+\sqrt{z}\,\sqrt{1-z}}, \qquad z \geqslant 0, \tag{11.14}$$

$$A = \begin{cases} \Gamma\Delta/2\delta, & \Gamma \gg \delta/\Delta, \\ (\Gamma\Delta/\pi\delta)^{1/2}, & \Gamma \ll \delta/\Delta; \quad \delta = (V - 2\Delta) \ll \Delta. \end{cases} \tag{11.15}$$

In spite of the fact that $\Gamma \sim \nu/\lambda$ is assumed to be small, the narrowness of the injection source may drive the distribution function to saturation (11.11) if $\Gamma \gg \delta/\Delta$. As noted above even in the saturation state there is no phonon instability. We may easily see that accounting for electron-electron inelastic collisions (in effective approximation (11.9)) also does not change this result.

Numerical calculations of the absorption coefficient accounting for the branch imbalance given in the preceding chapter confirm the conclusions here.

---

[5] In principle the situation may change at the final temperatures if we account for the contribution of inelastic electron-electron collisions together with the Eliashberg mechanism [35]. Unfortunately this problem has not been investigated in sufficient detail.

12. Asymmetrical SiS'-junction

This situation changes radically in the case of an *SiS'*-junction. This case was first examined in the study by Genkin and Protogenov [31] where it was discovered that the function n_ε in the region of ε over the gap may be close to unity. Bearing in mind the fact that this examination is not comprehensive,[6] we will provide a solution here of equations (11.1)-(11.4) in a more exact approximation.

As indicated by the results from Chapter 1 the initial equations for the *SiS'*-junction appear to be analogous to the case of a symmetrical *SiS*-junction with the difference appearing only in that the distribution function entering into (11.12) for the source $Q(\varepsilon)$ with deviating arguments relates to a bulk superconductor, while the factors $U_\pm$ and V_0 are

$$U_\pm = \frac{\varepsilon(\varepsilon+V)}{\sqrt{\varepsilon^2-\Delta^2}\sqrt{(\varepsilon\pm V)^2-\Delta'^2}}, \qquad V_0 = \frac{\varepsilon(V-\varepsilon)}{\sqrt{\varepsilon^2-\Delta^2}\sqrt{(V-\varepsilon)^2-\Delta'^2}} \tag{12.1}$$

(Δ' is the gap of the bulk injector).

If the width of the injector is much greater than the width of the thin film which in turn does not exceed the diffusion length of the quasi particles, the electron subsystem of the bulk superconductor may be considered to be unperturbed even when the thin film is significantly out of equilibrium. Assuming "narrowness" of the resulting highly-nonequilibrium ($n_\varepsilon \sim 1$) distribution of quasi particles with $T = 0$ the solution of equations (11.2)-(11.4) subject to (11.6), (11.8), (11.9), (11.12), and (12.1) is represented as

$$n_z = \frac{B\theta(1-z)}{B+\sqrt{1-z}}, \qquad z \geqslant 0, \tag{12.2}$$

where

$$B = \frac{A}{\bar{n}(1+c)}, \quad A = \sqrt{\frac{\Delta'}{2\delta}}\,\Gamma, \quad \Gamma = \Gamma_0/\Delta^3, \quad \Gamma_0 = \nu\omega_D^2/4\pi\lambda, \quad \Delta = \Delta_0 \exp(-2\bar{n}), \tag{12.3}$$

while the parameter $\bar{n}$ will be determined from the equation

$$\bar{n} = \sqrt{\frac{\delta}{2\Delta}}\, A \int_0^1 \frac{dz}{\sqrt{z}\,[A+\sqrt{1-z}\,(1+c)\,\bar{n}]}. \tag{12.4}$$

---

[6] Unfortunately study [31] did not carry out a detailed analysis of the kinetic equations accounting for the self-consistency equation and did not reveal the ambiguity of the solutions of the kinetic equations related to this. This approach is not sufficient for our purposes.

We remember that the quantities Δ, δ, c and A in (12.3) are dependent on $\bar{n}$ and equation (12.4) determines the values of B for given injection parameters. Henceforth it will be convenient to reduce (12.4) to the form

$$\bar{n}\left(\frac{\Delta}{2\delta}\right)^{1/2} = B\frac{\pi}{2} - \frac{B^2}{\sqrt{1-B^2}}\begin{cases} \ln\left|\dfrac{1+B+\sqrt{1+B^2}}{-1-B+\sqrt{1-B^2}}\right|, & B<1, \\ 2\,\mathrm{arctg}\sqrt{\dfrac{B-1}{B+1}}, & B>1. \end{cases} \tag{12.5}$$

Equation (12.5) allows an analytic investigation in the limiting cases (for simplicity we will now set $c = 0$).

a) Assume the external parameters V and Γ_0 are such that the self-consistent value of $\bar{n}$ results in $B \ll 1$. Then (12.5) subject to (12.3) is reduced to the form

$$\bar{n}\exp(-7\bar{n}) = \frac{\pi}{2}\sqrt{\frac{\Delta'}{\Delta_0}}\frac{\Gamma_0}{\Delta_0^3}. \tag{12.6}$$

We note that in this case the roots of equation (12.6) are independent of the parameter δ, although the range of applicability of equation (12.6) does depend on δ since because $B \ll 1$ we require the steady-state solution to satisfy the inequality

$$\Gamma_0 \ll \sqrt{2\delta/\Delta'}\,\Delta^3\bar{n}. \tag{12.7}$$

The left half of (12.6) has a maximum at $\bar{n} = 2/7$. This value of $\bar{n}$ is the root if

$$\Gamma_0 = \Gamma_0^* = \left(\frac{2}{7e}\right)^2\frac{\Delta_0^3}{\pi}\sqrt{\frac{\Delta_0}{\Delta'}} \tag{12.8}$$

(here e is the base of the natural logarithm). With larger values of Γ_0 there are, obviously, no solutions with small values of B, while with larger values of Γ_0 there are two roots and as Γ_0 drops the smallest of the roots behaves as

$$\bar{n} \approx \left[\frac{\pi}{2}\sqrt{\frac{\Delta'}{\Delta_0}}\frac{\Gamma_0}{\Delta_0^3}\right]^{1/2}, \tag{12.9}$$

while the second increases rather rapidly finally ending up outside the range of applicability of our examination (we remember that we are assuming that the value of $\bar{n}$ is small). Near the characteristic value $\Gamma_0 \approx \Gamma_0^*$ we have the following condition for δ:

$$\delta/\Delta_0 \gg 10^{-2}. \tag{12.10}$$

Thus, if $\Gamma_0 \leqslant \Gamma_0^*$, the self-consistent kinetic equations always allow two solutions with $B \ll 1$, since we may easily see that condition (12.10) is satisfied if only δ_0 is not too small (δ_0 is the initial level over the critical level, $\delta_0 = V - \Delta' - \Delta$; for small values of $\bar{n}$ we have $\delta - \delta_0 \approx 2n$).

b) In the opposite limiting case $B \gg 1$ the right half of (12.5) approaches unity and hence (12.5) has a small root only with small values of δ. Using relation (11.7) we obtain from (12.5) the following expression for the small root:

$$\bar{n} \approx 2\left[1-\sqrt{1+\frac{1}{2}\frac{\delta_0}{\Delta_0}}\right], \tag{12.11}$$

which becomes meaningless when $\delta_0 > 0$. Thus, the self-consistent equation system with negative initial displacement ($\delta_0 < 0$) has a solution with

$$\bar{n} \approx -{}^1/_2\ \delta_0, \tag{12.12}$$

here $B \gg 1$ if $|\delta_0|$ is sufficiently small.

c) The subject of greatest interest is the intermediate case $b \approx 1$, since it is precisely with these values that the phonon instability is achieved. In this case the right half of (12.5) is equal to $(1/2\pi - 1)$ and for the small root the same value of (12.12) is obtained. Thus the roots of interest to us in cases b) and c) coincide and differ only in the quantity B determined by the external parameters δ_0 and Γ_0.

A numerical analysis of transcendental equation (12.5) confirms the concepts outlined here. Thus, for the parameters $\Gamma_0 = 10^{-3}$, $\gamma_0 = -0.02$, $c_0 = 0$, $\Delta' = 2$ (in units of Δ_0) we have three roots of equation (12.5): $\bar{n}_1 \approx 0.01$, $(B \approx 14)$, $\bar{n}_2 \approx 0.04$ $(B \approx 0.1)$; $\bar{n}_3 \approx 0.95$ $(B \approx 0.3)$. In the case $\Gamma_0 > \Gamma_0^*$, for example, $\tau_0 - 0.05$, we have (with $\delta_0 = -0.3$, $\Delta' - 0.5$) only a single solution $\bar{n}_1 \approx 0.2$ (here $B \approx 2.9$; $\delta \approx 0.02$; $\Delta \approx 0.67$) for the case $c_0 = 0$. With growth of c_0 the value of B drops ($B \approx 1.9$ when $c_0 = 1$) and at sufficiently large values of c_0 (for example, $c_0 = 10$) there are generally no roots. Thus, the behavior of the tunnel SiS'-junction is extraordinarily sensitive to the initial injection parameters and the parameters of the junction itself.

The situation we are now dealing with has become the subject of intense study in recent years in nonequilibrium superconductivity theory. Specifically, in SiS-injection there is also a multitude of solutions (see, for example, the results from study [8] as well as Chapter 2). We note that in addition to the solutions we derived $\bar{n}_{1-3}$ where $\delta_0 < 0$ there also exists the solution $\bar{n}_0 = 0$ corresponding to the situation where there are no excess quasi particles as well as the solution corresponding to the normal state ($\Delta = 0$). If we avoid solutions with large values of $\bar{n}$[7] in the general case we will ascertain the existence of three solutions where the case $b \sim 1$ corresponds to the intermediate value of $\bar{n}_1$. Proceeding analogous to the SiS-junction we may expect that this state is not stable, while the E-I char-

---

[7] Consideration of the cases corresponding to large values of $\bar{n}$ requires going beyond the approximations used here. It is important to bear in mind that states with large values of $\bar{n}$ are separated from states with small values of $\bar{n}$ by a large energy barrier and transitions between them are highly improbable [21].

acteristic of the SiS'-junction is S-shaped and a spatially-inhomogeneous state is achieved in the superconducting film. The possibility of achieving the acoustical quantum-mechanical oscillator state in this case requires additional analysis. However, we will not examine this issue here, since as noted above, when $\Gamma_0 > \Gamma_0^*$ and $\delta_0 < 0$ (see, specifically, the cited case with $\Gamma_0 = 0.05$) there exists only one solution corresponding to the superconducting state with excess electron excitations and here the S-shape of the E-I characteristic vanishes.

The vanishing of the S-shape as Γ_0 increases, i.e., as the degree of homogeneity in the tunnel junction increases, continuing the analogy to study [14] between the nonlinear behavior of a nonequilibrium superconductor and the behavior of a nonideal Van der Waals gas. It is specifically an increase in the degree of inhomogeneity in the junction that is analogous to an increase in the temperature of a nonideal gas causing the Van der Waals isotherm to assume a monotonic form. This fact significantly simplifies consideration of the stability problem when achieving the critical state in an acoustic quantum-mechanical oscillator.

13. The role of fluctuations

An analysis of the role of fluctuations is extraordinarily important for solving the possibility of achieving an acoustical quantum-mechanical oscillator. In nonequilibrium superconductors there are many factors that influence the stability of the steady state (including, for example, fluctuations in the superfluidity velocity [30], high-frequency fluctuations in the order parameter [29] and an electromagnetic field [21], etc.).

We will begin with stability with respect to high-frequency fluctuations in Δ. The corresponding criterion of instability was obtained by Aronov and Gurevich and is related to the change in sign of the damping of the modes of the natural collective oscillations of the electron subsystem in the superconductor. The stable boundary value of Δ is determined by the change in the sign of the expression

$$P_i(\Delta)=\int_{\Delta}^{\infty}\frac{(1-2n_\varepsilon)\,d\varepsilon}{\sqrt{\varepsilon^2-\Delta^2}\sqrt{\varepsilon^2-\omega_i^2/4}}, \qquad (13.1)$$

where the frequency of the i-mode is determined by the equation

$$n(\omega_i/2) = 1/2. \qquad (13.2)$$

Moreover, the mode $\omega_0 = 0$ may also be unstable. In the case $B \gg 1$ which we will be using, the expression (13.1) is written (for a narrow source) in the form[8]

$$P_0(\Delta) = (1 - 2\bar{n})/\Delta^2, \tag{13.3}$$

and, consequently, the rapid "collapse" of the Cooper pairs may occur even when $\bar{n} \geqslant 1/2$. In other words with small values of $\bar{n}$ no instability occurs.

We will now consider instability with respect to fluctuations in the superfluidity velocity v_s [30] associated with the fact that the response of a nonequilibrium superconductor to an external magnetic field becomes paramagnetic [36] and here the sign of the "density of the superfluid component" N_s changes:

$$N_s = 1 + 2\int_{\Delta}^{\infty} \frac{\partial n}{\partial \varepsilon} \frac{\varepsilon}{\sqrt{\varepsilon^2 - \Delta^2}}\, d\varepsilon = 1 + \sqrt{\frac{2\Delta}{\delta}} \int_0^1 \frac{\partial n}{\partial z} \frac{dz}{\sqrt{z}}. \tag{13.4}$$

As indicated from (13.4) with the small values of δ and $B \sim 1$ we employ here, N_s may change sign. (We note that in fact a change in the sign of N_s does not mean the development of instability.) A strict analysis based on a solution of Maxwell-London equations carried out in study [31] has demonstrated that the existence of a junction with a bulk superconductor stabilizes the situation in a thin film if

$$\lambda' d < |\lambda^2|. \tag{13.5}$$

here λ' and λ are the depths of penetration of the magnetic field related to the London relation $\lambda^{-2} = \lambda_L^{-2} N_s$.

However the phonon instability occurs earlier than N_s becomes negative. In the case $\delta \ll \Delta$ this may be observed rather easily. We may drop the 1 in expression (13.4) while we may drop the second integral in (10.12) and set $\omega_q = V$ in the first (the least stable mode). Then by using explicit expression (12.2) for n_ε and the fact that the remaining integrals are expressed in quadrature, we may see that N_s changes sign when $b \geqslant 1$, while the phonon instability arises when $B \geqslant 0.54$. Thus, with an appropriate selection of parameters the instability related to the reversal of the sign of N_s will not be critical.

Slow spatial and temporal fluctuations in the excitation distribution function and order parameter, accounting for the multiple solutions of kinetic equations, may cause the appearance of a spatially-inhomogeneous state. When there is a unique small value of $\bar{n} \neq 0$ for the distribution of the type in (12.1), the problem is considered in study [31]. The results from study [31] indicate stability of homogeneous distribution (12.2) with an arbitrary value of B.

---

[8] In the case $B > 1$ no instability occurs when $i \neq 0$ (compare to [29]).

Regarding the stability of the situation with regard to high-frequency electromagnetic field fluctuations then, as noted in study [31] (see also [16]) an instability with respect to electromagnetic wave generation may occur in London superconductors. We should, however, bear in mind that such an "instability" does not yet signify breakdown of the homogeneous superconducting state. This is due to the fact [21] that the conductivity has a significant imaginary part that suppresses possible instability. In spite of this a precise system could operate as an amplifier of electromagnetic waves [16] by reflection.

Finally we note that fluctuations of the longitudinal electrical field related to the branch imbalance in the electron excitation spectrum are also possible in superconductors. However in our case of $T = 0$ (and in "narrow" distributions) the imbalance is insignificant and we may, evidently, assume that no instability develops with respect to fluctuations in the longitudinal field.

Clearly the investigation outlined in this section cannot be considered exhaustive, since it is quite difficult to consider all possible paths of development of an instability. We consider the problem of an acoustical quantum-mechanical oscillator to be sufficiently interesting and important and hence together with the following theoretical analysis it would be desirable to formulate the appropriate experiments. We will provide the numerical values of certain possible parameters. Experiment has indicated that nonequilibrium states are achieved most easily and most often in aluminum films as a result of the relatively long lifetime of the electron excitations. Characteristic values of γ for aluminum are estimated at $\gamma \sim 10^7$–10^8 s^{-1} (see table) and if $\nu \sim 0.01\gamma$, then for aluminum with $d \sim 10^{-4}$ cm we have from (2.15) the estimate: $RS \sim 10^{-5}$ ohms$\cdot$cm^2. Such low-resistance junctions may be implemented in practice (see, for example, [37, 38]). Another area requiring attention is the need to satisfy inequality (13.5) that imposes additional conditions on the pair comprising the SiS'-junction.

We note that the speed of sound in a superconductor is significantly less than the speed of light, while the dimensionless electron-phonon interaction constant exceeds the corresponding constant of electromagnetic interaction. As a result the gain of the acoustical quantum-mechanical oscillator is quite high. Using operator (10.8) we may obtain the following expression for the gain

$$K(\omega_q) = \frac{\pi\lambda}{2} \frac{\omega_D}{\varepsilon_F} \frac{\Delta}{u} I^0(\omega_q), \tag{13.6}$$

where $I^0(\omega_q)$ is the quantity in braces in (10.12), and for characteristic metals (see table) we have from (13.6) $K(\omega_q) \sim 10^3$ cm^{-3} at pumping corresponding to $I^0(\omega_q) \sim 1$. This is higher than the working gain of gas, solid-state and superconductor lasers. Hence, unlike a regular laser, the acoustic quantum-mechanical oscillator state in principle may be achieved without using an oscillator and the coherent phonon flux will propagate along the thin film. Moreover, when necessary it is possible to take advantage of the fact that at the metal-liquid helium interface the phonons experience total internal reflection and a ring laser scheme is achieved.

And, finally, we wish to identify one additional fact. The wavelength of phonons emitted by the acoustical quantum-mechanical oscillator are in the order to 10^{-5} cm at the same time that the energy of the optical channel with this same wavelength is five orders of magnitude greater. This energy feature of the acoustical quantum-mechanical oscillator based on a superconductor may make it quite interesting from the practical viewpoint.

CONCLUSION

We will briefly outline the results obtained in this study. In our view the most important element is investigating the possibility for an acoustical quantum-mechanical oscillator based on a nonequilibrium superconductor as discussed in the last chapter. As the natural basis of the concluding section, the preceding chapters are of independent interest. In particular we derived a specific expression for the tunnel source that accounts for macroscopic phase coherence in the tunnel junction and the branch imbalance of electron-hole excitations. We also derived the canonical form of the electron-electron collision operator that is valid when the imbalance noted above exists together with an expression for the current through the tunnel junction in nonequilibrium conditions. Even as early as the first stage of using these results we noted such a phenomenon as quantum oscillations in the distribution function of electrons in a nonequilibrium tunnel superconductor junction whereby the predicted "vibration" of the chemical potential and satellite formation in the emission scattered by the junction appear. We focused on the kinetics of the phonon subsystem. We also derived the electron-phonon collision operator by analytic continuation and on this basis described the canonical form of the collision integral which made it possible, by incorporating results of a numerical analysis of the kinetics of electron-hole excitations, to derive the spectral dependence of the phonon fluxes from a tunnel junction in nonequilibrium conditions. It turned out that the phonon deficit effect predicted previously appears even when there are excess quasi particles generated by the field from the condensate in the superconductors.

In this study we dealt with only a few aspects of the physics of nonequilibrium superconductivity. Without a doubt further research in this field will produce new interesting results.

BIBLIOGRAPHY

1. Eliashberg, G.M. "Inelastic electron collisions and nonequilibrium steady states in superconductors" ZhETF, 1971, V. 61, No. 3(9), p. 1254-1271.

2. Gulian, A.M., Zharkov, G.F. "The kinetics of nonequilibrium phonons and electrons in superconductors in a high-frequency electromagnetic field" Tr. FIAN, 1983, V. 148, p. 83-137.

3. Clarke, J. "Experimental observation of pair-quasiparticle potential difference in nonequilibrium superconductors" PHYS. REV. LETT., 1972, V. 28, No. 21, p. 1363-1366.

4. Tinkham, M. "Tunneling generation, relaxation and tunneling detection of hole-electron imbalance in superconductors" PHYS. REV. B, 1972, V. 6, No. 5, p. 1747-1756.

5. Kirichenko, I.K., Seminozhenko, V.P. "The distribution of nonequilibrium excitations and phonon generation in superconducting junctions" FNT, 1977, V. 3, No. 8, p. 986-1000.

6. Lapkin, A.I., Ovchinnikov, Yu.N. "The tunnel effect between superconductors in an alternating field" ZhETF, 1966, V. 51, No. 5, p. 1536-1543.

7. Bulyzhenkov, I.E., Ieolev, B.I. "Nonequilibrium phenomena in superconductor junctions" ZhETF, 1978, V. 74, No. 1, p. 224-235.

8. Elesin, V.F., Kopaev, Yu.V. "Superconductors with excess quasi particles" UFN, 1981, V. 133, No. 21, p. 259-307.

9. Keldysh, L.V. "The diagram technique for nonequilibrium processes" ZhETF, 1964, V. 4, No. 4(10), 1515-1527.

10. Fizika fononov bol'shikh energii [The physics of high energy phonons] Edited by I.B. Levinson, Moscow: Mir, 1976, 266 p.

11. Berberich, P., Buemann, R., Kinder, H. "Monochromatic phonon generation by the Josephson effect" PHYS. REV. LETT., 1982, V. 40, No. 20, p. 1500-1503.

12. Valeev, V.G., Kukharenko, Yu.A. "Kinetic equations for superconductors with a collision integral that conserves the number of particles" Prepr. FIAN, No. 111, Moscow, 1983, 25 p.

13. Lapkin, A.I., Obchinnikov, Yu.N. "Nonlinear effects from vortex motion in superconductors" ZhETF, 1977, V. 73, No. 1(7), p. 299-312.

14. Volkov, A.F. "Nonequilibrium states in superconducting tunnel structures" ZhETF, 1975, V. 68, No. 2, p. 756-765.

15. Abrikosov, A.A., Gor'kov, A.P., Dzyaloshinskiy, I.E. "Metody kvantovoy teorii polya v statisticheskoy fizike" [Methods from quantum field theory in statistical physics] Moscow: Fizmatgiz, 1962, 443 p.

16. Ivlev, V.I., Lisitsyn, S.G., Eliashberg, G.M. "Nonequilibrium excitations in superconductors in high-frequency fields" J. LOW TEMP. PHYS., 1973, V. 10, No. 3/4, p. 449.

17. Landau, L.D., Lifshits, Ye.M. "Elektrodinamika sploshnykh sred." [Continuum electrodynamics] Moscow: Nauka, 1982.

18. Golovashkin, A.I., Elenskiy, V.G., Likharev, K.K. "Effekt dzhozefsona i ego primenenie: Bibliogr. ukaz. [The Josephson effect and its application] 1962-1980, Moscow: Nauka, 1983, 222 p.

19. Chang, J.J., Scalapino, D.J. "Nonequilibrium superconductivity" J. LOW. TEMP. PHYS., 1978, V. 31, No. 1/2, p. 1-32.

20. Otschik, P., Eschrig, H., Lange, F. "Sound wave amplification in superconductors" J. LOW TEMP. PHYS., 1981, V. 43, No. 3/4, p. 397-408.

21. Aronov, A.G., Spivak, B.Z. "The properties of nonequilibrium excitation superconductors" FTT, 1976, V. 18, No. 2, p. 541-553.

22. Gulian, A.M., Zharkov, G.F. "The kinetics of phonons in nonequilibrium superconductors in an external magnetic field" ZhETF, 1981, V. 80, No. 1, p. 303-325.

23. Gor'kov, L.P., Eliashberg, G.M. "The generalization of Ginzburg-Landau theory for nonsteady-state problems in the case of alloys containing paramagnetic dopants" ZhETF, 1968, V. 54, No. 2, p. 612-626.

24. Bolkov, A.F., Kogan, Sh.M. "Collisionless relaxation of the energy gap in superconductors" ZhEFT, 1973, V. 65, V. 5(11), p. 2038-2045.

25. Chang, J.J., Scalapino, D.J. "Kinetic-equation approach to nonequilibrium superconductivity" PHYS. REV., B, 1977, V. 15, No. 5, p. 2651-2670.

26. Bulyzhenkov, I.E. "Amplification of hypersound in nonequilibrium superconductors" FNT, 1979, V. 5, No. 12, p. 1386-1390.

27. Chang, J.J., Scalapino, D.J. "Transport properties of nonequilibrium superconductors" PHYS. REV. B, 1974, V. 10, No. 9, p. 4047-4057.

28. Chang, J.J., Scalapino, D.J. "New instabilities in superconductors under external dynamic pair breaking" PHYS. REV. B, 1974, V. 10, No. 9, p. 4047-4757.

29. Aronov, A.G., Gurevich, V.L. "The stability of nonequilibrium Fermi distributions with respect to Cooper pairing" ZhETF 1973, V. 65, No. 3(9), p. 1111.

30. Baru, V.G., Sukhanov, A.A. "New types of instability in nonequilibrium superconductor excitation" PIS'MA V ZhETF 1975, V. 21, No. 4, p. 209-212.

31. Genkin, V.M., Protogenov, A.P. "Nonequilibrium states in tunneling in superconductors" FTT, 1976, V. 18, No. 1, p. 24-32.

32. Eckern, V., Schmid, A., Schmutz, M., Shön, G. "Stability of superconducting states out of thermal equilibrium" J. LOW TEMP. PHYS., 1979, V. 36, No. 5/6, p. 643-687.

33. Vardanyan, R.A., Ivlev, B.I. "The influence of laser emission in superconductivity" ZhETF, 1973, V. 65, No. 6(12), p. 2315-2336.

34. Elesin, V.F. "The nonequilibrium state of superconductors with optical excitation of quasi particles" ZhETF, 1974, V. 66, No. 5, p. 1755-1761.

35. Gulian, A.M., Zharkov, G.F. "Nonequilibrium kinetics of electrons and phonons in superconductors in intense UHF-fields" J. LOW TEMP. PHYS., 1982, V. 48, No. 1/2.

36. Aronov, A.G. "Paramagnetic effects in superconductors" PIS'MA V ZhETF 1973, V. 18, No. 6, p. 387-390.

37. Gray, K.E., Willemsen, H.W. "Inhomogeneous state of superconductors by intense tunnel injection of quasi particles" J. LOW TEMP. PHYS., 1978, V. 31, No. 5/6, p. 911.

38. Iguchi, J. "Nonequilibrium gap instability and phase transition under intense tunnel injection in superconducting lead films" J. LOW TEMP. PHYS., 1978, V. 33, No. 5/6.

39. Gulian, A.M., Zharkov, G.F. "Quantum oscillations of the nonequilibrium chemical potential in Josephson junction" PHYS. LETT. A, 1984, V. 103, No. 5, p. 283-285.

40. Gulian, A.M., Zharkov, G.F. "Electromagnetic emission scattering by a nonequilibrium Josephson junction" PIS'MA V ZhETF, 1984, V. 40, No. 4, p. 134-136.

41. Kurichenko, I.K., Seminozhenko, V.P. "The distribution of nonequilibrium excitations and phonon generation in superconducting tunnel junctions" FNT, 1977, V. 3, No. 8, p. 987-1001.

42. Gulian, A.M., Zharkov, G.F. "The 'phonon deficit' effect in superconductors induced by UHF-radiation" PHYS. LETT. A, 1980, V. 80, No. 1, p. 79-80.

43. Eliashberg, G.M. "High-frequency field-induced superconductivity of films" PIS'MA V ZhETF 1970, V. 11, No. 3, p. 186-188.

Equilibrium and Nonequilibrium Phenomena in Inhomogeneous and Weakly-Coupled Superconductors

A.D. Zaikin

Abstract: In this chapter we formulate a microscopic theory of the order parameter suppression effect in a superconductor near the N-S-boundary when $T \ll T_C$ and $T \sim T_C$. The Meissner effect in a normal metal layer making contact with the superconductor is investigated. A microscopic theory of the stationary Josephson effect in various types of *SNS* - junctions is formulated. Nonstationary and nonequilibrium properties of *SNS*-junctions with direct conductivity are investigated together with the same properties of *SNS* - junctions containing a dielectric interlayer. A microscopic theory of the nonstationary Josephson effect in such systems is formulated. The enchancement of supercurrent of such systems in an external microwave field is investigated together with a number of other effects. An expression is derived for the spectrum of minor modes in a system of Josephson junctions in granular superconductors that is not accompanied by a deviation of the quasi particle distribution function from equilibrium.

INTRODUCTION

To date there has been significant progress in theoretical in-estigations of inhomogeneous superconductors which relies primarily n the derivation and utilization of microscopic equations for energy-ntegrated quasi-classical Green-Gorkov functions first introduced in tudy [1]. In superconducting systems with an inhomogeneous con-iguration of the order parameter a critical role is played by struc-ures consisting of different metals, one of which is in a supercon-ucting state while the other is in a normal state; such metals are in irect contact. Such a situation is achieved in so-called SNS-junc-ions (superconductor-normal metal-superconductor junctions); alterna-ion of the *N*- and *S*-regions is also used in type-I superconductors in he intermediate state, as well as in layered and granular supercon-ucting structures, etc. The extensive investigation of such systems s due to the multiple and various interesting physical phenomena that ccur near the boundary between the superconducting metal and the nor-al metal; it is necessary to understand such phenomena to develop the hysics of the superconducting state which is even more important in iew of the significant practical importance of such systems. In this egard one of the most elegant and interesting phenomena in supercon-ictivity is the Josephson effect that is extensively employed in odern electronics and computer technology. Josephson junctions have been used in high-sensitivity voltmeters and magnetometers as well a devices for detecting and generating electromagnetic and acoustica waves and computer memory components. We should note that the mos important Josephson junctions for practical application (as well a *SIS* and *SCS* junctions, where *I* is the dielectric interlayer, and *C* i geometrical constriction) are SNS-junctions.

The foundation of the theoretical analysis of normal metal-su perconductor systems (*NS*) was established by De Gennes [2] who inves tigated the configuration of the order parameter Δ of such a syste and demonstrated by means of Ginzburg-Landau equations that $\Delta(x)$ in superconductor near the critical temperature T_c changes at distance $\xi(T) \sim \xi_0[T_c/(T_c - T)]^{1/2}$, which is much greater than the characteris tic dimensions of the pair $\xi_0(\xi_0 = \upsilon_F/2\pi T_c$ is the coherence length, υ is the Fermi velocity). He also identified the important role o boundary conditions in the consideration of *NS*-systems.

The next step was by Zaytsev [3-5] who showed that the effec tive boundary condition for Ginzburg-Landau equations may be foun from an integral equation for Green-Gorkov functions of microscopi theory, which is linearized in the region $\xi_0 \ll x \ll \xi(T)$. The cor responding solution of Ginzburg-Landau equations is a hyperbolic tan gent solution, the random constant is found from boundary condition of the type $\Delta'(0) = \Delta(0)/\beta$, and β is determined from microscopi theory.

Such ideas were subsequently investigated in studies [6-8] However the Gorkov equations [9] containing information on the be havior of the Green's functions at interatomic distances, are rathe complex for solution in many inhomogeneous situations. Moreover, suc information is often extraneous in many problems of inhomogeneous su perconductivity. In this regard it is convenient to carry out Fourier transform of the Green's functions in terms of the coordinat difference which allows us to identify slow changes in such functior against a background of rapid oscillations in the interatomic distan ces. Likewise carrying out an energy integration process we may ob tain functions that from the mathematical viewpoint are significantl simpler than the Green's functions. At the same time the equatior for these functions are microscopic equations and may be used at ar bitrary temperatures to describe the properties of a system at distan ces much greater than the interatomic distances. Such equations wer derived by Eilenberger [1] as well as Larkin and Ovchinnikov [20] The quasi-classical approach has turned out to be quite useful in th investigation of *NS*-systems. For systems with a short free pat length $l \ll \xi_0$ (the dirty limit) further simplifications of the Eilen berger equations result in the Usadel equations [11] that were used i studies [12-14] to describe the proximity effect in dirty *NS*-system when $T \sim T_c$. Thus, the ideology for investigating the proximity ef fect in *NS*-systems may be considered well-developed and sufficientl incorporated into practice (at least with respect to the dirty case today.

It is important, however, to note at this point that the entire ;cheme for examining the proximity effect is valid only near the :ritical temperature of the superconductor $T \sim T_c$. At lower tempera- :ures the Ginzburg-Landau equations may no longer be used and in this :ase it is necessary to solve the microscopic equations in all space in a self-consistent manner which has not yet been done. Moreover, letermining the configuration of the order parameter and the exact ;reen's functions of the *NS*-system at low temperatures is an impor- :ant problem that is of interest in and of itself and from the view- oint of the use of the Josephson effect in various types of *SNS*-junc- ions and a number of other properties if various systems consisting f *N*- and *S*-metals.

The stationary Josephson effect [15] (i.e., the nondissipative urrent flow with zero voltage across the junction) in *SNS*-sandwiches t $T \sim T_c$ has been investigated in many studies [2, 7, 8, 12-14] and t was demonstrated that suppression of the order parameter of the uperconductor near the *NS*-boundary has a significant influence on the evel of Josephson current. Kulik [16] began the investigation of the tationary Josephson effect in pure *SNS*-junctions with $T \ll T_c$. Kulik emonstrated that the expression for the energy of the discrete spec- ral levels in such junctions (the Andreev levels [17, 18]) depends on he phase difference of the order parameters of two superconductors r, in other words, on the current flow. Studies [19] (for the case T 0 only) and [20] found the Green-Gorkov functions of pure *SNS*-junc- ions and the stationary Josephson current. It turned out that at low emperatures $T \lesssim v_F/d$ (d is the thickness of the *N*-layer) the "cur- ent/phase" relation deviates significantly from the sinusoidal law haracteristic of the *SIS* tunnel junctions [15], and when $T = 0$ there s a characteristic sawtooth form [19]. A detailed discussion of the hysical essence of this effect was provided in studies [21, 22].

The influence of a small quantity of nonmagnetic impurities $\xi_0 \ll$ $< d$ on the expression for the current was investigated in study 23]. However in these studies [16, 19, 23] as well as many other tudies devoted to an investigation of *SNS*-junctions, when $T \ll T_c$ a odel with a "stepwise" order parameter was employed, i.e., the order arameter at any point in the *S*-metal was considered to be equal to ts equilibrium value in a homogeneous superconductor, while in the *N*- etal it was assumed to be equal to zero. Such a model is convenient r calculations although it does not account for suppression of the rder parameter near the *N*-layer. Hence calculations of the current 1 such a model, strictly speaking, are not self-consistent. In order) calculate the influence of suppression of Δ on the Josephson cur- ent, additional research is required.

One additional issue is the problem of the influence of an ex- rnal magnetic field on the spectrum of the Andreev levels and the osephson effect in *SNS*-junctions. Galayko [24] and subsequently gadze and Kulik [25] demonstrated that when a magnetic field is ap- ied the discrete spectrum becomes unstable and is replaced by a uasi-continuous spectrum. A significant role in this process is played by the cited dependence of the energy of the Andreev levels o the gradient-invariant phase difference, which, generally speaking, i the presence of a magnetic field is a function of the coordinate [position]. The expression for the Josephson current in these conditions is significantly complicated [26]. There exist, however, systems in which the phase of the order parameter has, in general, n influence on the spectrum of the Andreev levels (*NS*- or *INS*-systems) or such influence is negligible (an *SNINS*-system). The issue of th behavior of the discrete spectrum in a magnetic field as well as th features of the Josephson effect in such conditions have not been considered in the literature.

However, if the primary regularities of the stationary equilibrium behavior of *NS*- and *SNS*-systems have already been identifie in research to date (in spite of the many cited and other importan issues that still require resolution), many nonstationary propertie of such systems (including the nonstationary Josephson effect) hav been investigated in only isolated cases or not at all.

The theory of the nonstationary Josephson effect (i.e., the pas sage of current through the junction with applied voltage) in tunne *SIS*-junctions was formulated as early as 1966 by Werthamer [27] Larkin and Ovchinnikov [28]. In this case the high resistance of th dielectric layer makes it possible to calculate the current by per turbation theory in the small parameter of the transparency of such a interlayer and to limit the calculations in the first order wit respect to this parameter. Here the entire voltage impacts the di electric layer, so the nonequilibrium effects associated with curren flow and penetration of the electrical field into the superconducto are small and may be manifest only in the next order of the expansio in terms of transparency. The situation in other types of Josephso junctions (*SCS*, *SNS*) is significantly more complex, since the nonequi librium nature of the excitation distribution function caused by th current flow may be significant. The simplest theory of superconduc tivity - the BCS theory [Bardeen-Cooper-Schrieffer]-Gorkov, which doe not account for the relaxation processes, is insufficient for describ ing the behavior of the system and in this case requires incorporatio of kinetic methods. The situation is made more complex by the stron inhomogeneity of the system.

All these difficulties long made it impossible to develo methods for a consistent microscopic solution of many issues as sociated with the description of various nonequilibrium states of su perconductors (including the nonstationary Josephson effect in *SCS* and *SNS*-junctions).

Significant progress in the investigation of nonequilibrium su perconductors was made in the study by Gorkov and Eliashberg [29] i which nonstationary Ginzburg-Landau equations were derived for gaples superconductors. The next important step in developing a theory o nonequilibrium superconductivity was the study by Eliashberg [30] Eliashberg derived general equations describing the kinetics of super onductors and in deriving such equations employed a self-consistent pproach in which the order parameter exists and conserves its meaning n nonequilibrium cases as well. A certain simplification of the examination is related to the possibility of carrying out energy integration of the Green's functions. In this regard the equations erived in study [30] may in some sense be considered a generalization f the Eilenberger equations to the nonequilibrium case. However, the umber of unknown functions in the Eliashberg equations increases significantly compared to the equilibrium situation (the matrix functions , G^R, G^A) related to the need to determine the very state (in the bsence of equilibrium) over which the statistical averaging should be arried out. The kinetic examination was used to solve a number of mportant and interesting problems of nonequilibrium superconductivity see, for example, [31-39]).

In studies [38, 39] the Eliashberg equations were used to formulate a microscopic theory of the nonstationary Josephson effect in *CS*-junctions. The direct generalization of the approach first used n the equilibrium case [40] made it possible to derive general expressions for the nonstationary current in narrow *SCS*-bridges as well s to investigate these expressions comprehensively in two voltage anges $eV \ll \Delta$ and $eV \gg \Delta$. It turned out that many characteristics of he nonstationary Josephson effect differ significantly from the nalogous characteristics of a simple resistive model [41]. An expression was derived for the so-called excess current whose existence n the current-voltage characteristic of the junction when $eV \gg \Delta$ for the case of gapless superconductors) was first noted in study 42].

A number of studies [43, 33, 34, 44, 45] have considered the roblem of electrical field penetration into the superconductor. This ffect is a significantly nonequilibrium effect. It was demonstrated n study [43] that current flow breaks the symmetry between the ranches of the electron and hole distribution (the so-called branch mbalance) which causes the electrical field to penetrate the superonductor to a significant distance $\sim \sqrt{D\tau_Q} \gg \xi(T)$ (D is the diffusion oefficient, τ_Q is the characteristic vanishing time of this imalance). This causes additional resistance to appear in the *NS*-sysem near T_c [44, 45] which in turn produces certain features of the E- characteristic of the *SNS*-junctions [46] ($T \sim T_c$). We should, howver, note that a consistent theory of the nonstationary Josephson ffect in *SNS*-junctions (this applies primarily to pure *SNS*-junctions) as not yet been formulated to date. At the same time such systems ave many specific features capable of producing a variety of nomalies in the nonstationary and nonequilibrium behavior of these unctions. This study is devoted to an investigation of these and any other issues in the theory of inhomogeneous and weakly-coupled uperconductor systems.

Chapter 1 investigates certain phenomena that occur between the oundary between the normal metal and the superconductor. In Section we consider the proximity effect in a pure SNS-system. It is demonstrated that at $T \ll T_c$ the order parameter of the superconductor i the case of equal Fermi velocities of both metals is approximately 1/ its value Δ_0 in the homogeneous case near the *NS*-boundary and change as it penetrates into the superconductor at the characteristic lengt $\sim\xi_0$. The case $T \sim T_c$ is also discussed. In Section 2 we formulate microscopic theory of the Meissner effect in *NS*-systems; in such sys tems the Meissner effect has many specific features.

In study [2] stationary Josephson current is investigated i various types of *SNS*-junctions. In Section 3 it is demonstrated o the basis of the formulated theory of the proximity effect that when $\ll T_c$ *and* $d \gg \xi_0$ suppression of the order parameter in an *SNS*-sandwic does not reduce the Josephson current compared to the calculated valu in the "stepwise" model [20]. In Section 4 we formulate a microscopi theory of the stationary Josephson effect in dirty *SNS*-junctions whic is valid for sandwiches when $T \ll T_c$ and for bridge structures at ran dom temperatures. In Section 5 the stationary Josephson current i investigated in *SNS*-sandwiches containing the dielectric interlayer i the *N*-layer. The derived boundary conditions make it possible in for mulating the theory to employ quasi-classical Eilenberger and Usade equations. The influence of an external magnetic field and impuritie on the Josephson current is investigated.

In Chapter 3 we formulate a theory of the nonstationary Joseph son effect in pure *SNS*-junctions. Section 6 provides the primar equations describing nonequilibrium processes in superconductors a well as general relations for the *G*-matrices that will be employe later. Section 7 derives the theory of the nonstationary Josephso effect in *SNINS*-systems with random voltages. The discrete spectru produces a number of special features in the E-I characteristic. I is demonstrated that a microwave field may cause a sharp increase i the critical current. Section 8 considers the nonstationary Josephso effect in pure *SNS*-bridges of variable thickness. The effect is in vestigated at both small and large voltages together with the respons to a low a.c. and stimulation of supercurrent in a microwave field.

Certain issues of the theory of vortex states in extende Josephson junctions and systems of such junctions are examined i Chapter 4. In establishing the current versus phase relation and th expression for critical current in microscopic theory, distances i the order of ξ_0 are significant (referring to the transverse dimen sions of the junction). At the same time the characteristic distanc over which the magnetic field changes in the junction (and its relate current) is in the order of λ_J, which is much greater than ξ_0. Thi makes it possible to ignore the current field when formulating th microscopic theory, and in investigating the dependence of the curren and the field on the position in the extended junction the $j(\varphi)$ rela tion may be assumed to be local (the familiar "scale separation", see for example [47, 48]). The *SIS* and *SNS*-junctions may have a length i the order of or greater then λJ (normally $\lambda_J \sim 0.1$ mm). In this cas the "self-action" of the current is significant. We know that th magnetic field may penetrate into the transition in the form of vor ices [49-51]. In Section 9 we derive an expression for the function hat plays the role of the free energy of such vortices in the presence of current and we briefly discuss the stability conditions of vortex states in a Josephson junction that to some degree models actual magnetometers. In Section 10 we derive the expression for the spectrum of minor vortex oscillations in a system of Josephson junction and granular superconductors. The existence of such oscillations n granular superconductors unlike bulk superconductors is not related o the branch imbalance.

Chapter 1

PHENOMENA NEAR THE NORMAL METAL-SUPERCONDUCTOR BOUNDARY

1. Proximity effect in the normal metal-superconductor system

As noted previously the most effective approach in the theory of he inhomogeneous superconducting state is the use of quasi-classical ilenberger equations [1] that may be written in the following form:

$$\mathbf{v}_F \nabla \hat{g}_\omega (\mathbf{v}_F, \mathbf{r}) + \left[(\omega + 2ie\mathbf{v}_F \mathbf{A}) \hat{\tau}_3 + \hat{\Delta} - \frac{v_F}{2l} \langle \hat{g}_\omega (\mathbf{v}_F, \mathbf{r}) \rangle, \hat{g}_\omega (\mathbf{v}_F, \mathbf{r}) \right] = 0, \tag{1.1}$$

$$\hat{g}_\omega = \begin{pmatrix} g_\omega & f_\omega \\ f_\omega^+ & -g_\omega \end{pmatrix}, \quad \hat{\Delta} = \begin{pmatrix} 0 & \Delta \\ -\Delta^* & 0 \end{pmatrix}, \quad g_\omega^2 + f_\omega f_\omega^+ = 1.$$

ere $\hat{g}_\omega$ is the Green's function integrated with respect to the energy ariables ξ_p:

$$\hat{g}_\omega (\mathbf{v}_F, \mathbf{r}) = \frac{i}{\pi} \int_{-\infty}^{\infty} d\xi_p \int \hat{G}_\omega (\mathbf{r}_1, \mathbf{r}_2) e^{-i\mathbf{p}(\mathbf{r}_1 - \mathbf{r}_2)} d(\mathbf{r}_1 - \mathbf{r}_2),$$

here $\mathbf{r} = (\mathbf{r}_1 + \mathbf{r}_2)/2$, $\xi_p = v_F (p - p_F)$; $p_F = m\mathbf{v}_F$ is the Fermi momentum; $\hat{G}_\omega$ is the regular matrix temperature Green-Gorkov function; $\omega = \pi T(2n + 1)$ is the Mössbauer frequency. Here and henceforth we utilize the auli matrix system:

$$\hat{1} = \begin{pmatrix} 1 & 0 \\ 0 & 1 \end{pmatrix}, \quad \hat{\tau}_1 = \begin{pmatrix} 0 & 1 \\ 1 & 0 \end{pmatrix}, \quad \hat{\tau}_2 = \begin{pmatrix} 0 & -i \\ i & 0 \end{pmatrix}, \quad \hat{\tau}_3 = \begin{pmatrix} 1 & 0 \\ 0 & -1 \end{pmatrix},$$

nd the brackets in (1.1) designate the commutator, while the angle rackets designate averaging over the directions of the vector $\mathbf{v}_F$; A s the vector potential. The current and order parameter of the superconductor are found by the formulae

$$\mathbf{j}(\mathbf{r}) = -\frac{iep_F}{\pi} T \sum_{\omega>0} \mathrm{Sp}\langle \mathbf{p}_F \hat{\tau}_3 \hat{g}_\omega(\mathbf{v}_F, \mathbf{r})\rangle, \tag{1.2}$$

$$\Delta(\mathbf{r}) = \lambda N(0)\, 2\pi T \sum_{\omega>0} \langle f_\omega(\mathbf{v}_F, \mathbf{r})\rangle, \qquad N(0) = \frac{mp_F}{2\pi^2}, \tag{1.3}$$

where λ is the BCS interaction constant ($\lambda > 0$ designates the existence of effective attraction between the electrons).

We will consider the following situation. Assume the superconductor occupies the half-space $x > 0$, while a normal metal (which we will assume is "truly" normal, i.e., $\lambda_N = 0$) occupies the half-space $x < 0$, and the metals are in good electrical contact. Electrons may travel from one half space to another and as a result the density of the Cooper pairs is nonzero in the N-metal, while the order parameter of the superconductor near the NS-boundary becomes less than its value in the homogeneous case.

We will assume that there is no current in the system ($A = 0$) and we will select the order parameter of the superconductor Δ and the f-function to be real. We will introduce the $h^{\pm}_\omega$ functions such that

$$h^{\pm}_\omega(\mathbf{v}_F, \mathbf{r}) = \frac{1}{2}(f_\omega(\mathbf{v}_F, \mathbf{r}) \pm f_\omega(-\mathbf{v}_F, \mathbf{r})). \tag{1.4}$$

Subject to (1.4) the Eilenberger equations may be rewritten as

$$2\omega h^{+}_\omega + \mathbf{v}_F \nabla h^{-}_\omega = 2\Delta g_\omega + \frac{v_F}{l}(g_\omega \langle h^{+}_\omega\rangle - h^{+}_\omega \langle g_\omega\rangle), \tag{1.5a}$$

$$\mathbf{v}_F \nabla h^{+}_\omega + \left(2\omega + \frac{v_F}{l}\langle g_\omega\rangle\right) h^{-}_\omega = 0, \tag{1.5b}$$

$$\mathbf{v}_F \nabla g_\omega - \left(2\Delta + \frac{v_F}{l}\langle h^{+}_\omega\rangle\right) h^{-}_\omega = 0, \tag{1.5c}$$

$$\Delta(\mathbf{r}) = \lambda N(0)\, 2\pi T \sum_{\omega>0} \langle h^{+}_\omega(\mathbf{v}_F, \mathbf{r})\rangle. \tag{1.6}$$

We will consider the case of few impurities $l \gg \xi_0$. Then after some simple transformations we have from equations (1.5):

$$v_x^2 \frac{d^2 h^{+}_\omega}{dx^2} - 4\Omega^2 h^{+}_\omega + 4\Delta\left(\int^x \frac{d\Delta}{dx'} h^{+}_\omega\, dx' + C\right) = 0, \qquad \Omega^2 = \omega^2 + \Delta^2. \tag{1.7}$$

Here we utilized the fact that the Eilenberger functions are dependent only on x and the component of the Fermi velocity v_x due to the homogeneity of the problem with respect to y and z. The random integration constant in (1.7) is determined from the condition

$$h^{+}_\omega(x \to \infty) = \Delta_0/\Omega_0, \qquad \Omega_0^2 = \Delta_0^2 + \omega^2. \tag{1.8}$$

Now representing $\Delta(x)$ and $h^{+}_\omega(x)$ as

$$\Delta(x) = \Delta_0 - \delta\Delta(x), \qquad h_\omega^+(v_x, x) = \frac{\Delta_0}{\Omega_0} - \delta h_\omega^+(v_x, x),$$

$$\delta\Delta = \lambda N(0)\, 2\pi T \sum_{\omega>0} \langle \delta h_\omega^+ \rangle, \tag{1.9}$$

we will have for the superconducting half-space

$$v_x^2 \frac{d^2 \delta h_\omega^+}{dx^2} - 4\Omega^2 \delta h_\omega^+ + 4\omega^2 \frac{\delta\Delta}{\Omega_0} - 4(\Delta_0 - \delta\Delta) \int^x \frac{d\delta\Delta}{dx'}\, \delta h_\omega^+\, dx' = 0. \tag{1.7a}$$

Above all we are interested in the low temperature range $T \ll T_C$. The investigation of the proximity effect in this range is most complex since it is necessary to solve microscopic equations in the entire superconductor. Moreover, we will also discuss the case $T \sim T_C$.

a) Low temperatures $T \ll T_C$. Above all we will establish the nature of the decay of the quantity $\Delta(x)$ far from the NS-boundary. When $T \neq 0$ we may always identify the range of values x $\left(x \gg \xi_0 \ln \frac{\Delta_0}{T}\right)$, in which equation (1.7a) is linearized for all values of ω (for a given T):

$$v_x^2 \frac{d^2 \delta h_\omega^+}{dx^2} - 4\Omega_0^2 \delta h_\omega^+ + 4\frac{\omega^2}{\Omega_0} \delta\Delta = 0. \tag{1.10}$$

The solution of this equation results in the relation

$$\delta h_{\omega p}^+ = \frac{\omega^2}{\Omega_0} \frac{\delta\Delta_p}{\Omega_0^2 - \left(\frac{v_x p}{2}\right)^2}, \tag{1.11}$$

where $\delta h_{\omega p}^+$ and $\delta\Delta_p$ are the Laplacian transforms of the quantities δh_ω^+ and $\delta\Delta$. Together with the self-consistency equation, (1.11) yields

$$1 = \lambda N(0) 2\pi T \sum_{\omega>0} \int_0^1 d\alpha \frac{\omega^2}{\Omega_0 [\Omega_0^2 - (v_F \alpha p/2)^2]}. \tag{1.12}$$

Equation (1.12) is the dispersion equation for p. Let its solution be $p = 0$. This means that in the expression for $\delta\Delta_p$ (and, consequently, for $\delta h_{\omega p}^+$) there is a pole, i.e., $\delta\Delta_p = A/(p - p_0)$, from which (subject to the requirement of the diminishing of $\delta\Delta(x)$ at infinity) we obtain

$$\delta\Delta(x) = A e^{-|p_0| x}, \tag{1.13}$$

where A is a constant. Equation (1.12) has the root $p_0^2 = \frac{4\Delta_0^2}{v_F^2 \alpha^2}$, which, however, must be dropped since p is not dependent on α. The solution of (1.12) is given in Appendix 1. Here we will provide only the final result. When $T \ll T_C$ in the principal approximation we have

$$p_0^2 = \frac{4(\Delta_0^2 + \gamma^2)}{v_F^2}, \qquad \gamma = T\left(\pi - 2\exp\left\{-\frac{\Delta_0^3}{\pi^3 T^3} \sum_{n=2}^{\infty} \frac{2^{n-1}(n-1)!}{(4n^2-1)(2n-1)!!}\right\}\right) \simeq \pi T, \tag{1.14}$$

i.e., $|p_0| \sim 2\Delta_0/v_F \sim \xi_0^{-1}$. We emphasize one additional fact: we will not allow the formal transformation to the case $T = 0$ in (1.12) since when $T = 0$ this equation is valid only at infinity and has no solutions (aside from a trivial solution). Obviously it does not follow from here that the law of decay of $\delta\Delta(x)$ for the case of large values of x in the case $T = 0$ will differ significantly from (1.13)-(1.14). With small values of x in order to determine $\delta\Delta(x)$ and $\delta h_\omega(x)$ we must solve exact equation (1.7a). In the high frequency range $\omega \gg \Delta_0$ it is rather easy to find the solution for $\delta \overset{+}{h}_\omega$. In this frequency range (for all values of x) (1.7a) becomes (1.10) and from here we have

$$\delta h_\omega^+(v_x, x) = \frac{\delta\Delta(x)}{\omega} + \dot{B}_\omega \exp\left\{-\frac{2\Omega_0 x}{v_x}\right\}. \quad (1.15)$$

Let the Fermi velocities of the N- and S-metals be equal. The continuity condition of the functions g_ω and f_ω on the transit trajectories of the electrons incorporates the continuity of the functions $\overset{+}{h}_\omega$ from which in turn we have the continuity of $\nabla \overset{+}{h}_\omega$ (see, for example (1.5b)). Joining the continuity $\overset{+}{h}_\omega$ and $\nabla \overset{+}{h}_\omega$ at the point $x = 0$ yields

$$|v_x| \frac{d\delta h_\omega^+}{dx}\bigg|_{x=0} = 2\omega\left(\delta h_\omega^+(0) - \frac{\Delta_0}{\Omega_0}\right). \quad (1.16)$$

At low frequencies $\omega \ll \Delta_0$ the function $\delta \overset{+}{h}_\omega(x)$ is small (when $\omega = 0$ $\delta h^+(x) \equiv 0$). In the frequency range $\omega \sim \Delta_0$ it is not possible to analytically solve equation (1.7a) and it is necessary to rely on numerical calculations. However before providing the results of these calculations we will make the following comment. Generally speaking the order parameter Δ sharply drops to zero near the NS-boundary at distances in the order of the interatomic distances and is not of special interest. Further, within the scope of the BCS model we cannot calculate $\Delta(x)$ with $x \lesssim v_F/\omega_D$ (ω_D is the Debye frequency). This is due to the fact that at such distances it is necessary to account for the delay effect of electron-phonon interaction which, obviously, cannot be accounted for in a model with four-fermion interaction. Moreover the length $v_F/\omega_D \sim 10^{-6}$ cm $\ll \xi_0$, so calculations of the order parameter at such short distances is also not of primary interest (the characteristic distance at which $\Delta(x)$ changes when $D \ll T_c$ is in the order of ξ_0). Accounting for this we must solve equation (1.7a) with boundary conditions (1.8) and (1.6) in the range $x \geqslant 0$ and the solution found must also satisfy the self-consistency equation in the range $x \gg v_F/\omega_D$. We will assume that $\delta\Delta(0)$ (and generally $\delta\Delta(x)$, $x \lesssim v_F/\omega_D$) is equal to $\delta\Delta(x)$ when $v_F/\omega_D \ll x \ll \xi_0$. Then we directly obtain from (1.15) and (1.16)

$$B_\omega = \frac{\Delta_0 - \delta\Delta(0)}{2\omega}, \quad \omega \gg \Delta_0. \quad (1.17)$$

Thus we know the solution of (1.7a) when $\omega \gg \Delta_0$ and it is given by formulae (1.15), (1.17). In this regard the self-consistency equation when $x \gg v_F/\omega_D$ may be rewritten in the form ($T \ll T_c$):

$$2\pi T \sum_{\omega>0}^{\omega_m} < \delta h^{+}_{\omega} (v_x, x) > = \delta\Delta (x) \ln \frac{2\omega_m}{\Delta_0} . \tag{1.18}$$

The quantity ω_m must satisfy the conditions

$$\omega_m \gg \Delta_0, \qquad \int_0^1 d\alpha \exp\left\{- \frac{2\omega_m x}{v_F \alpha}\right\} \ll \frac{\delta\Delta (x)}{\delta\Delta (0)} , \tag{1.19}$$

although it cannot be fixed exactly since ω_m, obviously, will not enter into the final result. In the range of frequencies $\omega < \omega_m$ the function $\delta h^{+}_{\omega}(v_x, x)$ satisfying the Eilenberger equations with the corresponding boundary conditions as well as equation (1.18) was found numerically. The results from numerical calculations are given in Fig. 1a and 2. The order parameter near the *NS*-boundary has the value $\Delta(0) \simeq 0.5\ \Delta_0$ and at distances of $\sim\xi_0$ approaches its equilibrium value. Fig. 2 gives results from numerical calculations of the function $\delta h^{+}_{\omega}(x)$ for several values of x. It is clear that when $\omega \ll \Delta_0$ the function $<\delta h^{+}_{\omega}>$ is small, and then increases and reaches its maximum value when $\omega \simeq 0.8\ \Delta_0$, after which it begins to decay, rather rapidly approaching $1/\omega$.

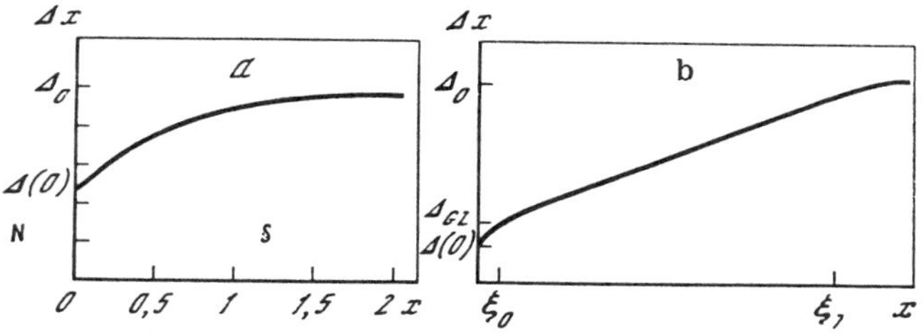

Fig. 1. The order parameter $\Delta(x)$ in an *NS*-system with $T \ll T_c$ (a) (the coordinate x is measured in units of length $v_F/2\Delta_0$) and with $T \sim T_c$ (b)

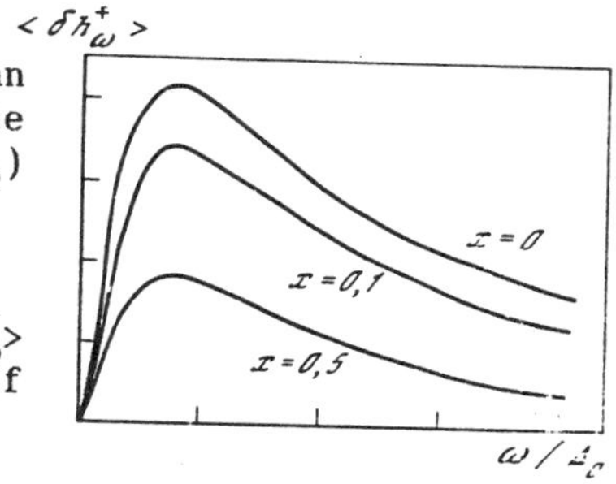

Fig. 2. Numerical calculations of $<\delta h^{+}_{\omega}>$ as a function of ω for various values of x ($T \ll T_c$)

We note that the behavior of $\Delta(x)$ and the f-function in *NS*-systems with a small free path length $l \ll \xi_0$ was investigated in study [55] in which, specifically, it was demonstrated that when $T \ll T_c$ the value of $\Delta(x) \simeq 0.4\ \Delta_0$. The characteristic distance over which $\Delta(x)$ changes in this case is equal to $\sqrt{\xi_0 l}$. Thus, the behavior of $\Delta(x)$ in pure and dirty *NS*-systems is of the same nature.

We note that the value of $\Delta(x)$ when $T \ll T_c$ is only slightly dependent on temperature. The reduction in $\Delta(0)$ at the final (yet low) temperature (compare to the case $T = 0$) is in the order of $\Delta_0 T/T_c$.

Study [56] used the following method to find the form of $\Delta(x$ near the NS-boundary. The Green-Gorkov functions of the NS-syste were calculated with a constant order parameter in the superconducto (i.e., $\Delta(x) = \Delta_0$, $x > 0$) and a zero value of the order parameter i the N-metal. Then the expression derived for the F-function was sub stituted into the right half of the self-consistency equation and $\Delta(x$ was calculated. Here it was assumed in study [56] that the functio $\tilde{\Delta}(x)$ found in this manner was only slightly different from the exac function $\Delta(x)$. This statement, however, is incorrect. In order t demonstrate this it is sufficient to calculate the Green's function by means of Gorkov equations by substituting into these equations $\tilde{\Delta}(x$ and then to again substitute the F-function into the self-consistenc equation. The $\tilde{\tilde{\Delta}}(x)$ value obtained in this manner near the NS-boundar is significantly different from $\tilde{\Delta}(x)$, i.e., the described iteratio process is not convergent and hence in this case cannot be used. Thi error was also repeated in later studies [57, 58]. Moreover, a stated above, self-consistency equation (1.6) should not be used fo calculating $\Delta(x)$ in direct proximity of the NS-boundary ($x \leqslant v_F/\omega_D$) which also reveals that results from studies [56-58] are in error.

b) High temperatures $T \sim T_C$. It is well known that at suc temperatures the order parameter of a superconductor sufficiently fa from the NS-boundary ($x \gg \xi_0$) is the solution of a Ginzburg-Landa equation of the type

$$\Delta_{GL}(x) = \Delta_0 \operatorname{th}\frac{x+\beta}{\sqrt{2}\,\xi(T)}, \qquad \xi^2(T) = \xi_0^2\,\frac{7\zeta(3)}{12}\,\frac{T_c}{T_c - T}, \tag{1.20}$$

where $\zeta(n)$ is the Riemann zeta-function. Equation (1.7) for $h_\omega^+(x$ when $T \sim T_C$ is significantly simplified:

$$4\omega^2 h_\omega^+ - v_x^2\,\frac{d^2 h_\omega^+}{dx^2} = 4\Delta\omega. \tag{1.21}$$

The self-consistency equation may be written in the following manner

$$2\pi T\sum_{\omega>0}\left(\frac{\Delta}{\omega} - \langle h_\omega^+\rangle\right) = 0. \tag{1.22}$$

At the point $x = 0$ the function h_ω^+ satisfies the boundary conditio resulting from the continuity of the functions h_ω^+ and $h_\omega^{+\prime}$:

$$h_\omega^+(0) = \frac{|v_x|}{2\omega}\,h_\omega^{+\prime}(0). \tag{1.23}$$

The exact solution of equation system (1.21)-(1.23) is quite complex Here we will provide an approximate solution. Subject to (1.22) th solution of equation (1.21) in a superconductor may be represented a

$$h_\omega^+(x) = \frac{\Delta(x)}{\omega} + \left(\frac{|v_x|\,\Delta'(0)}{4\omega^2} - \frac{\Delta(0)}{2\omega}\right)\exp\left\{-\frac{2\omega x}{|v_x|}\right\}. \tag{1.24}$$

It should be emphasized that $\Delta(x)$ when $x \leqslant \xi_0$ does not coincide wit $\Delta_{GL}(x)$ (1.20). Specifically, as will be clear from the discussio below, $\Delta(0)$ is significantly different from $\Delta_{GL}(0)$. Thus, we have

$$\Delta(0)/\Delta'(0) = \delta, \quad \Delta(x)/\Delta'(x) = x + \beta, \quad \xi_0 \ll x \ll \xi(T), \tag{1.25}$$

and it follows from this discussion that, generally speaking, $\Delta \neq \beta$.

Similar to the case of dirty systems [14] it is possible to express β through $h^+_\omega(0)$ and $h^{+\prime}_\omega(0)$. For this we multiply (1.21) by ω^{-2}, and average over the directions of v_F and sum with respect to ω. Using (1.22) we obtain

$$\sum_{\omega>0} \frac{1}{\omega^2} \frac{d^2}{dx^2} \langle v_x^2 h_\omega^+ \rangle = 0, \tag{1.26}$$

which yields

$$\sum_{\omega>0} \frac{1}{\omega^2} \langle v_x^2 h_\omega^+(x) \rangle = \sum_{\omega>0} \frac{1}{\omega^2} \langle v_x^2 h_\omega^+(0) \rangle + x \sum_{\omega>0} \frac{1}{\omega^2} \langle v_x^2 h_\omega^{+\prime}(0) \rangle. \tag{1.27}$$

We now multiply equation (1.21) by v_x^2/ω^4 and then again carry out averaging over the directions and sum over the frequencies. Then subject to (1.27) we will have

$$\frac{\Delta(x)}{3} v_F^2 \sum_{\omega>0} \omega^{-3} = \sum_{\omega>0} \frac{1}{\omega^2} \langle v_x^2 h_\omega^+(0) \rangle +$$
$$+ x \sum_{\omega>0} \frac{1}{\omega^2} \langle v_x^2 h_x^{+\prime}(0) \rangle - \sum_{\omega>0} \frac{1}{4\omega^4} \langle v_x^4 h_\omega^{+\prime\prime}(x) \rangle. \tag{1.28}$$

Further, as is normally done, we assume that $\Delta(x) \ll \Delta_0$ when $x \ll \xi(T)$. Then from (1.20) and (1.28) we have

$$\beta = \Big(\sum_{\omega>0} \frac{1}{\omega^2} \langle v_x^2 h_\omega^+(0) \rangle\Big) \Big/ \Big(\sum_{\omega>0} \frac{1}{\omega^2} \langle v_x^2 h_\omega^{+\prime}(0) \rangle\Big). \tag{1.29}$$

In deriving (1.29) we utilized the fact that when $x \gg \xi_0$ the third term in the right half of (1.28) is much less than the other two. However this neglect is no longer valid in the range $x \lesssim \xi_0$. Moreover, in order to establish the form of $\Delta(x)$ it is precisely this range of values of x that is the most significant.

We will substitute the approximate expression for $h^+_\omega(x)$ (1.24) into exact relation (1.28). Carrying out integration with respect to the directions v_F and summation with respect to the frequencies, we obtain

$$\Delta''(x) + a\xi_0^{-2}\Delta(x) = \Big(\frac{a}{2}\xi_0^{-2}\Delta(0) + b\xi_0^{-1}\Delta'(0)\Big)\Big(1 + \frac{x}{\beta}\Big) +$$
$$+ c\xi_0^{-2}\Delta(0) \sum_{n=0}^{\infty} \frac{1}{(2n+1)^3} \int_0^1 \alpha^2 \, d\alpha \exp\Big\{-\frac{(2n+1)x}{\xi_0 \alpha}\Big\} -$$
$$- c\xi_0^{-1}\Delta'(0) \sum_{n=0}^{\infty} \frac{1}{(2n+1)^4} \int_0^1 d^3 \, d\alpha \exp\Big\{-\frac{(2n+1)x}{\xi_0 \alpha}\Big\}, \tag{1.30}$$

where $a = \dfrac{35\zeta(3)}{93\zeta(5)}, \quad b = \dfrac{5\pi^4}{2976\zeta(5)}, \quad c = \dfrac{80}{31\zeta(5)}.$

Thus we have obtained an equation for $\Delta(x)$ valid with all x. We emphasize that (1.30) is not the Ginzburg-Landau equation and will not become this equation when $x \gg \xi_0$. This equation is a form of auxiliary equation that allows us to determine $\Delta(x)$ when $x \leqslant \xi_0$. In the vicinity $x \gg \xi_0$ the form of $\Delta(x)$ is known and is given by formula (1.20).

The solution of (1.30) that becomes (1.20) when $x \gg \xi_0$ takes the form

$$\Delta(x)=\left(\frac{\Delta(0)}{2}+\frac{b\xi_0\Delta'(0)}{a}\right)\left(1+\frac{x}{\beta}\right)+c\sum_{n=0}^{\infty}\int_0^1\alpha^4 d\alpha \times$$

$$\times \exp\left\{\frac{(2n+1)x}{\xi_0\alpha}\right\}(a\alpha^2+(2n+1)^2)^{-1}\left[\frac{\Delta(0)}{(2n+1)^3}-\frac{\alpha\xi_0\Delta'(0)}{(2n+1)^4}\right]. \tag{1.31}$$

Now we must set $x = 0$ in (1.31) in order to determine $\Delta(0)/\Delta'(0)$. Finally we obtain

$$\delta=\xi_0\left[\frac{b}{a}-c\sum_{n=0}^{\infty}\int_0^1\frac{\alpha^5 d\alpha}{(a\alpha^2+(2n+1)^2)(2n+1)^4}\right]\left[0{,}5-\right.$$

$$\left.-c\sum_{n=0}^{\infty}\int_0^1\frac{\alpha^4 d\alpha}{(a\alpha^2+(2n+1)^2)(2n+1)^3}\right]^{-1}\simeq 0{,}40\xi_0. \tag{1.32}$$

It is easy to establish the relation between β and δ by means of (1.24) and (1.29):

$$\beta=\frac{\delta\frac{7\zeta(3)}{12}+\xi_0\frac{\pi^4}{192}}{\delta\pi^2/8\xi_0+\frac{7\zeta(3)}{12}}. \tag{1.33}$$

Substituting (1.32) into (1.33) we have $\beta \simeq 0.66\xi_0$, which is somewhat different from the result obtained by Zaytsev [3, 4] ($\beta = 0.73\xi_0$). It is also easy to determine the value of $\Delta(0)$:

$$\Delta(0)=2\Delta_0\sqrt{\frac{7}{7\zeta(3)}\left(1-\frac{T}{T_c}\right)}\left[\frac{3\pi^2}{14\zeta(3)}+\frac{\xi_0}{\delta}\right]^{-1}\simeq$$

$$\simeq 0{,}395\Delta_0\sqrt{1-\frac{T}{T_c}}\cong 1{,}21(T_c-T). \tag{1.34}$$

Formulae (1.20), (1.31)-(1.33) completely determine the form of the superconductor parameter Δ for all values of x. Finally we have

$$\Delta(x)=\Delta_{GL}(x)+\frac{80\Delta(0)}{31\zeta(5)}\sum_{n=0}^{\infty}\int_0^1\alpha^4 d\alpha\exp\left\{-\frac{(2n+1)x}{\xi_0\alpha}\right\}\times$$

$$\times\left[\frac{1}{(2n+1)^3}-\frac{\alpha\xi_0/\delta}{(2n+)^4}\right]\left[\frac{35\zeta(3)}{93\zeta(5)}\alpha^2+(2n+1)^2\right]^{-1}. \tag{1.35}$$

The function $\Delta(x)$ is shown in Fig. 1b. It is clear that in the range $x \leqslant \xi_0$ there is an additional reduction in $\Delta(x)$ (compared to $\Delta_{GL}(x)$) so $\Delta(0) \simeq 0.7\ \Delta_{GL}(0)$. Determining the behavior of $\Delta(x)$ in immediate proximity of the *NS*-boundary is also important in order to establish the exact form of $\Delta(x)$ far from the boundary (when $x \gg \xi_0$), since the quantity β is related to δ by relation (.133). And although the function $\beta(\delta)$ is sufficiently smooth ($\beta(0) \simeq 0.72\xi_0$, $\beta(\beta) \simeq$

$0.63\xi_0$, $\beta(\infty) \simeq 0.57\xi_0$) we cannot use expression (1.20) with small values of $x \lesssim \xi_0$ or for determining β. Thus, when $T \sim T_c$ there are two different scales of variation in $\Delta(x)$: significant changes occur at distances of $\xi(T) \gg \xi_0$ and, moreover, there is an additional suppression of $\Delta(x)$ in immediate proximity to the *NS*-boundary ($x \lesssim \xi_0$). When $T \ll T_c \xi(T) \sim \xi_0$, i.e., in this case both characteristic distances are of the same order and, consequently, there is only one scale of variation in $\Delta(x)$.

2. The Meissner effect and external magnetic field action on the discrete spectrum in the NS-system

If the superconducting properties of an *S*-metal near the *NS*-boundary are damped, the *N*-metal acquires such properties upon contact with the superconductor. One manifestation of these properties is the Meissner effect. When a superconductor/normal metal system is exposed to an external magnetic field, an undamped current is excited in this system that screens the field; this current will flow not only in the superconductor but also in the normal metal layer. The Meissner effect in such systems, as we will see, has a number of interesting features. Assume a magnetic field is in the system. For a homogeneous superconductor at moderate field intensities ($ev_F A/T_c \ll 1$) from (1.1) we have

$$g_\omega = \frac{\omega}{\Omega_0} + iev_F\mathbf{A}\frac{\Delta_0^2}{\Omega_0^2\left(\Omega_0 + \frac{v_F}{2l}\right)}, \qquad f_\omega = \frac{\Delta_0}{\Omega_0} + \left(\frac{\omega}{\Omega_0} - g_\omega\right)\frac{\omega}{\Delta_0}, \tag{1.36}$$

and from here by means of (1.2) we obtain the familiar relation between the current and vector potential for a type II homogeneous superconductor [9]:

$$\mathbf{j} = -\mathbf{A}\frac{2e^2p_F^3}{3\pi m}T\sum_{\omega>0}\frac{\Delta_0^2}{\Omega_0^2\,(\Omega_0 + v_F/2l)}. \tag{1.37}$$

In this case it will be convenient for us to place the superconductor in the left half space $x < 0$; the *N*-metal occupies the region $0 < x < d$. We will carry out the calculations in the model of a "stepwise" order parameter: $\Delta(x) = \Delta_0\theta(-x)$, $\theta(x)$ is the heaviside function. Such a configuration of $\Delta(x)$ exists when there is a significant differential between the Fermi velocities of the *N*- and *S'*-metals $v_{FN} \ll v_{FS}$ (in the dirty case: with a significant difference in the conductances ($\sigma_N \ll \sigma_S$). When $v_{FN} \simeq v_{FS}$, as we have already determined, the actual order parameter differs significantly from the "step." Nonetheless all expressions for current obtained in this simple model are valid exactly (and not within an order of magnitude) for any types of *NS*- and *SNS*-systems even when $v_{FN} = v_{FS}$ in conditions $T \ll T_c$ and $d \gg \xi_0$ (in the dirty case $d \gg \sqrt{\xi_0 l}$). This statement will be substantiated in detail in the following chapter. Here we will assume that the quantities v_F and l coincide in the *N*- and *S*-metals.

We will select the following gauge of the vector potential: $\mathbf{A} = \{0, A(x), 0\}$. Such a selection is possible in view of the homogeneity of the system on the y and z axes. We will consider $A(x)$ to be equal to zero if there is no magnetic field. We will assume that specular reflection of the electrons exists at the boundary between the normal metal and the vacuum. Then

$$\hat{g}_\omega (v_x, d) = \hat{g}_\omega (-v_x, d). \tag{1.38}$$

Such reflection also exists when there is a thin oxide film on the metal boundary [40]. The expression for the screening current in the system will be significantly dependent on the free path length l. We will consider the two limiting cases $l \gg \xi_0$ and $l \ll \xi_0$.

a) Pure NS-systems. The solution of equations (1.1) in a superconductor ($x < 0$) takes the form

$$\hat{g}_\omega = \hat{C}_1 + \hat{C}_2 \exp \{(2\Omega_0 + v_x/l)\, x/| v_x |\}. \tag{1.39}$$

Here $\hat{C}_1$ is the solution of (1.1) in a homogeneous superconductor which is given by formulae (1.36). We will now assume that there are no impurities. In this case the solution of the Eilenberger equations in the normal metal layer will be rather simple

$$\hat{g}_\omega = g_\omega (v_x, v_y)\, \hat{\tau}_3 + f_\omega (v_x, v_y)\, [\mathrm{ch}\, \chi (x)\, \hat{\tau}_1 - \mathrm{sh}\, \chi (x)\, i\hat{\tau}_2],$$

$$\chi (x) = \frac{r\omega x}{v_x} + 2ie \frac{v_y}{v_x} \int_0^x A (x')\, dx', \tag{1.40}$$

where $g_\omega(v_x, v_y)$ and $f_\omega(v_x, v_y)$ are independent of the coordinates. Using (1.38) and joining (1.39) and (1.41) at the NS-boundary, we may easily find the matrix $\hat{g}_\omega$ for this system. For example, we may write the expression for $g_\omega(v_x, v_y)$:

$$g_\omega (v_x, v_y) = \frac{\omega \,\mathrm{sign}\, v_x\, \mathrm{ch}\, \chi (d) + \Omega_0\, \mathrm{sh}\, \chi (d)}{\omega_0 \,\mathrm{sign}\, v_x\, \mathrm{ch}\, \chi (d) + \omega\, \mathrm{sh}\, \chi (d)}. \tag{1.41}$$

From (1.41) using (1.2) we may directly derive the expression for the screening current in a normal metal layer:

$$j = -\frac{e p_F^2 \Delta_0^2}{2\pi^2} T \sum_{\omega>0} \int_0^{2\pi} d\varphi \int_0^1 d(\cos\theta) \sin\theta \cos\varphi \sin\left\{4e\, \mathrm{tg}\, \theta \cos\varphi \int_0^d A (x)\, dx\right\} \times$$
$$\times \left\{\Delta_0^2 \cos^2 \left[2e\, \mathrm{tg}\, \theta \cos\varphi \int_0^d A (x)\, dx\right] + \left[\Omega_0\, \mathrm{sh}\, \frac{2\omega d}{v_F \cos\theta} + \omega\, \mathrm{ch}\, \frac{2\omega d}{v_F \cos\theta}\right]^2\right\}^{-1}. \tag{1.42}$$

It is clear that the dependence of the current on the vector potential in the N-layer is nonlocal. Moreover, the function g_ω and, consequently, current (1.42) in the normal metal layer are independent of the coordinate x in the absence of impurities. Physically this fact may easily be understood from the following. In this case the excitation spectrum in the normal metal layer at energies below Δ_0 is discrete [18]. Expression (1.41) which is considered a function of the complex variable ω has poles on the imaginary axis that determine the corresponding values of the excitation energies. When $E \ll \Delta_0$ we have

$$E_n = \frac{\pi v_x}{2d}(n + 1/2) - e\frac{v_y}{d}\int_0^d A(x)\,dx. \tag{1.43}$$

Thus, the existence of a magnetic field in the N-layer shifts the Andreev levels and this shift is proportional to the integral of the vector potential as well as the excitation velocity component v_y. Such a shift (obviously uniform for the entire N-layer) also produces the screening current independent of x. The physical meaning of result (1.43) is well known. Indeed the reason for the discreteness of the spectrum is the interference effect. When an external magnetic field is present in the N-layer, the interfering excitations acquire an additional phase shift that produces the second term in the right half of (1.43). At first glance an analogous situation exists in SNS-systems in which the phase difference of the order parameters existing with current passing through the system also produces a shift of the Andreev levels [16]. However when the external magnetic field is switched on the spectrum of the SNS-system is significantly modified [24, 25] so that the discrete energy levels are replaced by zones. The reason for this reformulation lies in the dependence of the energy on the gradient-invariant phase difference that depends on the coordinate y when there is a magnetic field. In an NS (or INS)-system the discrete spectrum, on the other hand, remains stable when the magnetic field is switched on, which only shifts the energy levels. However, we will return to an investigation of the screening properties of the NS-system. When $T = 0$ and $d \gg \xi_0$ (1.42) becomes

$$j = -\frac{ep_F^3}{8\pi^3 md}\int_0^{2\pi} d\varphi \int_0^1 d(\cos\theta)\sin 2\theta\cos\varphi\,\mathrm{arctg}\left\{\mathrm{tg}\left[2e\,\mathrm{tg}\,\theta\cos\varphi\int_0^d A(x)\,dx\right]\right\}. \tag{1.44}$$

when $e\int_e^d A(x)\,dx \ll 1$ (and when $T \gg v_F/d$ with arbitrary values of A) we will have

$$j = -ej_s(T)\int_0^d A(x)\,dx,$$

$$j_s = \begin{cases} \dfrac{ep_F^3}{3\pi^2 md} & T = 0, \\[2ex] \dfrac{e}{\pi T}\left(\dfrac{p_F v_F}{\pi d}\right)^2 \exp\left\{-\dfrac{4\pi T d}{v_F}\right\} & T > \dfrac{v_F}{d}. \end{cases} \tag{1.45}$$

Let $H|_{x=d} = H_0$. From (1.45) and Maxwell's equations we may easily obtain

$$B(x) = H_0 + 4\pi j(d - x), \qquad j = -3eH_0 j_s d^2 [6 + 8\pi e j_s d^3]^{-1}, \qquad 0 \leqslant x \leqslant d. \tag{1.46}$$

At low temperatures $T \ll v_F/d$ for standard values of $v_F \sim 10^8$ cm/s, $d \sim 10^{-3}$-10-4 cm we have $ej_s d^3 \gg 1$ and from here $j = -3H_0/8\pi d$. In this case when $x < d/3$ the sign of the magnetic field will be opposite the sign of H_0 (specifically, $B(0) = -H_0/2$). Thus in the systems in certain conditions a "rescreening" effect of the external magnetic field is possible. We note that an analogous effect exists in type-I superconductors. The nonlocal relation between the current and the vector potential in this case also changes the sign of the magnetic field penetrating the superconductor [59]. Such an effect was observed in study [60].

This examination is valid in the condition $l > d$ and with homogeneity of the N-layer boundaries. With small values of l the screening current j begins to depend on x. We will not provide the corresponding expressions in view of their cumbersome quantity. We note only that in the case $\xi_0 \ll l \ll d$ when $x \gg l$ current j, even when $T = 0$, is proportional to the small factor $\exp\{-x/l\}$. We will now consider the case $l \ll \xi_0$ which evidently is most often achieved in experiment.

b) Dirty NS-systems. When $l \ll \xi_0$ we have

$$g_\omega(\mathbf{v}_F, \mathbf{r}) = G_\omega(\mathbf{r}) + \frac{\mathbf{v}_F}{v_F}\mathbf{G}_\omega(\mathbf{r}), \qquad f_\omega(\mathbf{v}_F, \mathbf{r}) = F_\omega(\mathbf{r}) + \frac{\mathbf{v}_F}{v_F}\mathbf{F}_\omega(\mathbf{r}), \tag{1.47}$$

where the quantities $\mathbf{G}_\omega$ and $\mathbf{F}_\omega$ are small. In this case equations (1.1) become the Usadel equation [11] ($D = v_F l/3$, $\hat{\nabla} = \nabla + 2ie\mathbf{A}$):

$$2\omega F_\omega(\mathbf{r}) - 2\Delta G_\omega(\mathbf{r}) - D\hat{\nabla}\left[G_\omega(\mathbf{r})\hat{\nabla}F_\omega(\mathbf{r}) + \frac{F_\omega(\mathbf{r})}{2G_\omega(\mathbf{r})}\nabla|F_\omega(\mathbf{r})|^2\right] = 0, \tag{1.48}$$

while formula (1.2) takes the form

$$\mathbf{j}(\mathbf{r}) = -\frac{iemp_F}{\pi}DT\sum_{\omega>0}(F_\omega^*\hat{\nabla}F_\omega - F_\omega\hat{\nabla}F_\omega^*). \tag{1.49}$$

In the absence of a magnetic field ($\mathbf{A} = 0$) equation (1.48) in the N-layer by substituting $F_\omega(x) = \sin\alpha_\omega(x)$ is reduced to the equation

$$\frac{d^2\alpha_\omega}{dx^2} = \frac{2\omega}{D}\sin\alpha_\omega. \tag{1.50}$$

The solution of (1.50) with boundary conditions

$$\alpha_\omega(0) = \arcsin\frac{\Delta_0}{\Omega_0}, \qquad \left.\frac{d\alpha_\omega}{dx}\right|_{x=d} = 0 \tag{1.51}$$

in implicit form is given by the relation

$$\left(\frac{2\omega}{D}\right)^{1/2}(d-x) = K(a) - F(b,a), \qquad a = \cos\left\{\frac{\alpha_\omega(x-d)}{2}\right\},$$
$$b = \arccos\{[\cos\alpha_\omega(d) - \cos\alpha_\omega(x-d)]^{1/2}/\sqrt{2}\cos\alpha_\omega(x-d)\}. \tag{1.52}$$

Here $F(b, a)$ and $K(a)$ are the incomplete and complete first order elliptical integrals, respectively [61].

With large d ($d^2 \gg D/T$) (1.52) becomes

$$F_\omega(x) = \frac{2\,\mathrm{sh}\,\eta(x)}{\mathrm{ch}^2\eta(x)}, \qquad \eta(x) = x\left(\frac{2\omega}{D}\right)^{1/2} + \eta_0, \qquad \mathrm{sh}\,\eta_0 = \frac{\Omega_0 + \omega}{\Delta_0}. \tag{1.53}$$

Now assume that an external magnetic field exists in the system, where $ev_F lA/T \ll 1$. For magnetic field values of $H \sim$ 1-10 Gauss and $l \sim 10^{-7}$ cm this inequality is not satisfied even at temperatures $T \gtrsim 10^{-2}$-10^{-4} K. From (1.48) and (1.49) in the principal approximation we have

$$\mathbf{j}(x) = -\mathbf{A}(x)\frac{4e^2 p_F^2 l}{3\pi} T \sum_{\omega>0} F_\omega^2(x), \tag{1.54}$$

where $F_\omega(x)$ is determined from (1.52). Thus, in this case the magnetic field in the normal metal is screened only at distances of $\sim\xi_N = (D/2\pi T)^{1/2}$ from the *NS*-boundary.

Thus, the nature of magnetic field screening in a normal metal layer differs significantly in pure and dirty *NS*-systems. In the case of only a few impurities the screening current is determined by the shift of the levels of the discrete spectrum (1.43) due to the magnetic field in the *N*-layer. Here the current is homogeneous throughout such a layer, while the dependence of j on the vector potential is nonlocal. At low temperatures $T \lesssim v_F/d$ there may be a sign reversal of the magnetic field penetrating the system which is due to such nonlocality. We note that in this case the *NS*-structure is an interesting example of a system in which the Fourier-transform of the magnetic susceptibility $\mu(k)$ may become negative with certain values of the wave vector $\mathbf{k}$. Thus, when $ej_s d^3 \gg 1$ in the principal approximation

$$\mu(k_x) = \frac{6}{dk_x^2}\sin^2\frac{k_x d}{2} - \frac{\sin k_x d}{k_x}. \tag{1.55}$$

The behavior of a pure *NS*-system in a magnetic field in some sense is similar to the behavior of a type-I superconductor (the nonlocal relation between $\mathbf{j}$ and $\mathbf{A}$ and the possibility for rescreening of the external field; the role of the inverse coherence length for the *N*-layer is played by the quantity $1/\xi_T + 1/l$, $\xi_T = v_F/2\pi T$). In the dirty case the $\mathbf{j}(\mathbf{A})$ relation manifests a local nature (1.54), while the value of the screening current is determined by the value of the f-function which in the *N*-layer drops at distances in the order of $\xi_N \sim (l\xi_T)^{1/2}$, since pair diffusion to the dirty normal metal occurs at precisely such distances. This result is consistent with experimental results [62]. Thus, a *NS*-system with a short free path length appears to be like a type-II superconductor.

We note that all results were obtained assuming that the BCS interaction constant in the *N*-metal is equal to zero. If it is nonzero, so the transition temperature $T_{c/N} > 0$, when $T > T_{c/N}$ the order parameter in the *N*-layer near the *NS*-boundary is nonzero, which influences the nature of magnetic field screening. Specifically this may cause a sharp jump in the dependence of the effective screening area on the external magnetic field H [62], since such "induced" superconductivity may break down with certain values of H. It was significant that even when $\Delta = 0$, significant screening currents may flow in the *N*-layer; these are associated with the penetration of the Cooper pairs from the superconductor into this layer.

Chapter 2

THE STATIONARY JOSEPHSON EFFECT IN SUPERCONDUCTOR-NORMAL METAL-SUPERCONDUCTOR JUNCTIONS

The penetration of Cooper pairs from the superconductor into the normal metal is behind a number of other interesting phenomena (in addition to the Meissner effect). One such phenomenon is the stationary Josephson effect in *SNS*-junctions, i.e,. nondissipative current flow through the *N*-layer in the absence of voltage. In this chapter we will investigate a number of issues associated with such current flow through various types of *SNS*-junctions.

3. The influence of order parameter suppression on critical current in *SNS*-junctions

The order parameter of a superconductor, as noted above, near the *NS*-boundary becomes less than its value in a homogeneous superconductor. This fact, obviously, must be accounted for in formulating a theory of the Josephson effect in *SNS*-junctions. The theory of the proximity effect has been sufficiently well developed near T_c, which allows us to find an exact microscopic expression for the stationary Josephson current of *SNS*-junctions in this temperature range. Such calculations were carried out in a number of studies [12-14].

A number of studies have investigated the Josephson effect in the temperature range $T \ll T_c$ [16, 19-23] using the "stepwise" model noted above for the order parameter. This model is valid for bridge structures and random temperatures as well as for *SNS*-sandwiches when $v_{FN} \ll v_{FS}$ [63] (in the dirty case $\Gamma_N/\Gamma_S \ll 1$, $\Gamma^2 = N^2(0)D$) also at virtually all temperatures. However when $v_{FN} \simeq v_{FS}$ even when $T \ll T_c$ there is a noticeable drop in Δ near the *N*-layer (see Section 1). We will investigate the influence of this reduction on the Josephson current. We will limit our case to pure *SNS'*-junctions with an *N*-layer width of $d \gg \xi_0$.

Let the boundaries of the *N*-layer with the superconductors be homogeneous and located at the points $x = \pm d/2$. When $d \gg \xi_0$ the influence of one superconductor on the order parameter of the other superconductor is small, since the behavior of $\delta(x)$ in the superconductors does not differ from the case examined in Section 1 ($T \ll T_c$, see Fig. 1a). Let current flow through the *SNS*-junctions so that the phase difference of the order parameters of the superconducting borders is equal to φ_0. We will represent the solution of (1.1) in the superconducting regions in the form

$$f_j = \frac{\Delta_0}{\Omega_0} \exp\{(-1)^j i\varphi_0/2\} + \delta f_j(v_x, x), \qquad g_j = \frac{\omega}{\Omega_0} + \delta g_j(v_x, x), \qquad j = 1, 2. \tag{2.1}$$

Here and henceforth the index 1 refers to the left superconductor ($x < -d/2$) while index 2 refers to the right superconductor ($x > d/2$). The solution of (1.1) in the N-domain is trivial. As a result we obtain

$$\delta g(v_x) = \delta g_1(v_x, -d/2) = \delta g_2(v_x, d/2), \quad \left(\frac{\Delta_0}{\Omega_0} e^{-i\frac{\varphi_0}{2}} + \delta f_1\right) e^{-\frac{2\omega d}{v_x}} = = \frac{\Delta_0}{\Omega_0} e^{i\frac{\varphi_0}{2}} + \delta f_2. \tag{2.2}$$

From equations (1.1) when $\omega \ll |\Delta(\pm d/2)|$ we have the relations

$$\delta f_1(v_x, -d/2) = \delta g(v_x) e^{-i\frac{\varphi_0}{2}} \operatorname{sign} v_x, \qquad \delta f_2(v_x, d/2) = -\delta g(v_x) e^{i\frac{\varphi_0}{2}} \operatorname{sign} v_x. \tag{2.3}$$

From (2.2) and (2.3) we may easily obtain

$$\delta g_\omega(v_x) = \operatorname{sign} v_x \operatorname{th}\left\{i\frac{\varphi_0}{2} + \frac{\omega d}{v_x}\right\}, \qquad \omega \ll |\Delta(\pm d/2)|. \tag{2.4}$$

When $v_{FN} \simeq v_{FS}$ $|\Delta(\pm d/2)| \approx \Delta_0/2$, i.e., expression (2.4) is valid when $\omega \ll \Delta_0$. In the case $T \lesssim v_F/d$ and $d \gg \xi_0$ the primary contribution to the sum over the frequencies (1.2) comes from the region $\omega \lesssim v_F/d$, while when $T \gg v_F/d$ (although $T \ll T_c$) we may ignore all terms aside from the first term in this sum. As a result we have

$$j = \frac{2ep_F^2}{\pi} \int_0^1 \alpha\, d\alpha\, T \sum_{\omega>0} \frac{\sin\varphi_0}{\cos\varphi_0 + \operatorname{ch}\frac{2\omega d}{v_F\alpha}}. \tag{2.5}$$

This expression is well known. It was derived in a number of studies [19-22] ignoring suppression of the order parameter. Thus, in the range of applicability of formula (2.5) the critical current in the SNS-sandwiches with similar values of the Fermi velocities of the N- and S-metals is not suppressed, in spite of the suppression of Δ. Indeed, if the primary contribution to the current comes from low frequencies, then in determining Δ from self-consistency equation (1.2) high frequencies are significant, and the contribution of the region $\omega \ll \Delta_0$ is insignificant (see Fig. 2). The function f_ω for frequencies in the order of (or greater than (Δ_0) drops in a self-consistent manner with a drop in Δ. With low frequencies $\omega \ll \Delta_0$ it turned out that the f-function was not suppressed thereby producing result (2.5). Again we will enumerate the cases in which we may use the "graduated" model for the modulus of the order parameter to find the exact expression for the Josephson current in SNS-junctions: a) bridge structures. Arbitrary values of T and d; b) SNS-sandwiches. Virtually any values of T and d, although $v_{FN} \ll v_{FS}$ [63]; c) SNS-sandwiches. Arbitrary values of v_{FN} and v_{FS}, although $v_F/|\Delta(\pm d/2)| \ll d \ll v_F/T$ (specifically when $v_{Fn}0 \approx v_{FS} \ll T_c$ and $d \gg \xi_0$).

Analogous conditions also exist in the dirty limit. If such conditions are not satisfied, the critical current becomes less than the value determined by formula (2.5). As a rule the value of such current may be determined numerically only.

In the range $T \sim T_C$ suppression of Δ in the majority of cases is quite strongly manifest in the critical current level. When $d \gg \xi_0$ it is possible to ignore the mutual influence of the superconducting borders. Equations (1.1) in this case are easily solvable. Using the solution and formula for current (1.2) we have

$$j = \frac{ep_F^2}{2\pi} T \sin\varphi_0 \int_{-1}^{1} \alpha\, d\alpha \exp\left\{-\frac{2\omega d}{v_F |\alpha|}\right\} \left[\left|f\left(-\frac{d}{2}, v_x\right)\right|^2 + \left|f\left(\frac{d}{2}, v_x\right)\right|^2\right], \tag{2.6}$$

where $|f(\pm d/2)|$ are the values of the modulus of the Eilenberger f-function at the NS-boundaries when $\omega = \pi T$. Using our proximity effect theory these values are easily found:

$$f\left(\pm\frac{d}{2}, v_x\right) \doteq \frac{\Delta(0)}{\pi T}\left[1 \mp \frac{v_x}{2\pi T\delta} + \left(\frac{|v_x|}{4\pi T\delta} - \frac{1}{2}\right)\left(1 \pm \frac{v_x}{|v_x|}\right)\right],$$

where $\Delta(0)$ and δ are determined from (1.32), (1.34). Substituting this expression into (2.6), finally we have

$$j_c = Q\,\frac{ep_F^2 v_F}{d}\left(1 - \frac{T}{T_c}\right)^2 \exp\{-d/\xi_0\},$$

$$Q = \frac{96}{49\pi^2\zeta^2(3)}\left[\frac{1 + \xi_0/\delta}{3\pi^2/14\zeta(3) + \xi_0/\delta}\right] \simeq 0{,}093.$$

The numerical coefficient Q is somewhat greater than the calculated value in study [17] ($Q \simeq 0.084$); the authors of this study assumed that Δ at the superconducting borders obeys law (1.20) up through the NS-boundary, which as we have seen, is not entirely true. Study [97] attempted to account for the influence of the proximity effect on the critical current of SNS-junctions. These results were incorrect, since they employed improper integral equations for $\Delta(x)$.

Thus, the behavior of $\Delta(x)$ in the superconductor region immediately adjacent to the N-metal ($x \lesssim \xi_0$) is also significant for determining the critical current of the SNS-junction. The numerical coefficient Q in the formula for j_c, as we have seen, is dependent only on $\Delta(0)/\Delta'(0)$.

4. The stationary Josephson effect in wide *SNS*-junctions containing impurities

We will now investigate the properties of SNS'-junctions consisting of metals with a high quantity of nonmagnetic impurities: $l \ll \xi_0$. Such dirty alloys, as noted above, may be described by Usadel equations (1.40). We will consider wide SNS-junctions: $d \gg \sqrt{l\xi_0}$. When $T \sim T_C$ the stationary Josephson effect in dirty SNS-junctions has been investigated in sufficient detail [12-14]. Here we will deal primarily with the low temperature range $T \ll T_c$.

The Usadel equation in the N-metal will take the form

$$\frac{d^2\alpha_\omega}{dx^2} = \frac{2\omega}{D}\sin\alpha_\omega + \frac{1}{2}\left(\frac{d\chi_\omega}{dx}\right)^2 \sin 2\alpha_\omega,$$
$$F_\omega = e^{i\chi_\omega}\sin\alpha_\omega. \tag{2.7}$$

In deriving (2.7) from (1.48) we used the property of conservation of the quantity $j_\omega = \sin^2\alpha_\omega d\chi_\omega/dx$ in the N-metal (as before we consider the case $T_{cN} = 0$). Generally in order to find the Josephson current we must still solve the Usadel equations in the S-regions and then join the quantities F_ω and $DN(0)F'_\omega$ at the points $x = \pm d/2$. In a number of cases, however, we may use the boundary conditions

$$F_\omega(\pm d/2) = \frac{\Delta_0}{\Omega_0}\exp\{\pm i\varphi_0/2\} \tag{2.8}$$

and solve equations (2.7) in the N-metal only. Below we will identify the range of applicability of this approach.

Thus, we must find the solution of equation (2.7) with boundary conditions (2.8). Equation (2.7) has the first integral

$$\left(\frac{d\alpha_\omega}{dx}\right)^2 + \frac{4\omega}{D}\cos\alpha_\omega + \frac{j_\omega^2}{\sin^2\alpha_\omega} = C(\omega). \tag{2.9}$$

The quantity $\sin\,\alpha_\omega$ is an even function of x and from here

$$C(\omega) = \frac{4\omega}{D}\cos\alpha_\omega(0) + \frac{j_\omega^2}{\sin^2\alpha_\omega(0)}\,. \tag{2.10}$$

As a result we obtain

$$|x| = \int\limits_{\cos\alpha_\omega(x)}^{\cos\alpha_\omega(0)} dy R^{-1}(y), \qquad |\chi_\omega(x)| = \int\limits_{\cos\alpha_\omega(x)}^{\cos\alpha_\omega(0)} dy\,[R(y)(1-y^2)]^{-1}; \tag{2.11}$$

$$R^2(y) = (1-y)^2\left[\frac{4\omega}{D}(\cos\alpha_\omega(0) - y) + \frac{j_\omega^2}{\sin^2\alpha_\omega(0)}\right] - j_\omega^2. \tag{2.12}$$

Relations (2.11), (2.12) subject to boundary conditions (2.8) yield the solution of equation (2.7) in implicit form. Integrals (2.11) allow representation through the first and third order elliptical integrals. Here, however, it is not necessary to provide the corresponding rather cumbersome expressions.

In the temperature range $T \gg D/d^2$ calculation of the quantity j is significantly simplified. In this case we may ignore the mutual influence of the superconducting borders. Then

$$F_\omega(x) = e^{i\frac{\varphi_0}{2}}\sin\alpha_\omega^+(x) + e^{-i\frac{\varphi_0}{2}}\sin\alpha_\omega^-(x), \tag{2.13}$$

where the functions $\alpha_\omega^\pm(x)$ satisfy equation (1.50). It is interesting that such an equation is widely employed in weak superconductivity theory for solving completely different problems (it describes the magnetic field and current distribution, and the vortex structures in a wide Josephson junction; we will use this equation for this purpose in Chapter 4). In this case equation (1.50) may be solved analytically:

$$\alpha_{\omega}^{\pm} = 4\,\mathrm{arctg}\,\exp\left\{\pm x\sqrt{\frac{2\omega}{D}} + A_{\omega}^{\pm}\right\}; \tag{2.14}$$

while the $A_{\omega}^{\pm}$ constants are found from boundary conditions (2.8).

When $T \gg D/d^2$ approximation (2.13) allows conservation of current within the normal metal layer with the exception of a small region in the order of $(D/T^2)^{1/2} \ll d$ near the NS-boundaries. Therefore it will be convenient for us to calculate the current in the center of the N-layer ($x = 0$). Using (1.49), (2.13), and (2.14) we derive an expression for the Josephson current density in wide dirty SNS-junctions [64]:

$$j = \frac{64\pi T}{eR_N}\,\frac{d}{\xi_N}\,\frac{\Delta_0^2 \exp\{-d/\xi_N\}\sin\varphi_0}{\{\pi T + S(T) + \sqrt{2}\,[S^2(T) + \pi T S(T)]^{1/2}\}^2}, \tag{2.15}$$

where $S^2 = \pi T^2 + \Delta_0^2$; $R_N = d/2DN(0)e^2$ is the resistance per unit of area of the N-layer. When $T \ll T_c$ we have from (2.15):

$$j = \frac{64\pi}{3 + 2\sqrt{2}}\,\frac{T}{eR_N}\,\frac{d}{\xi_N}\exp\{-d/\xi_N\}\sin\varphi_0. \tag{2.16}$$

The derived relations are in good agreement with existing experimental data [65]. Formula (2.16) is in fact valid up through temperatures of $T \gtrsim D/d^2$ (more precisely, $T \gg v_F l/6\pi d^2 \sim 10^{-2}$-$10^{-4}$ K for typical parameters of the problem). The "current versus phase" relation deviates significantly from the sinusoidal in the range $T \lesssim D/d^2$. At such temperatures, unfortunately, it is not possible to obtain an analytic solution of the problem and it is necessary to rely on numerical calculations. Fig. 3 gives the current density versus phase difference φ_0 relation for $T = 0$ and $T \gg D/d^2$. When $T = 0$ this relation is near linear in the range $\varphi_0 \lesssim \pi/2$. In the range of high values of φ_0 there is a certain deviation from the linear relation. As the temperature is increased the $j(\varphi_0)$ curve is deformed and becomes sinusoidal when $T \gg D/d^2$. We remember that the $j(\varphi_0)$ relation (2.5) for clean SNS-junctions is sinusoidal when $T \gg v_F/d$, while when $T = 0$ it is linear [19] ($-\pi < \varphi_0 < \pi$). It is clear that for dirty SNS-junctions the $j(\varphi_0)$ function when $T = 0$ is more complex although it is also significantly different from a sinusoid. Likharev [66] discovered

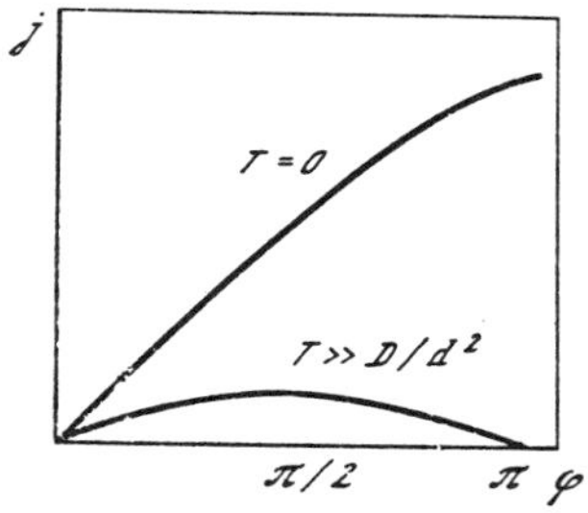

Fig. 3. The "current versus phase" relation in wide "dirty" SNS-junctions when $T = 0$.

the nonsinusoidal form of the $j(\varphi_0)$ relation at low temperatures (with small values of d) based on numerical calculations of the quantity $j_c(dj/d\varphi_0)^{-1}|_{\varphi_0=\pi}$.

The stationary Josephson effect in wide dirty *SNS*-junctions when $T \ll T_c$ was also investigated in study [67]. This investigation also utilized the Usadel equation. However the method of solving this equation used in study [67] is incorrect. Moreover, study [67] assumed that the drop in the *F*-function near the *NS*-boundaries occurs with a constant Δ. This, as we will see, is also incorrect, since the f-function of the superconductor at low frequencies $\omega \ll \Delta_0$ is not suppressed in spite of the suppression of Δ. Therefore the result is virtually an order of magnitude lower than our results (2.15), (2.16)

Our examination with "rigid" boundary conditions (2.8) is valid for bridge structures as well as sandwiches with $\Gamma_N \ll \Gamma_S$ even with arbitrary values of T and, moreover, when $T \ll T_c$ for sandwiches with significantly less stringent requirements on the relation between Γ_N and Γ_S (specifically, when $\gamma = \Gamma_N/\Gamma_S \sim 1$). Results (2.14), (2.15) were confirmed by numerical calculations given in study [55]. The $j_c(T)$ relations were calculated for various values of the parameter Γ. It was demonstrated that when $\gamma \ll 1$ the $j_c(T)$ relation coincides with (2.15), and when $\gamma \gtrsim 1$ this relation coincides with (2.16) in the limit $T \ll T_c$, i.e., expression (2.16) when $T \ll T_c$ is in any case valid in the most realistic situation of close values of v_F and l in *N*- and *S*- metals. We note, however, that the $T \ll T_c$ inequality must be rather strong. Indeed, as indicated from the boundary conditions, for validity of (2.16) the condition $\gamma \ll (T_c/T)^{12}$ must be satisfied, i.e., when $\gamma \simeq 1$, smallness of the quantity $\sqrt{T/T_c}$ is required.

Thus, the dependence of the stationary Josephson current in wide *SNS*-junctions on the phase difference and impurity concentration is sufficiently nontrivial. Indeed, the existence of impurities becomes manifest as early as $l \lesssim d$, and when $\xi_0 \ll l \ll d$ the value of the critical current is exponentially small even when $T = 0$ (we remember the results from study [23] obtained when $l_N = l_S = l$), while $j(\varphi_0)$ is deformed to a sinusoidal relation. The role of the "coherence length" of the *N*-layer in this case is played by the quantity $\xi_T(l) = (1/\xi_T + 1/l)^{-1}$. With a further increase in the impurity concentration ($l \ll \xi_0$) when $T \ll D/d^2$ the critical current (as in our pure case when $T \ll v_F/d$) does not include exponential smallness, while the function $j(\varphi_0)$ is not significantly different from a linear function (Fig. 3), i.e., again it is significantly nonsinusoidal. Thus, the behavior of *SNS*-junctions at low temperatures with a reduction in l first changes significantly ($\xi_0 \ll l \ll d$) and then ($l \ll \xi_0$) it again appears to be similar to the pure (clean) case (obviously in a qualitative sense only). The physical interpretation of this process is as follows. When $l > d$ and $T \ll v_F/d$ the superconducting correlation of electrons transiting the *N*-layer is not broken, since they do not experience collisions. In this case the critical current $j_c \sim 1/d$ and does not contain exponential smallness. Any electron collision with an impurity breaks the correlation in the velocity space, which leads to the "detachment" of the scattered electron from the superconducting current transport process. When $l \ll d$ (although $l \gg \xi_0$) the number of electrons that have not experienced scattering during their transit in the N-layer and, consequently, conserve the superconducting correlation, is exponentially small, i.e., $j_c \sim \exp\{-d/l\}$. When $l \ll \xi_0$ the electron experiences many collisions at distances in the order of the Cooper pair. In this case a diffuse superconducting current transport mechanism predominates. If the temperature is much less than the inverse diffusion time $T \ll D/d^2$, the superconducting correlation of the diffusive electrons will not be broken and j_c will have a power (rather than an exponential) dependence on the thickness. One manifestation of the physical differentiation of these two mechanisms is the difference in the expressions determining the effective "coherence length" of the N-metal (compare $\xi_T(l)$ and ξ_N).

We should note that the concept of "coherence length" of the N-metal itself is valid only if the superconductor makes contact with the metal (i.e., the "source" of the Cooper pairs). Hence this length is unavoidably a characteristic of processes occurring both in normal and in superconducting metals. When $l_s \ll \xi_0$ the Eilenberger functions of the superconductor are anisotropic with respect to velocity, since any scattering act in the N-metal will cause a jump in the phase coherence. In this case a determinant role is played by the quantity $\xi_T(l_N)$. If $l_s \ll \xi_0$, the F-functions are isotropized in the S-metals. When $l > d$ the primary role is played by ξ_T, as before. When $l < d$ the situation changes. Thus, when $l \ll \xi_T \ll d$ the quantity ξ_N already becomes important. These differences in the current transport mechanisms in SNS-junctions are manifest in experiment (see the results from studies [65 and 68, 69]).

5. The stationary Josephson effect in SNINS-junctions

So far we have examined SNS-junctions with transparency of the NS-boundaries equal to unity. In addition there have been no irregularities in the N-layer itself. Moreover, if there is an oxide film on the boundary between the normal and superconducting metals, the transparency of this boundary will be determined by the potential barrier generated by the film and in any case will be less than unity. We may also consider the Josephson effect in two NS-systems configured so that the dielectric film divides the two N-layers. Such systems are of both theoretical and experimental interest (see, for example, [69-72]).

In this section we will consider the stationary Josephson effect in $SNINS$-junctions where the position of the oxide film within the N-layer (on the NS-boundary) is given by the coordinate x_0, which allows us to describe the various situations within the scope of a unified approach.

In order to determine the matrix Green-Gorkov functions of *SNINS*-system we must solve a diagram equation of the following type:

 (2.17)

where the thin line designates the matrix Green-Gorkov function of a pure (clean) *SNS*-junction [20], the thick line corresponds to the desired matrix for an *SNINS*-junction averaged over the position of the impurities, and the wavy line represents the vertex $U(x)\hat{\tau}_3$. The existence of the potential $U(x)$ is related to the oxide film in the system. Since the size of this film is comparable to the interatomic distances, $U(x)$ may be modeled by the δ-function $U(x) = V^0\delta(x - x_0)$.

Equations (2.17) may be solved rather easily only in the pure limit and when impurities are present with smallness of V_0^{-1} (Appendix 2). However, as in other cases, the examination may be significantly simplified by using quasi-classical Eilenberger equations (1.1). The situation is somewhat complicated by the fact that the potential $U(x)$ should not be included in these equations, since they cannot be used to correctly describe the system at distances in the order of the interatomic distances. The existence of a dielectric interlayer may be accounted for only by means of effective boundary conditions.

The electron tunneling process through the dielectric barrier is rather complex and hence an electron, generally speaking, in this case, ceases to be a quasi-classical electron, since the concept of "trajectory" becomes meaningless during tunneling. It is well known that such a process cannot be adequately described by means of the continuity conditions of the Eilenberger functions on the transit trajectories which are used in examining pure systems with nontunnel conductivity. For this process we require boundary conditions of a different type that may be represented in the following form for the case of contacting metals with identical Fermi velocities [73]:

$$\hat{g}_\omega(v_x, x_0+0) - \hat{g}_\omega(-v_x, x_0+0) = \hat{g}_\omega(v_x, x_0-0) - \hat{g}_\omega(-v_x, x_0-0) =$$
$$= i\frac{v_x}{V_0}[\hat{g}_\omega(v_x, x_0+0) - \hat{g}_\omega(v_x, x_0-0)]. \quad (2.18)$$

When $V_0 = 0$ we have continuity of $\hat{g}_\omega$ at the point X_0 from (2.18), while when $V_0 = \infty$ we obtain condition (1.38) on the intact *NS*-boundary. We should note that an important condition of the applicability of (2.18) is the condition $\Delta(+\infty) = \Delta(-\infty)$.

Conditions (2.18) are valid for an arbitrary concentration of impurities in the system. In the case $l \ll \xi_0$ we may easily obtain from equations (1.1) [11]

$$2\omega \mathrm{F}_\omega + v_F\nabla F_\omega = 2\Delta \mathrm{G}_\omega + \frac{v_F}{l}[\mathrm{G}_\omega F_\omega - \mathrm{F}'_\omega G_\omega]. \quad (2.19)$$

We will now integrate boundary conditions (2.18) with respect to the directions of the vector v_F. Then subject to (1.47) we will have

$$\frac{iv_x}{2V_0}(F_\omega(x_0+0)-F_\omega(x_0-0))=F_{1\omega}(x_0+0)=F_{1\omega}(x_0-0) \quad (2.20)$$

and analogous relations for the *G*-functions. Here we have accounted for the fact that in this problem only the *x*-components of the vectors $\mathbf{F}_\omega$ and $\mathbf{G}_\omega$ are nonzero ($F_{1\omega}$ and $G_{1\omega}$). It follows from (2.19), (2.20) when $l \ll \xi_0$ that

$$\left.\frac{\partial F_\omega}{\partial x}\right|_{x=x_0+0}=\left.\frac{\partial F_\omega}{\partial x}\right|_{x=x_0-0}=$$
$$=\frac{iv_F}{2V_0 l}\left[G_\omega(x_0+0)F_\omega(x_0-0)-G_\omega(x_0-0)F_\omega(x_0+0)\right]. \quad (2.21)$$

When $V_0 = 0$ the regular continuity condition of Usadel functions at the point x_0 derives from (2.21), and when $V_0 = \infty$ we obtain the condition at the boundary with the nonpermeable dielectric [11] $\partial F_\omega/\partial x = 0$.

Thus relations (2.18) and (2.21) are effective boundary conditions on the Eilenberger and Usadel equations for describing these superconducting systems with a tunnel barrier and are valid for an arbitrary value of V_0.

1. Pure (clean) SNINS-junctions. Here and henceforth all calculations will be carried out in the model

$$\Delta(x)=\Delta_0\exp\{i\varphi_0/2\,\mathrm{sign}\,x\}\,[\theta(x-d/2)+\theta(-x-d/2)], \quad (2.22)$$

whose applicability for calculating the current in *SNS*-junctions has already been discussed in Section 3. For simplicity we will also assume that the quantities v_F and l are identical in the entire system. It is clear that all our results will also be valid in a number of other cases (see Section 3). Henceforth, however, we will not provide any special stipulations in this regard.

First assume the free path length l is large compared to all characteristic parameters of the problem (the pure (clean) limit). In this case it is rather easy to solve equations (1.1) with boundary conditions (2.18). Finally we obtain by means of the formula for the current (1.2):

$$j=\frac{2ep_F^2\Delta_0^2}{\pi}\sin\varphi_0 T\sum_{\omega>0}\int_0^1\alpha\,d\alpha\left\{\Delta_0^2\left[\cos\varphi_0+\frac{V_0^2}{v_F^2\alpha^2}\,\mathrm{ch}\,\frac{4\omega x_0}{v_F\alpha}\right]+\right.$$
$$\left.+\left(1+\frac{V_0^2}{v_F^2\alpha^2}\right)\left[(\omega^2+\Omega_0^2)\,\mathrm{ch}\,\frac{2\omega d}{v_F\alpha}+2\omega\Omega_0\,\mathrm{sh}\,\frac{2\omega d}{v_F\alpha}\right]\right\}^{-1}. \quad (2.23)$$

This result, of course, may also be obtained (after some significantly more complex calculations) by means of expressions for the exact Green's functions of the *SNINS*-system (Appendix 2). We note that formula (2.23) describes the stationary Josephson effect in undoped superconducting systems with weak links of various type with arbitrary values of V_0. Indeed, first assume $d = 0$. Then the current density will take the form

$$j = \frac{e p_F^2 \Delta_0^2 \sin \varphi_0}{4\pi} \int_0^1 \alpha \, d\alpha \frac{D(\alpha)}{R(\alpha)} \operatorname{th}\left\{\frac{\Delta_0}{2T} R(\alpha)\right\}, \qquad (2.24)$$

$$R(\alpha) = \left\{1 - D(\alpha) \sin^2\left(\frac{\varphi_0}{2}\right)\right\}^{1/2}, \qquad D(\alpha) = \left\{1 + \frac{V_0^2}{v_F^2 \alpha^2}\right\}^{-1}.$$

With large values of V_0 we have directly from (2.24) the Ambegaokar-Baratoff formula [74]. Formula (2.24) is also suitable for describing the current state in narrow superconducting bridges. When multiplying the current density by the hole area we directly reproduce the results from study [40] ($V_0 = 0$) and [75] ($V_0 \neq 0$).

Now assume $d \neq 0$ (and in fact $d \gg \xi_0$). Then when $V_0 = 0$ the formula yields the expression for the current density in the *SNS*-junction [20]. If the resistance of the dielectric interlayer is high: $R \gg 8\pi^2/\to^2 p_F^2$ ($R = 2\pi^2 V_0^2/e^2\mu^2$ [47], μ is the chemical potential), we obtain the result [52]

$$j = \frac{8\pi\Delta_0^2}{eR} T \sum_\omega \int_0^1 \alpha^3 \, d\alpha \sin \varphi_0 \left[(\Omega_0^2 + \omega^2) \operatorname{ch} \frac{2\omega d}{v_F \alpha} + \right.$$
$$\left. + \Delta_0^2 \operatorname{ch} \frac{4\omega x_0}{v_F \alpha} + 2\omega\Omega_0 \operatorname{sh} \frac{2\omega d}{v_E \alpha}\right]^{-1}. \qquad (2.25)$$

For $x_0 = \pm d/2$ (*SINS*-junction) (2.26) coincides with the result from study [76] obtained previously by another method.

When $T \ll T_c$ and $d \gg \xi_0$ this expression is simplified:

$$j = \frac{16\pi}{eR} T \sum_{\omega > 0} \int_0^1 \alpha^3 \, d\alpha \left[\operatorname{ch} \frac{2\omega d}{v_F \alpha} + \operatorname{ch} \frac{4\omega x_0}{v_F \alpha}\right]^{-1} \sin \varphi_0. \qquad (2.26)$$

When $T = 0$ we may easily obtain from (2.26):

$$j(x_0 = 0) = \frac{4}{5} \frac{v_F}{eRa} \sin \varphi_0, \qquad j\left(x_0 = \pm \frac{d}{2}\right) = \frac{\pi}{5} \frac{v_F}{eRd} \sin \varphi_0, \qquad (2.27)$$

i.e., the critical current of the *SNINS*-junction is somewhat greater than the analogous current of the *SINS*-junction. It follows from formula (2.26) (or (2.27)) that the critical current at any temperature as a function of x_0 is maximized when $x_0 = 0$ (i.e., for the symmetrical *SNINS*-junction). In fact this result is rather clear. Indeed, according to Bardeen and Johnson [21], the superconducting properties of the system containing a wide ($d \gg \xi_0$) normal metal layer is determined by the gap in the spectrum of quasi-particle excitations of this system. We will compare, for example, a symmetrical *SINS*-junction and an *SNINS*-junction with identical parameters. In the first case a gap exists in the excitation spectrum with a fixed value of v_x which is equal to (Fig. 4a) $\Delta_e = \pi v_x/2d$. In the second case the value of this gap in the spectrum is half as large (Fig. 4b) and hence the superconducting properties of this system will be weaker, i.e., the critical current through the *SINS*-junction will be less than the current through the *SNINS*-junction, which is confirmed by exact calculation.

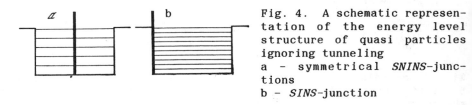

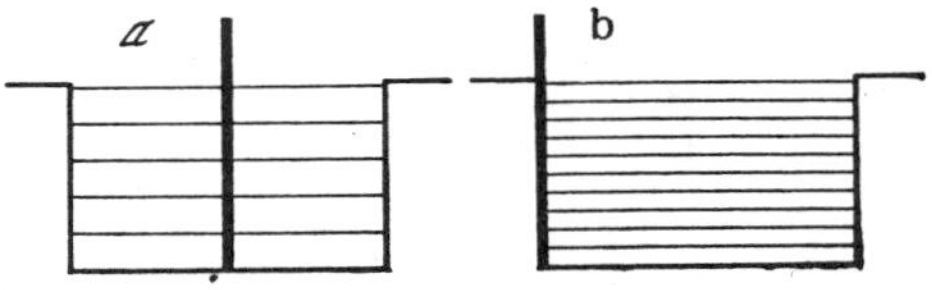

Fig. 4. A schematic representation of the energy level structure of quasi particles ignoring tunneling
a - symmetrical *SNINS*-junctions
b - *SINS*-junction

When $v_F/d \ll T \ll T_c$ we obtain for the *SNINS*-junction

$$j = \frac{8v_F}{eRd} \exp\{-d/\xi_T\} \sin \varphi_0. \tag{2.28}$$

This formula is valid over a rather broad range of values of the parameter x_0: $d/2 - |x_0| \gg \xi_T$. When $|x_0| \sim (d - \xi_T)/2$ the tunnel current in the system begins to drop somewhat and when $x_0 = \pm d/2$ we arrive at the expression for current in the *SINS*-junction [76], which is one-half the value of expression (2.28) at the same temperatures.

In the temperature range $T \sim T_c$ $\Delta_0(T) \ll T_c$ (2.25) yields

$$j = \frac{2v_F}{\pi^2 eRd} \left(\frac{\Delta_0}{T_c}\right)^2 \exp\{-d/\xi_T\} \sin \varphi_0. \tag{2.29}$$

Result (2.29) is independent of x_0, since when $T \sim T_c$ there is in fact no discrete spectrum. Here the specific value of the current is determined solely by the width of the *N*-layer *d*, although not by the position of the dielectric layer.

Formula (2.29) is valid in the case of bridges as well as *SNS*-sandwiches with $v_{FN} \ll v_{FS}$, and here $j_c \sim T_c - T$.

2. The influence of impurities. We will see that the various characteristics of the stationary Josephson effect in *SNINS*-junctions are common to both *SIS*-junctions ($j \propto 1/R$, $j \propto \sin \varphi_0$ with any value of T) and to *SNS*-junctions ($j \sim 1/d$ when $T = 0$ and $j \sim \exp\{-d/\xi_T\}$ when $T \gg v_F/d$, $j \sim (T_c - T)^2$ and when $T \sim T_c$). The issue of the influence of impurities on the critical current in the *SNINS*-junctions is interesting. Indeed, the existence of nonmagnetic impurities in the *SIS*-junctions generally has no influence on this current, while in *SNS*-junctions, as we determined above, this influence is significant. How do the *SNINS*-junctions behave?

We will first consider the case of a low impurity concentration $l \gg \xi_0$. Here the Eilenberger functions at the superconducting borders are anisotropic with respect to velocity and may be represented as (1.39). Equations (1.1) may be solved by means of perturbation theory in the parameter V_0^{-1}. In the zeroth order the half-spaces $x < x_0$ and $x > x_0$ are isolated. The solution of (1.1) for these half-spaces may be obtained rather easily. Thus, the functions $g_\omega^{(0)}(v_x, x_0 \pm 0)$ have the same form as in the pure limit if in place of the Mössbauer frequency we substitute its renormalized value ($T \ll T_c$):

$$\omega_{\pm}^{*} = \omega + \frac{v_F}{2l} \int_0^1 d\alpha \, \mathrm{th} \frac{\omega_{\pm}^{*}(d \mp 2x_0)}{v_F \alpha} . \tag{2.30}$$

We may also obtain formula (2.31) by using the regular averaging technique [9] of the exact Green's functions of the system (Appendix 2).

Strictly speaking it is also necessary to account for the renormalization of the quantity Δ. When $l \gg \xi_0$ it is easily demonstrated [23] that we may ignore the contribution of this renormalization. Using conditions (2.18) and the formula for the current (1.2) we obtain the following result ($T \ll T_c$, $d \gg \xi_0$):

$$j = \frac{8\pi \sin \varphi_0}{eR} T \sum_{\omega>0} \int_0^1 \alpha^3 \, d\alpha \left[\mathrm{ch} \frac{\omega_{+}^{*}(d - 2x_0)}{v_F \alpha} \, \mathrm{ch} \frac{\omega_{-}^{*}(d + 2x_0)}{v_F \alpha} \right] . \tag{2.31}$$

It is clear that when $l > d$ the impurities have little influence on the current value. In the opposite case, $l \ll d$ we will have from (2.30) and (2.31):

$$j = \frac{16\pi}{eR} \sin \varphi_0 T \int_0^1 \alpha \, d\alpha \left[\exp\left(\frac{d}{l\alpha}\right) \mathrm{sh} \frac{d}{\xi_T \alpha} \right]^{-1} . \tag{2.32}$$

At temperatures of $v_F/d \ll T \ll T_c$ and with random values of l we obtain from (2.30) (2.31):

$$j = \frac{8 v_F}{eRd} \frac{T \sin \varphi_0}{T + 2\pi v_F/l} \exp\{-d/\xi_T - d/l\}. \tag{2.33}$$

In the case $l \ll d$ even when $T = 0$ expression (2.32) for the current is exponentially small:

$$j = \frac{8 v_F}{eRd} \frac{l}{d} \sin \varphi_0 \exp\{-d/l\}. \tag{2.34}$$

Results (2.32)-(2.34) are valid for $d - 2\,|x_z \gg \xi_T$. When $|x_0| \sim (d - \xi_T)/2$ the tunnel current begins to drop somewhat and when $x_0 = \pm d/2$ (*SINS*-junction). The corresponding expressions are half the value of the expressions (2.33)-(2.35) (as in the case of the pure limit).

Thus, the physical cause for the influence of impurities on the Josephson current in *SNINS*-junctions (and in *SNS*-junctions [23]) is blurring of the discrete spectrum of levels due to scattering of electron excitations by the impurities localized in the normal metal layer. The specific difference from the case of *SNS*-systems (with an analogous condition $l > \xi_0$) derives from the existence of two systems of levels which results in different contributions to frequency renormalization from both sides of the dielectric interlayer.

We will now consider the reverse case: $l \ll \xi_0$. In the case of large values of V_0 of interest to us, the problem may also be solved by perturbation theory in the parameter V_0^{-1}. In the zeroth approximation we may represent the F-function in the form $F_{\omega j}^{(0)} = \sin \alpha_\omega^j \exp-\{(1)^j i\varphi_0/2\}$, where α_ω^j satisfies equation (1.50) and the boundary conditions

$$\alpha_\omega^{1,2}(\pm d/2) = \arcsin \frac{\Delta_0}{\Omega_0}, \qquad \left.\frac{d\alpha_\omega^j}{dx}\right|_{x=x_0} = 0, \qquad j = 1, 2. \tag{2.35}$$

Solutions of equation (1.50) with analogous boundary conditions have already been given in Section 2. For the case of interest to us $\alpha_\omega^j(x)$ are determined from the equations

$$\left(\frac{2\omega}{D}\right)^{1/2}(x - x_0) = (-1)^j \left[K\left(\cos\frac{\alpha_\omega^j(x)}{2}\right) - F\left(\chi^j, \cos\frac{\alpha_\omega^j(x)}{2}\right)\right], \tag{2.36}$$

Assume now $F_\omega^j = F_\omega^{(0)j} + F_\omega^{(1)j}$, where $F_\omega^{(1)j} = \beta_\omega^j \cos\alpha_\omega^j \exp\{-1)^j i\varphi_0/2\}$ is the first approximation for the Usadel function in the parameter V_0^{-1}. The β_ω^j functions in the normal metal layer satisfy the equation

$$\frac{d^2\beta_\omega^j}{dx^2} = \frac{2\omega}{D}\beta_\omega^j \cos\alpha_\omega^j, \qquad |x| \leqslant d/2. \tag{2.37}$$

In the superconducting regions the equations for β_ω^j assume a very simple form:

$$\frac{d^2\beta_\omega^j}{dx^2} - \frac{2\Omega_0}{D_i}\beta_\omega^j = 0. \tag{2.38}$$

At infinity the β_ω^j functions will approach zero. Using this fact and joining $F_\omega^{(1)j}$ and their derivatives at the points $x = \mp d/2$, we obtain

$$\left.\frac{d\beta_\omega^j}{dx}\right|_{x=(-1)^j d/2} + \left[2\Delta_0\left(\frac{\Omega_0 - \omega}{2\omega\Omega_0}\right)^{1/2} + (-1)^j\left(\frac{2\Omega_0}{D}\right)^{1/2}\right]\beta_\omega^j = 0, \quad x = (-1)^j d/2. \tag{2.39}$$

The general solution of (2.37) takes the form

$$\beta_\omega^j(x) = A_j h_j(x) + B_j h_j(x)\int^x \frac{dx}{h_j^2}, \qquad h_j(x) = \frac{d\alpha_\omega^j}{dx}. \tag{2.40}$$

By substituting (2.40) into the boundary conditions we may determine the random constants A_j and B_j.

These relations allow us to calculate the Josephson current at random temperatures. The final result for the current density is expressed through the elliptical functions. When $T \gg D/d^2$ this consideration is significantly simplified. Assume initially $x_0 = -d/2$ (the *SINS*-junction). The quantity $\alpha_\omega^j(x_0)$ is easily calculated from relations (2.36):

$$\alpha_\omega^j(x_0) = \frac{8\Delta_0}{\Omega_0 + \omega[2\Omega_0(\Omega_0 + \omega)]^{1/2}} \exp\left\{-d\left(\frac{2\omega}{D}\right)^{1/2}\right\}, \qquad j = 2. \tag{2.41}$$

The solution of (2.37) satisfying boundary condition (2.21) is represented as (here we assume $\sin\alpha^j_\omega = \Delta_0/\Omega_0,\ j = 1$):

$$\beta^j_\omega = -\frac{iv_F}{2V_0 l}\left(\frac{D}{2\omega}\right)^{1/2}\left[\frac{\omega}{\Omega_0}\frac{d\alpha^j_\omega}{dx} + \frac{\Delta_0}{\Omega_0}e^{-i\varphi_0}\exp\left\{-d\left(\frac{2\omega}{D}\right)^{1/2}\right\}\right], \quad j = 2, \tag{2.42}$$

where this expression is valid at distances that are not too close to the point $x = d/2$ (when $d/2 - x \gg (D/\omega)^{1/2}$). We directly derive from (1.49) and (2.42) the result for the Josephson current density in a dirty *SINS*-junction

$$j = \frac{16\pi^3 d\,(2\pi T D)^{1/2}}{e^3 R R_N p_F^2 l^2}\frac{\Delta_0^2 \sin\varphi_0}{\pi^2 T^2 + \Delta_0^2}\frac{T}{W}\exp\{-d/\xi_N\},$$

$$W = S(T) + \pi T + [2S^2(T) + 2\pi T S(T)]^{1/2}, \quad S^2(T) = \pi^2 T^2 + \Delta_0^2. \tag{2.43}$$

When $T \ll T_c$ general formula (2.43) becomes the equality

$$j = 16\pi^3(2 - \sqrt{2})\frac{T}{\Delta_0}\frac{d\,(\pi T D)^{1/2}\sin\varphi_0}{e^3 R R_N p_F^2 l^2}\exp\{-d/\xi_N\}. \tag{2.44}$$

Now assume $d/2 - |x_0| \gg (D/T)^{1/2}$ (*SNINS*-junction). Carrying out the calculations in this case we have

$$j = \frac{16\pi^2 d\,(2\pi T D)^{1/2}\Delta_0^2\sin\varphi}{e^3 R R_N p_F^2 l^2 W^2}\exp\{-d/\xi_N\}. \tag{2.45}$$

When $T \ll T_c$ we obtain

$$j = 16\pi^2(3\sqrt{2} - 4)\frac{d\,(\pi T D)^{1/2}\sin\varphi_0}{e^3 R R_N p_F^2 l^2}\exp\{-d/\xi_N\}. \tag{2.46}$$

We see that in the case $l \ll \xi_0$ the Josephson current (as in the case of a low concentration of impurities) depends on the position of the dielectric interlayer within the normal metal layer. Thus, the current density in the *SNINS*-junction is significantly greater than the analogous value for the *SIS*-junction (compare (2.44) and (2.46)).

Thus, the existence of impurities in the *SNINS*-junctions has a significant influence on the critical current level, and such influence (as in the case of *SNS*-junctions) is qualitatively different in the cases $l \gg \xi_0$ and $l \ll \xi_0$.

3. The influence of an external magnetic field. The spectrum of pure *SNINS*-junctions, as may be determined rather easily, takes the form ($E \ll \Delta_0$)

$$\cos\frac{2Ed}{v_x} + \frac{\cos(4Ex_0/v_x)}{1 + v_x^2/V_0^2} + \frac{\cos\varphi_0}{1 + V_0^2/v_x^2} = 0. \tag{2.47}$$

Now assume that an external magnetic field exists in the system. Here we will assume that the field is uniform in the *N*-metal layer and will avoid the screening effect (Section 2). When it must be accounted for in all our calculations we must carry out the substitution

$$\frac{Hx^2}{2} \rightarrow \int_0^x A(x')\,dx'.$$

By means of (1.1) and (2.20) we may easily find the dispersio equation determining the excitation energy in such conditions. It takes the form ($E \ll \Delta_0$)

$$\cos\left[\frac{2Ed}{v_x} - \gamma_+ + \gamma_-\right] + \cos\left(\frac{4Ex_0}{v_x} + \gamma_+ + \gamma_-\right)\left\{1 + \frac{v_x^2}{V_0^2}\right\}^{-1} + \frac{\cos\tilde{\varphi}}{1 + \frac{V_0^2}{v_x^2}} = 0,$$

$$\gamma_\pm = \frac{ev_yH}{v_x}\left(\frac{d}{2} \mp x_0\right)^2, \qquad \tilde{\varphi} = \varphi_0 - 2eHy\,(d + 2\lambda_L), \tag{2.48}$$

where λ_L is the depth of penetration of the magnetic field into th superconductor. Strictly speaking expression (2.48) is contradictory since the intrinsic energy E is independent of the coordinates. I fact the instability of the spectrum in the *SNS*-junctions is relate to this [24, 25]. However if the transparency of the dielectric i small, then in the principal approximation it follows from (2.48) tha

$$\cos\left[\frac{2Ed}{v_x} - \gamma_+ + \gamma_-\right] + \cos\left[\frac{4Ex_0}{v_x} + \gamma_+ + \gamma_-\right] = 0. \tag{2.49}$$

As we see blurring does not occur in this case but rather the level shift by a quantity proportional to the magnetic field H and th derived result is valid for arbitrary values of H.

The significant differences in the behavior of the spectrum o *SNS*- and *SNINS*-junctions in an external magnetic field will also de termine the different influences of this field on the stationar Josephson effect in such junctions. In the case of an *SNINS*-system i is rather easy to account for the magnetic field. Level shift (2.49 results in renormalization of the Mössbauer frequency. Carrying ou the substitution of E by $-i\omega$ in (2.49) we directly obtain

$$\tilde{\omega}_\pm = \omega \mp \frac{1}{4}\, ieHv_F\,(d \mp 2x_0)\sin\theta\cos\chi, \quad x \gtrless x_0. \tag{2.50}$$

Here $v_x = v_F\cos\theta$, $v_y = v_F\sin\theta\cos\chi$. Thus, the magnetic fiel "shifts" the frequency on the imaginary axis (unlike the impuritie whose presence causes the frequency to shift on the real axis) Further calculations were carried out analogous to the case of deriv ing formula (2.30). Finally we will have ($T \ll T_c$)

$$j = \frac{2\sin\tilde{\varphi}}{eR}\, T \sum_\omega \int_0^{2\pi} d\chi \int_0^1 \cos^3\theta\, d(\cos\theta)\left[\mathrm{ch}\,\frac{\tilde{\omega}_+ (d - 2x_0)}{v_F\cos\theta}\,\mathrm{ch}\,\frac{\omega_- (d + 2x_0)}{v_F\cos\theta}\right]^{-1} \tag{2.51}$$

When $T = 0$ for a symmetrical *SNINS*-junction ($x_0 = 0$) expression (2.51 becomes the equality:

$$j=\frac{2v_F\sin\tilde{\varphi}}{\pi eRd}\int\limits_0^{2\pi}d\chi\int\limits_0^1\cos^4\theta d(\cos\theta)\frac{\arcsin Q(H)}{Q(H)}, \tag{2.52}$$

where $Q(H)=\sin\left\{\frac{eHd^2}{2}\operatorname{tg}\theta\cos\chi\right\}$.

The dependence of the critical current density on the magnetic field is given in Fig. 5. It is clear that the magnetic field causes a certain increase in j_c as well as oscillations in this quantity. We emphasize that here we refer specifically to current density. The total current through the junction with dimensions in the order of (or greater than) the Josephson depth of penetration will, of course, be significantly reduced due to the interferential effect.

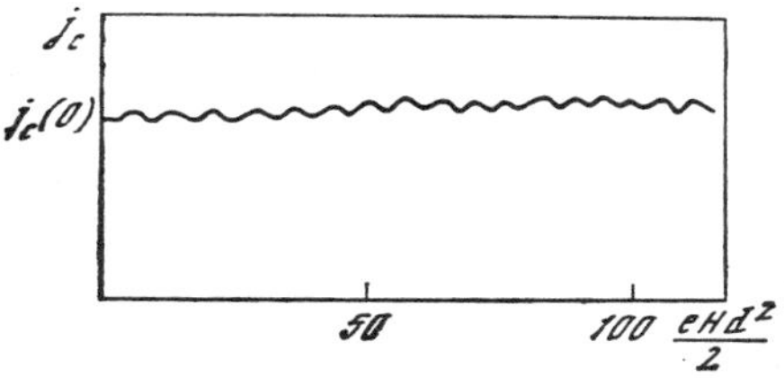

Fig. 5. The $j_c(H)$ relation in an *SNINS*-junction at $T \ll \upsilon_F/d$

For an *SNINS*-junction (2.51) (T = 0) yields $j=\frac{\pi}{5}\frac{v_F}{eRd}\sin\tilde{\varphi}$, i.e., in this case the magnetic field has no influence on the amplitude of the Josephson current. When $T \gg \upsilon_F/d$ the contribution to current comes only from excitations whose velocity component υ_x is similar to $\upsilon_F(\theta \lesssim (\upsilon_F/Td)^{1/2})$. For such excitations the shift in levels is small and, consequently, the existence of a magnetic field in the normal metal layer is not manifest at all in the critical density of tunnel current at such temperatures, and this is valid for any values of x_0.

We see that the Josephson behavior of the *SNINS*-system in an external magnetic field is rather nontrivial and differs significantly from the analogous behavior of an *SNS*-junction. In the case of *SNS*-junctions the dependence of the Josephson current on the magnetic field is rather complex [26]. For *SNS*-junctions the amplitude of the tunnel current depends on the magnetic field only at low temperatures: $T \lesssim \upsilon_F/d$; this dependence varies significantly for various values of x_0. The phase dependence of this current remains sinusoidal at all temperatures.

Thus, the discrete spectrum of quasi particles in the *N*-layer may be noticeably manifest in the properties of *SNS*-junctions. We emphasize that the Andreev levels will exist in an *SNS*-junction only when tunnel conductivity predominates at at least one of the *NS*-boundaries (the *SNS, SINS, SNINS* junctions). In this sense these junctions are fundamentally different from *SINIS*-junctions [46]. There is no discrete spectrum in these junctions, and the supercurrent transport mechanism in this case (opposite the case of junctions with direct conductivity of the *NS*-boundaries) is not related to Andreev reflection.

Chapter 3

THE NONSTATIONARY JOSEPHSON EFFECT AND NONEQUILIBRIUM PROPERTIES OF PURE SUPERCONDUCTOR-NORMAL METAL-SUPERCONDUCTOR JUNCTIONS

The consideration of *SNS*- and *SNINS*-systems given above referred to the stationary equilibrium case. The general picture of the behavior of various weakly-coupled superconducting systems in this case may be considered to be sufficiently well-explored to date (see, for example, survey [79]). The situation is different with nonstationary and nonequilibrium properties of such systems. If the theory of the nonstationary Josephson effect in *SIS*-junctions was completely formulated long ago [27, 28] , significant progress in the investigation of superconducting weak links of other types (primarily *SCS*-junctions [38, 39, 80, 81]) has been made only in recent years. In this chapter we will investigate the nonstationary and nonequilibrium properties of pure *SNS*-junctions.

6. Fundamental equations and general relations

The most common method of microscopic analysis of nonequilibrium systems is the Keldysh method [82]. The Gorkov equations in this method take the form [35] (in this section we consider the case $\mathbf{A} = 0$):

$$\left\{\left(i\hat{\tau}_3 \frac{\partial}{\partial t} + \frac{1}{2m}\frac{\partial^2}{\partial r^2} + \hat{\Delta} - e\Phi + \mu\right)\check{1} - \check{\Sigma}\right\}\check{G} = \check{1}\delta(t-t')\,\delta(r-r'). \tag{3.1}$$

Here Φ is the scalar potential,

$$\check{G} = \check{G}(\mathbf{r}, \mathbf{r}'; t, t') = \begin{pmatrix} \hat{G}^R & \hat{G} \\ \hat{0} & \hat{G}^A \end{pmatrix}, \qquad \check{\Sigma} = \begin{pmatrix} \hat{\Sigma}^R & \hat{\Sigma} \\ \hat{0} & \hat{\Sigma}^A \end{pmatrix}.$$

We may use (3.1) to obtain quasi-classical equations for the Green's functions that are integrated with respect to energy and in the stationary case become (1.1). Such equations were first derived by Eliashberg [30]. They may be represented here in the form

$$\mathbf{v}_F \frac{\partial \check{G}}{\partial \mathbf{R}} + \hat{\tau}_3 \frac{\partial \check{G}}{\partial t} + \frac{\partial \check{G}}{\partial t'}\hat{\tau}_3 + i\{(e\Phi(t) - \hat{\Delta}(t))\check{G} - \check{G}(e\Phi(t') - \hat{\Delta}(t'))\} + i\int_{-\infty}^{\infty} dt_1 \{\check{\Sigma}_{tt_1}\check{G}_{t_1t'} - \check{G}_{tt_1}\check{\Sigma}_{t_1t'}\} = 0, \qquad \check{G} = \check{G}(\mathbf{v}_F, \mathbf{R}; t, t'). \tag{3.2}$$

The function $\check{G}$ satisfies the condition [35]

$$\int_{-\infty}^{\infty} dt_1 \check{G}_{tt_1}\check{G}_{t_1t'} = \check{1}\delta(t-t'). \tag{3.3}$$

The self-energy part of Σ takes the form

$$\check{\Sigma} = \frac{v_F}{2l} \langle \check{G}(\mathbf{v}_F, \mathbf{R}; t, t') \rangle + \check{\Sigma}_{ph}, \tag{3.4}$$

where Σ_{ph} describes the electron-phonon collisions. When a dielectric interplay exists in the system we must add the term $V_0 \delta(x - x_0) \check{1}$ to the right half of (3.4). With the same assumptions as in the stationary case we obtain the boundary conditions from (3.1) for $\check{G}$ at the point x_0:

$$\check{G}(v_x, x_0 + 0) - \check{G}(-v_x, x_0 + 0) = \check{G}(v_x, x_0 - 0) - \check{G}(-v_x, x_0 - 0) =$$
$$= i \frac{v_x}{V_0} \{\check{G}(v_x, x_0 + 0) - \check{G}(v_x x_0 - 0)\}. \tag{3.5}$$

The $\check{G}$ functions in (3.5) are taken with coincident time arguments.

The current density in the system is given by the formula

$$\mathbf{j}(t) = \frac{ep_F}{4\pi} \langle \mathrm{Sp}\, \hat{\tau}_3 \mathbf{p}_F \hat{G}(t, t) \rangle. \tag{3.6}$$

As before we will assume that the superconductors are located in the range $|x| > d/2$, while the N-metal is in the range $|x| < d/2$ and we will use the "stepwise" model for $\Delta(x)$. The solutions of equations (3.2) in the superconductors are sought as [39]

$$\check{G}_j(v_x, x) = \check{A}_j + \exp\{-\check{K}_j(x - (-1)^j d/2)\} \check{B}_j(v_x) \exp\{\check{K}_j(x - (-1)^j d/2\}. \tag{3.7}$$

The matrices $\check{A}_j$ are known:

$$\check{A}_j = \begin{pmatrix} \hat{A}_j^R & \hat{A}_j \\ \hat{0} & \hat{A}_j^A \end{pmatrix}, \qquad \hat{A}_j^{R(A)} = \hat{S}_j(t) \int \hat{g}^{R(A)}(\varepsilon)\, e^{-i\varepsilon(t-t')} \frac{d\varepsilon}{2\pi} \hat{S}_j^+(t'),$$

$$\hat{g}^{R(A)} = g^{R(A)}(\varepsilon)\hat{\tau}_3 + f^{R(A)}(\varepsilon)\, i\hat{\tau}_2 = \frac{(\varepsilon + i\gamma_1^{R(A)})\hat{\tau}_3 + (\Delta_0 + i\gamma_2^{R(A)})\, i\hat{\tau}_2}{[(\varepsilon + i\gamma_1^{R(A)})^2 - (\Delta_0 + i\gamma_2^{R(A)})^2]^{1/2}}, \tag{3.8}$$

$$\hat{S}_j(t) = \cos\frac{\varphi(t)}{4} \hat{1} + i(-1)^j \sin\frac{\varphi(t)}{4} \hat{\tau}_3, \qquad \varphi(t) = \varphi_0 + 2e \int^t V(t_1)\, dt_1$$

($V(t)$ is the difference in potentials between the superconducting borders; the expressions for the quantities $\gamma_{1,2}^{R(A)}$ describing the electron-phonon relaxation are given in study [38]).

We may show that the matrices $\check{A}$ and $\check{B}$ are related by the relations [39]

$$\check{B}_j(v_x)\check{A}_j = -\check{A}_j\check{B}_j(v_x) = (-1)^j \check{B}_j(v_x) \operatorname{sign} v_x. \tag{3.9}$$

Moreover, the following conditions are valid

$$\hat{A}_j = \hat{A}_j^R \hat{n}_j - \hat{n}_j \hat{A}_j^A,$$

$$\hat{n}_j(t,t') = \hat{S}_j(t)\int \operatorname{th}\frac{\varepsilon}{2T}\, e^{-i\varepsilon(t-t')}\frac{d\varepsilon}{2\pi}\, S_j^+(t'). \tag{3.10}$$

We will consider two cases: an *SNINS*-junction and an *SNS*-bridge.

7. The nonstationary Josephson effect in *SNINS*-junctions for a random voltage

1. The general expression for current. We will assume that th resistance of the dielectric barrier is large so that all voltag crosses this barrier. Then the electrical potential distribution i space may be written as

$$\Phi(t) = \frac{1}{2} V(t)\operatorname{sign}(x - x_0).$$

The solution of equations (3.2) when $|x| < d/2$ (i.e., when $\Delta = 0$) i trivial. It yields (in the case of no impurities)

$$\check{G}_j(v_x, x_0) = \check{C}_j(v_x)(\check{A}_j + \check{B}_j(v_x))\hat{C}_j^+(v_x), \qquad \check{G}_{1(2)}(v_x, x_0) \equiv \check{G}(v_x, x_0 \mp 0),$$

$$\hat{C}_j(v_x; t, t') = \hat{S}_j(t)\int C_j(v_x, \varepsilon) e^{-i\varepsilon(t-t')}\frac{d\varepsilon}{2\pi}\hat{S}_j^+(t'),$$

$$C_j(v_x, \varepsilon) = \cos\alpha_j + i(-1)^{j+1}\sin\alpha_j, \qquad \alpha_j(v_x, \varepsilon) = \frac{\varepsilon(d/2 + (-1)^j x_0)}{v_x}. \tag{3.11}$$

We still must find the matrices $\check{B}_j(v_x)$ Using the smallness of th transparency of the dielectric barrier we may represent these matrice in the following form:

$$\check{B}_j = \check{B}_j^{(0)} + \frac{v_x}{V_0}\check{B}_j^{(1)} + \frac{v_x^2}{V_0^2}\check{B}_j^{(2)} + \ldots, \tag{3.12}$$

and, as is normally the case, in order to find the current we mus calculate the term of this expansion $\sim V_0^{-2}$. Using formulae (3.5) (3.12) we obtain the final result [73]:

$$j(t) = \operatorname{Im}\iint d\omega\, d\omega' \{W(\omega) W^*(\omega') e^{-i(\omega'-\omega)t} j_q(\omega') + e^{i\varphi_0} W(\omega) W(\omega') e^{i(\omega+\omega')t} j_p(\omega')\}, \tag{3.13}$$

$$\operatorname{Re} j_q(\omega) = \frac{i}{2eR}\int_0^1 \alpha^3 d\alpha \int_{-\infty}^{\infty} d\varepsilon \{T_+(\varepsilon)[X^{RR}Y^{RR} - X^{AA}Y^{AA}] + T_-(\varepsilon)[X^{RA}Y^{RA} - X^{AR}Y^{AR}]\}, \tag{3.14a}$$

$$\operatorname{Im} j_q(\omega) = \frac{1}{2eR}\int_0^1 \alpha^3 d\alpha \int_{-\infty}^{\infty} d\varepsilon\, T_-(\varepsilon)\{X^{RR}Y^{RR} - X^{AR}Y^{AR} + X^{AA}Y^{AA} - X^{RA}Y^{RA}\}, \tag{3.14b}$$

$$\operatorname{Re} j_p(\omega) = \frac{i}{eR}\int_0^1 \alpha^3 d\alpha \int_{-\infty}^{\infty} d\varepsilon \{T_+(\varepsilon)[f^R(\varepsilon - \omega) f^R(\varepsilon) Y^{RR} -$$

$$- f^A(\varepsilon-\omega) f^A(\varepsilon) Y^{AA}] + T_-(\varepsilon) [f^A(\varepsilon-\omega) f^R(\varepsilon) Y^{RA} - $$
$$- f^R(\varepsilon-\omega) f^A(\varepsilon) Y^{AR}]\}, \qquad (3.15a)$$

$$\operatorname{Im} j_p(\omega) = \frac{1}{eR} \int_0^1 \alpha^3 d\alpha\, T_-(\varepsilon) f^R \{(\varepsilon-\omega) [f^R(\varepsilon) Y^{RR} - f^A(\varepsilon) Y^{AR}] +$$
$$+ f^A(\varepsilon-\omega) [f^A(\varepsilon) Y^{AA} - f^R(\varepsilon) Y^{RA}]\}, \qquad (3.15b)$$

$$X^{ij} = [1 + g^i(\varepsilon) g^j(\varepsilon-\omega)] \cos\beta_+ - [1 - g^i(\varepsilon) g^j(\varepsilon-\omega)] \cos\beta_- +$$
$$+ i [g^i(\varepsilon) + g^j(\varepsilon-\omega)] \sin\beta_+ - i [g^i(\varepsilon) - g^j(\varepsilon-\omega)] \sin\beta_-,$$
$$Y^{ij} = \{[1 + g^i(\varepsilon) g^j(\varepsilon-\omega)] \cos\beta_+ + [1 - g^i(\varepsilon) g^j(\varepsilon-\omega)] \cos\beta_- +$$
$$+ i [g^i(\varepsilon) + g^j(\varepsilon-\omega)] \sin\beta_+ + i [g^i(\varepsilon) - g^j(\varepsilon-\omega)] \sin\beta_-\}^{-1}, \qquad (3.16)$$
$$i(j) = R, A,$$
$$T_\pm(\varepsilon) = \operatorname{th}\frac{\varepsilon}{2T} \pm \operatorname{th}\frac{\varepsilon-\omega}{2T}, \qquad \beta_\pm = \frac{\varepsilon(d-2x_0)}{v_F\alpha} \pm \frac{(\varepsilon-\omega)(d+2x_0)}{v_F\alpha}.$$

In the formula for current (3.13) we employed the regular representation [27])

$$\exp\left\{ie\int^t V(t_1)\,dt_1\right\} = \int_{-\infty}^{\infty} W(\omega) e^{i\omega t}. \qquad (3.17)$$

Formulae (3.13)-(3.17) describe the nonstationary Josephson effect in *SNINS*-junctions with a random voltage variation principle. In the case $d = x_0 = 0$ the derived result coincides with the familiar expression for current in an *SIS*-junction first obtained by the tunnel Hamiltonian method [27, 28].

2. The d.c. case. We will now proceed to investigate the derived expressions. First assume that the voltage across the barrier V is constant in time. Then $W(\omega) = \delta(\omega - eV)$. In this case j_q is time-independent while we represent $j_p(t)$ as

$$j_p(t) = j_1 \sin\varphi(t) + j_2 \cos\varphi(t).$$

When $T = 0$ after the necessary intermediate calculations we have

$$j_0 = \operatorname{Im} j_q(eV) = j_2 = 0, \quad eV \ll \Delta; \qquad (3.18)$$
$$j_1 = j_1^r + j_1^a; \qquad (3.19a)$$

$$j_1^r(x_0 = \pm d/2) = \begin{cases} \dfrac{\pi v_F}{10eRd}\left(1 + \dfrac{\Delta_0}{\sqrt{\Delta_0^2 - e^2V^2}}\right), & \dfrac{v_F}{d} \ll eV,\ \Delta_0 - eV \gg \dfrac{v_F}{d}, \\ \dfrac{2}{9eR}\sqrt{\dfrac{2\pi v_F\Delta_0}{d}} \sum\limits_{k=0} \dfrac{(-1)^k}{\sqrt{2k+1}}, & |\Delta_0 - eV| \ll v_F/d; \end{cases} \qquad (3.19b)$$

$$j_1^r(x_0 = 0) = \frac{4v_F}{eRd}\int_0^1 \alpha^4 d\alpha \frac{\operatorname{arctg}\left(\operatorname{tg}\left(\frac{Vd}{v_F\alpha} + \chi\right)\right)}{\sin\left(\frac{Vd}{v_F\alpha} + \chi\right)},$$
$$\chi = \arcsin(eVd/v_F), \quad v_F/d \ll eV, \quad \Delta_0 - eV \gg v_F/d; \qquad (3.19c)$$

$$j_1^a\left(x_0=\pm\frac{d}{2}\right)=\frac{2\pi v_F}{eRd}\int_0^1\alpha^4\,d\alpha\sum_{n=0}^{N}\frac{(\Delta_0^2-\varepsilon_n^2)^{1/2}}{\sin(2\varepsilon_n d/v_F\alpha)\,[\Delta_0^2-(\varepsilon_n-eV)^2]^{1/2}},$$

$$eV<\Delta_0; \tag{3.19d}$$

$$j_1^a(x_0=0)=\frac{4\pi v_F}{eRd}\int_0^1\alpha^4\,d\alpha\sum_{m=0}^{M}\frac{(\Delta_0^2-\varepsilon_m^2)^{1/2}}{\sin(\varepsilon_m d/v_F\alpha)}\Big\{[\Delta_0^2-(\varepsilon_m-eV)^2]^{1/2}\times$$

$$\times\cos\frac{(\varepsilon_m-eV)\,d}{v_F\alpha}+(eV-\varepsilon_m)\sin\frac{(\varepsilon_m-eV)\,d}{v_F\alpha}\Big\}^{-1},\quad eV<\Delta_0. \tag{3.19e}$$

Here ε_n and ε_m are positive roots of the equations

$$\operatorname{tg}\frac{2\varepsilon_n d}{v_F\alpha}=\frac{(\Delta_0^2-\varepsilon_n^2)^{1/2}}{\varepsilon_n},\qquad \operatorname{tg}\frac{\varepsilon_m d}{v_F\alpha}=\frac{(\Delta_0^2-\varepsilon_m^2)^{1/2}}{\varepsilon_m}, \tag{3.20}$$

while N and M are the maximum numbers such that ε_N, ε_M are less tha eV. When $v_F/d \ll eV \ll \Delta_0$ expressions (3.19d), (3.19e) take the form

$$j_1^a(x_0=\pm d/2)=\frac{2v_F^2}{e^2RVa^2}\sum_{n=0}^{\infty}\frac{(-1)^n}{(2n+1)^2}\sin\left[\frac{2eVd}{v_F}+(2n+1)\right], \tag{3.21a}$$

$$j_1^a(x_0=0)=\frac{2v_F}{eRd}\ln\left|\frac{1+\cos(eVd/v_F)}{1-\cos(eVd/v_F)}\right|. \tag{3.21b}$$

It is clear the E-I characteristic of the *SNINS*-systems is suf ficiently nontrivial even when $eV<\Delta_0$ and its form differs signifi cantly with different positions of the dielectric barrier. The norma part of the Josephson current j_1^r in the *SNINS*-system monotonicall grows with an increase in V and reaches a maximum when $eV=\Delta_0$. Th increase in the current when $eV\sim\Delta_0$ is due to the root singularity i the density of states of the superconductor. The anomalous componen j_1^a for such a system oscillates with period $\pi v_F/d$, since as the poten tial difference across the barrier increases the position of the sin gularities in the state density within the normal metal layer als changes. The current $j_1^r(x_0=0)$ remains virtually the same in mag nitude ($eV<\Delta_0$) and oscillates only slightly with a period $2\pi v_F/d$ The formal expression for the anomalous part of the Josephson curren in this case, as we see from (3.21b), contains logarithmic divergence at points $V=\pi v_F n/ed$. This fact is related to the existence of pol singularities of the Green's functions of the system that reflect th existence of discrete levels. Such behavior of the current $j_1(V)$ i the *SNINS*-system in some sense is similar to the behavior of analogou current in *SIS*-junctions whose expression logarithmically diverge when $eV=2\Delta_0$ [27, 28]. This feature in the dependence of the Joseph son current on the voltage across the *SIS*-junctions is due to the roo divergence in the density of states of the superconductor at energie close to the spectral gap. In an *SNINS*-system the existence of a dis crete spectrum on both sides of the dielectric barrier causes the di vergence in the expression for current j_1.

The current divergences in *SNINS*-systems (as in *SIS*-systems manifest a formal nature. Indeed the discrete levels will be somewha broadened due to scattering processes (both elastic and inelastic which may be the result of a variety of causes (for example, in homogeneity of the NS'-boundaries, the existence of a few impurities, phonons, etc.). In this regard expression (3.21b) is not valid in the vicinity of the divergence points. With logarithmic accuracy near such points it will be replaced by

$$j_1^{a}(x_0=0)=\pm\frac{4v_F}{eRa}\ln\frac{v_F}{\gamma_N d}, \tag{3.22}$$

where γ_N is the characteristic value of the broadening of the discrete spectral levels (obviously, $\gamma_N \ll v_F/d$).

Result (3.18) is quite clear. It is explained by the fact that when $T = 0$ there are no quasi-particle excitations in the superconductor. Hence an electron, having tunneled through the dielectric barrier from the surface of the Fermi layer of the normal metal may end up above the gap of the superconductor only when[1] $eV > \Delta_0$. For comparison we remember that a similar situation exists in the SIS-junction. The normal current j_0 and current j_2 (often called the interference current of the quasi particles and the pairs) in this case (and in the case $T = 0$) are nonzero at voltages $eV > 2\Delta_0$, since pair breaking in one superconductor and the transit of the electrons in tunneling to the continuous spectrum over the gap of the other superconductor are possible only in this condition. In addition we note that result (3.18) is valid with any position of the dielectric layer within the normal metal. We also emphasize that (3.18) is valid only in the absence of impurities ($l > d$). As we have already seen the normal metal layer in some sense achieves the property of the superconductor even when the BCS interaction constant is equal to zero, since there is nothing to break down the superconducting correlation of the Cooper pairs transiting the N-layer. However if we now introduce impurities into the normal metal layer, the situation changes significantly. Thus, when $l \ll d$ the existence of superconducting boundaries will not be manifest in the value of the state density in the normal metal near the dielectric interlayer. In this case the result for normal current in the SIS-system in the principal approximation will coincide with the familiar expression for the current in the SIN-junction

$$j_0=\frac{1}{2eR}\int_{-\infty}^{\infty}\left(\operatorname{th}\frac{\varepsilon+eV}{2T}-\operatorname{th}\frac{\varepsilon}{2T}\right)\frac{|\varepsilon|\,\theta(|\varepsilon|-\Delta_0)}{(\varepsilon^2-\Delta_0^2)^{1/2}}\,d\varepsilon,$$

while in the $SNINS$-system the E-I characteristic will be similar to Ohm's Law.

---

[1] We note that such a result is valid only in the first approximation of the expansion of the current in $1/R$. In subsequent orders of this expansion the normal current may appear due to the existence of the nonequilibrium mechanism of electrical field penetration into the superconductor. We may ignore the contribution of this mechanism by virtue of the large value of R.

However, we will return to the case of no impurities. We wil now provide expressions for the currents in an *SINS-system* ($x_0 = \pm d/2$ when $v_F/d \ll eV - \Delta_0 \ll \Delta_0$ ($T = 0$):

$$j_0 = \frac{2\,\mathrm{sign}\,V}{3eR}\,\frac{e|V| - \Delta_0}{e|V| + \Delta_0}\,(e|V| + 2\Delta_0) + \frac{V}{R\,(e|V| - \Delta_0)} \times$$

$$\times \left(\frac{\pi v_F^3}{2d^3\,(e|V| + \Delta_0)}\right)^{1/2} \sum_{n=1}^{\infty} \frac{(-1)^{n+1}}{n^{3/2}} \sin \frac{4nd\,(e|V| - \Delta_0)}{v_F}, \tag{3.23}$$

$$j_1^r = \frac{v_F^2}{6eRd^2}\,\frac{\Delta_0}{(e|V| - \Delta_0)\,(e^2V^2 - \Delta_0^2)^{1/2}}\,G, \tag{3.24a}$$

$$j_1^a = \frac{2\pi v_F}{eRd} \int_0^1 \alpha^4\,d\alpha \sum_{n=N_1}^{N_2} \frac{1}{\sin(2\epsilon_n d/v_F\alpha)} \left(\frac{\Delta_0^2 - \epsilon_n^2}{\Delta_0^2 - (\epsilon_n - eV)^2}\right)^{1/2} + j'(V), \tag{3.24b}$$

$$j_2 = \frac{2\Delta_0\,\mathrm{sign}\,V}{eR\,(e|V| - \Delta_0)} \left(\frac{\pi v_F^3}{d^3\,(e|V| + \Delta_0)}\right)^{1/2} \sum_{n=0}^{\infty} \frac{(-1)^n}{(2n+1)^{3/2}} \cos\left[\frac{2d\,(e|V| - \Delta_0)}{v_F}\,(2n+1)\right]. \tag{3.25}$$

Here G is Katalan's constant; e_n is determined from (3.20); N_1 is th minimum number such that $e_{N_1} \geqslant eV - \Delta_0$; N_2 is the maximum number suc that $e_{N_2} \leqslant \Delta_0$. The expression for the component $j'(V)$ in (3.24b which determines the contribution to the current from the continuou spectrum will be provided below.

Analogous expressions are also valid in the case $x_0 = 0$. W will not provide the corresponding formulae here for space considera tions. We will simply focus on the fact that when $eV > \Delta_0$ the expres sions for the currents j_0 and j_2 will contain components that oscil late as a function of V with a period of $\pi v_F/2d$ and $\pi v_F/d$, respective ly. The physical cause for such oscillations are the interferentia effect that produces spatial quantization of the excitation energy In the *SINS*-systems this effect was identified in experiments [70-72] The oscillating $j(V)$ relation was also identified in theoretica studies [83, 84]. The nonstationary Josephson effect in an *SINS*-sys tem was investigated in these studies. The Green's functions of th system were calculated by means of an expansion in terms of the eigen functions of the single-particle problem and then by solving th Gorkov equations.[2]

With a further increase in V the expressions for the currents j and j_2 become somewhat more complex. In this case the contribution t current will come from not only the lower but also the higher excita

---

[2] There is no quantitative correlation between our results fo the currents in an *SINS*-system and analogous results from studies [82 84]. It may be demonstrated that the final expressions given in thes studies are imprecise.

tion energy levels in the normal metal layer, and when $eV > 2\Delta_0 (T = 0)$ the continuous spectrum also makes a contribution. The derived formulae are of the same type. For example, we may write out the result for j_0 in an *SINS*-system with $eV > \Delta_0 (T \ll T_0)$:

$$j_0 = j_{01} + j_{02},$$

$$j_{01} = \frac{\pi v_F}{eRd} \int_0^1 \alpha^4 \, d\alpha \sum_{n=0}^{N} \left(\operatorname{th} \frac{\varepsilon_n}{2T} - \operatorname{th} \frac{\varepsilon_n - eV}{2T} \right) \frac{eV - \varepsilon_n}{[(eV - \varepsilon_n)^2 - \Delta_0^2]^{1/2}} . \qquad (3.26)$$

In formula (3.26) e_n is determined from (3.20), while in this case N is the maximum number such that $e_N \leqslant eV - \Delta_0$. When $e_n \ll \Delta_0$ and $T = 0$ (3.26) becomes (3.23). Further we have

$$j_{02} = \frac{2}{eR} \theta (eV - 2\Delta_0) \int_0^1 \alpha^3 \, d\alpha \int_{\Delta_0}^{eV - \Delta_0} d\varepsilon \left(\operatorname{th} \frac{\varepsilon - eV}{2T} - \operatorname{th} \frac{\varepsilon}{2T} \right) \times$$

$$\times \frac{\varepsilon (\varepsilon - eV)}{[(\varepsilon - eV)^2 - \Delta_0^2]^{1/2}} \frac{\varepsilon^2 - \Delta_0^2}{\varepsilon^2 - \Delta_0^2 \cos^2 (2d/v_F \alpha)} . \qquad (3.27)$$

When $eV \gg \Delta_0$ and $T = 0$

При $eV \gg \Delta_0$ и $T = 0$ we find

$$j_{01} \simeq \Delta_0 / eR. \qquad (3.28)$$

In order to investigate the integral j_{02} we will use the following fact. We may show that

$$\int_1^A f(x) \frac{dx}{x^2 - \cos^2 bx} = \int_1^A \frac{f(x)}{x} \frac{dx}{\sqrt{x^2 - 1}} + O\left(\frac{1}{B}\right), \qquad A > 1, \ B \gg 1.$$

Utilizing this equality for j_{02} we may easily obtain $(T = 0)$

$$j_0 = (e^2 V^2 - 2eV\Delta_0)^{1/2} / eR. \qquad (3.29)$$

When $eV \gg \Delta_0$ we have from (3.28) and (3.29):

$$j = \frac{V}{R} - \frac{\Delta_0^2}{2e^2 VR} . \qquad (3.30)$$

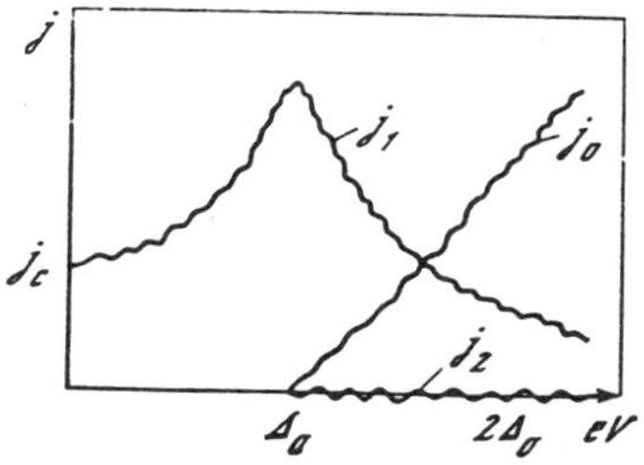

Fig. 6. The currents j_0, j_1 and j_2 plotted as a function of V for an *SINS*-junction with $T \ll v_F/d$

Thus, with large values of V the E-I characteristic of the *SINS*-systems is given by Ohm's Law. An analogous result may be obtained for *SNINS*-systems as well. In addition we note that in formulae (3.28)-(3.30) we have utilized the significant inequality $v_F/d \ll \Delta_0$ and have dropped the corresponding minor terms.

We will now provide the expression for the component $j_i(V)$ in formula (3.24b) which determines the contribution of the continuous spectrum to current j_1^a:

$$j'(V)=\frac{\theta(eV-\Delta_0)}{eR}\int_0^1\alpha^3\,d\alpha\int_M^{eV}d\varepsilon\left(\text{th}\frac{\varepsilon}{2T}-\text{th}\frac{\varepsilon-eV}{2T}\right)\times$$

$$\times\frac{\sqrt{\varepsilon^2-\Delta_0^2}}{\sqrt{\Delta_0^2-(\varepsilon-eV)^2}}\,\frac{\Delta_0^2\cos(2\varepsilon d/v_F\alpha)}{[\varepsilon^2-\Delta_0^2\cos^2(2\varepsilon d/v_F\alpha)]}. \tag{3.31}$$

Here $M = \max\{\Delta_0, eV - \Delta_0\}$. When $T = 0$ and $eV \gg \Delta_0$ (3.31) yields

$$j'(V)=\frac{\sqrt{\pi/2}}{e^3V^2R}\left(\frac{\Delta_0 v_F}{d}\right)^{3/2}\sin\frac{2(eV-\Delta_0)d}{v_F}. \tag{3.32}$$

The general form of the voltage dependence of the currents j_0, j_1, and j_2 in an *SINS*-system with $T = 0$ is shown in Fig. 6. The currents j_0 and j_2 in the *SNINS*-system behave in a like manner. The graphical representation of the $j_1(V)$ relation for the *SNINS*-systems is difficult to provide due to the significant irregularity of this relation. We note only that the logarithmic features of this current component (3.21) will exist when $eV < 2\Delta_0$. With large values of V the behavior of $j_1(V)$ is analogous in *SNINS*- and *SINS*-systems.

We will not provide certain formulae that determine the nonstationary Josephson effect in these systems in the temperature range $v_F/d \ll T \ll T_c$. When $eV < \Delta_0$ the currents j_0 and j_2 are exponentially small. For example, for j_0 in an *SINS*-system we have

$$j_0(x_0=\pm d/2)=\frac{\exp\{-(\Delta_0-eV)/T\}}{eR}\left\{\frac{1}{2}\sqrt{\pi T(2\Delta_0)+T}\right)+$$
$$+(\Delta_0-eV)^{-1}\sqrt{\frac{\pi\Delta_0 v_F^3}{2a^3}}\sum_{n=1}^{\infty}\frac{\sin\{4nd(\Delta_0-eV)/v_F+\pi/4\}}{n^{3/2}}\right\}. \tag{3.33}$$

When $eV > \Delta_0$ we have

$$j_{0,2}(T\gg v_F/d)=j_{0(2)}(T=0),\qquad 0<eV-\Delta_0\ll\Delta_0. \tag{3.34}$$

The pair current j_1^r when $T \gg v_F/d$ diminishes exponentially. For example, in the *SNINS*-system the expression for this current takes the form

$$j_1^r(x_0=0)=\frac{16\pi T}{eR\Delta_0 d}\,\frac{\exp\{-2d/\xi_T\}}{4\pi^2T^2+e^2V^2}\left\{[2\pi T\sqrt{\Delta_0^2-e^2V^2}-e^2V^2]\cos\frac{eVd}{v_F}-\right.$$
$$\left.-eV[2\pi T+\sqrt{\Delta_0^2-e^2V^2}]\sin\frac{eVd}{v_F}\right\},$$
$$v_F/d\ll eV\leqslant\Delta_0. \tag{3.35}$$

The anomalous component j_1^a has no multiplier that exponentially decays with temperature. Thus, when $v_F/d \ll eV \ll \Delta_0$ and $T \gg v_F/d$ the expression for j_1^a is in fact independent of T, i.e.,

$$j_1^a(T\gg v_F/d)=j_1^a(T=0). \tag{3.36}$$

The value $j_1^a(T = 0)$ is determined by formulae (3.21). If necessary it is easy to calculate the expressions for the currents in other particular cases as well. We may easily obtain the expression for the current in the *NINS*-system from general formulae (3.13)-(3.17) in the case where the parameter Δ_0 is equal to 0 at one of the borders. It is clear that only the normal current is nonzero in this case:

$$j_0 = \frac{1}{eR}\int_0^1 \alpha^3 d\alpha \int_{-\infty}^{\infty} d\varepsilon \left(\text{th}\frac{\varepsilon}{2T} - \text{th}\frac{\varepsilon - eV}{2T}\right)\left\{\frac{g^R(\varepsilon)c(\varepsilon) + is(\varepsilon)}{c(\varepsilon) + ig^R(\varepsilon)s(\varepsilon)} - \right.$$

$$\left. - \frac{g^A(\varepsilon)c(\varepsilon) + is(\varepsilon)}{c(\varepsilon) + ig^A(\varepsilon)s(\varepsilon)}\right\}, \quad c(\varepsilon) = \cos\left(\frac{2\varepsilon d}{v_F\alpha}\right), \quad s(\varepsilon) = \sin\left(\frac{2\varepsilon d}{v_F d}\right). \tag{3.37}$$

The E-I characteristic of such systems when $eV \gg v_F/d$ follows from regular Ohm's Law. In the inverse case $eV \ll v_F/d$ we have

$$j_0 = \frac{31}{20}\frac{V}{R}\left(\frac{4eVd}{\pi v_F}\right)^4 \zeta(5), \qquad T = 0. \tag{3.38}$$

At higher temperatures $T \gg v_F/d$ the $j_0(V)$ relation in the *NINS*-systems becomes resistive for any values of V.

3. The a.c. case. Supercurrent enhancement. We will now return to the investigation of *SNINS-systems*. *Assume now that in addition to the d.c. voltage across the barrier there is also a.c.:* $V(t) = V + V_1 \cos \omega_1 t$. Here

$$W(\omega) = \sum_{n=-\infty}^{\infty} J_n\left(\frac{eV_1}{\omega_1}\right)\delta(\omega - eV - n\omega_1),$$

where J_n is the Nth order Bessel function. The current will take the form

$$j(t) = \text{Im}\sum_{n=-\infty}^{\infty}\sum_{k=-\infty}^{\infty} J_n\left(\frac{eV_1}{\omega_1}\right)e^{ik\omega_1 t}\left\{J_{k+n}\left(\frac{eV_1}{\omega_1}\right)j_q(eV + n\omega_1) + \right.$$

$$\left. + \exp\{i\varphi_0 + 2iVt\}\, J_{k-n}\left(\frac{eV_1}{\omega_1}\right)j_p(eV + n\omega_1)\right\}. \tag{3.39}$$

Here $j_q(\omega)$, $j_p(\omega)$ are determined by formulae (3.14)-(3.16). It is clear that in this case there are current oscillations as a function of voltage as well as logarithmic peaks in the Josephson current component (in *SNINS*-systems). Moreover, such effects are also manifest as a function of the frequency ω_1 of the a.c. component.

When the condition $2eV = n\omega_1$ in these systems is satisfied the same effect as for *SIS*-junctions exists; in this case a d.c. current may flow through the junction. For *SINS*-junctions when $T = 0$ in the principal approximation (for simplicity we assume $eV \ll \Delta_0$, $\omega_1 \ll \Delta_0$) it is equal to

$$j_n = \frac{\pi v_F}{5eRd}(-1)^n J_n\left(\frac{2eV}{\omega_1}\right)\sin\varphi_0. \tag{3.40}$$

If there is no d.c. voltage across the barrier ($V = 0$) and the frequency and amplitude of the a.c. satisfy the condition $eV_1 \ll \omega_1 \ll \Delta_0$, the expression for the current in the system will take the following simple form:

$$j(t) = j_c \sin\varphi_0 + 2j_1(\omega_1)\frac{eV_1}{\omega_1}\sin\omega_1 t\cos\varphi_0. \tag{3.41}$$

Here j_c is the critical current of the *SNINS*-junction. In the *SINS* systems the expression yields ($T < \upsilon_F/d$):

$$j(t) = \frac{\pi v_F}{5eRd}\left\{\sin\varphi_0 + \frac{2eV_1}{\omega_1}\left[1 + \frac{\omega_1^2}{4\Delta_0^2} + \right.\right.$$

$$\left.\left. + \frac{10v_F}{\pi\omega_1 d}\sum_{n=0}^{\infty}\frac{(-1)^n}{(2n+1)^2}\sin\left(\frac{2\omega_1 d}{v_F}(2n+1)\right)\right]\sin\omega_1 t\cos\varphi_0\right\}. \tag{3.42}$$

An analogous expression (unfortunately, containing inaccuracies) was obtained by another method in study [76].

In the *SNINS*-system the expression for current $j(t)$ becomes logarithmically large if the frequency satisfies the condition $\omega_n = \pi\upsilon_F n/d$ (with the d.c. component: $eV + \omega_n = \pi\upsilon_F n/d$). Here we have resonant absorption of the photon with frequency ω_n by the tunneling electron and the conductivity of the junction becomes large.

We will now continue the expansion of the general expressions for current (3.39) up through the term proportional to the second power of the small parameter eV_1/ω_1. Averaging over time we obtain

$$j = \left[j_1^r(0) - \frac{e^2V_1^2}{2\omega_1^2}\left(j_1^r(0) - j_1^r(\omega_1) + j_1^r(\omega_1)\right)\right]\sin\varphi_0 - \frac{e^2V_1^2}{2\omega_1^2}j_2(\omega_1)\cos\varphi_0. \tag{3.43}$$

As we have seen when $T \gg \upsilon_F/d$ the current j_1^r is exponentially small, while the currents j_1^a and j_2 do not contain such smallness. Thus, the existence of an alternating electromagnetic field increases the superconducting properties of the *SNS*-junctions.

To some degree this effect is analogous to the familiar phenomenon of microwave enhancement of superconductivity [85]. Indeed, the role of the gap in the quasi-particle excitation spectrum of pure *SNS*-systems is played by the quantity $\Delta_e \sim \upsilon_F/d$. When $\omega_1 \gtrsim \Delta_e$ the quasi-particle distribution function in the *N*-layer is a nonequilibrium function and in the energy range near Δ_e there is a shortage of quasi particles. When $T \gg \upsilon_F/d$ the amplitude of the current flowing through the junction will also be determined largely by the nonequilibrium quasi particles with energies in the order of the reverse transit time through the *N*-layer $\upsilon_F/d \sim \Delta_e$, which results in effective amplification of the critical current (enhancement). Clearly this effect will exist for any type of *SNS*-junctions. Only the current - not the effect itself - depends on the position of the dielectric interlayer in the *N*-layer (and generally on its existence or absence).

The critical current enhancement effect also exists in superconducting bridge structures where (in certain conditions) the primary contribution to the nonhomogeneity of the quasi particles is not directly as a result of their electrical field-acceleration, but rather the "vibration" of the superconducting gap in the junction [80]. This effect has been observed experimentally.

There has also been experimental evidence of microwave-field enhancement of the critical current in *SNS*-junctions with a short free path length [86]. We again note that dirty *SNS*-bridges in a microwave field of low intensity have been considered theoretically in a recent study [87]. The critical current in these conditions is proportional to the electron-phonon relaxation time in the *N*-layer and is inversely proportional to d^4, i.e., it does not have a multiplier that decays exponentially with the thickness of the *N*-layer.

In *SNINS*-systems there will be additional supercurrent enhancement related to the existence of logarithmic peaks in the $j_1^a(\omega_1)$ relation. When $v_F/d \ll T$, $\omega_1 \ll \Delta_0$ and $d \gg \xi_0$ the critical current density in the junction is

$$j_c = \frac{e^2 V_1^2}{2\omega_1^2} \, | j_1^a(\omega_1) |, \tag{3.44}$$

where j_1^a is determined from (3.36), (3.21b). Accounting for the noted formal nature of the divergence of j_1^a at frequencies of $\omega_n = \pi v_F/d$ near these values we have, with logarithmic accuracy [77]

$$j_c = \frac{2 v_F e V_1^2}{R\, d\omega_1^2} \ln \frac{v_F}{\gamma_N d} . \tag{3.45}$$

Thus, critical current enhancement in pure *SNINS*-systems by a microwave field may occur for two reasons. Irradiation of the junction results in the formation of nonequilibrium electrons in the *N*-layer that freely travel through the junction and, moreover, produce a resonant (at certain frequencies) jump of the tunneling electrons from one energy level to another. There exists a type of "double" supercurrent enhancement effect. As a result the critical current, even at a low irradiation power, may have a significant magnitude, and varies in inverse proportion to the thickness of the *N*-layer (however, not exponentially in the equilibrium case).

All results obtained in this section are valid in the conditions

$$\nu_T \ll 1/\tau_{eN} \ll v_F/d, \qquad \nu_T \ll 1/\tau_{eS}^{*} \ll \Delta_0. \tag{3.46}$$

Here ν_T is the characteristic frequency of electron jumps through the tunnel barrier ($\nu_T \propto 1/R$), τ_{eN}, τ_{eS} are the inelastic relaxation times of excitations in the *N*- and *S*-metals. These inequalities make it possible to calculate the current by an expansion of the transparency of the dielectric in terms of the small parameter, using as the zeroth approximation the equilibrium Green's function of the borders in the absence of current, and moreover, makes it possible to identify the discrete spectrum in the *N*-metal and use expressions for the Green's functions obtained in the BCS model.

In addition we note that we cannot make a formal transition to the case $d \to \infty$ in our general formulae for current. Indeed, at sufficiently high values of d the broadening of the Andreev levels even at $l = \infty$ becomes in the order of the distance between them ($d \sim v_F\tau_{eN}$), so (3.6) is no longer satisfied. At the same time if we make Γ_{eN} as large as possible (as is clear from (3.46)) limited in the range of application of our calculations to small currents only ($\nu_T \to 0$, $R \to \infty$), since in these calculations we have not accounted for the nonequilibrium effects associated with the use of the distribution function due to current flow, these appear when $\nu_T\tau_{eN} \gtrsim 1$. In order to pass to the limit $d \to \infty$ it is necessary to account for the finiteness of the quantity l. Here, as noted above, we are led to Ohm's Law.

8. The nonstationary and nonequilibrium properties of *SNS*-bridges

In this section we will examine a bridge of variable thickness as our model of an *SNS*-bridge [79]: two bulk superconductors are connected by a thin *N*-metal filament. As before the problem is considered quasi one-dimensional. We will assume that the filament length is d with a cross-sectional area of $S = \pi a^2$.

1. General expression for current. In this model the electrical potential at the superconducting borders is independent of the coordinates. Assume the right superconductor has the potential $V(t)/2$,

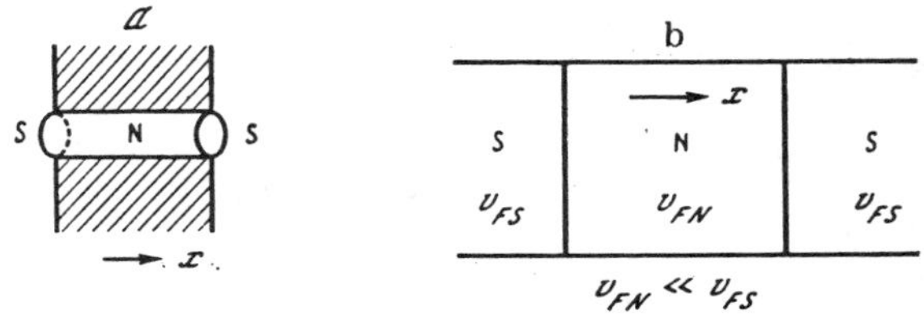

Fig. 7. Schematic representation of the *SNS*-structures in section 8
a - *SNS*-bridge, b - *SNS*-sandwich with $v_{FN} \ll v_{FS}$

while the left superconductor has a potential $-V(t)/2$. Then the solution of equations (3.2) in the superconductors on the transit trajectories will, as before, take the form of (3.7). The solution of (3.2) when $|x| < d/2$ may also be searched rather easily. We will introduce the matrix

$$\hat{C}(\varepsilon, \varepsilon'; v_x) = \left(\cos\frac{\varepsilon d}{2v_x}\hat{1} + i\sin\frac{\varepsilon d}{2v_x}\hat{\tau}_3\right)\delta(\varepsilon - \varepsilon'). \tag{3.47}$$

We may use (3.47) to easily find the expression for $\check{G}(x = 0) \equiv \check{G}_0(v_x)$. We have (for simplicity we drop the energy variables):

$$\begin{aligned} \check{A}_1 + \check{B}_1(v_x) &= \hat{C}(v_x)\check{G}_0(v_x)\hat{C}^+(v_x), \\ \check{A}_2 + \check{B}_2(v_x) &= \hat{C}^+(v_x)\check{G}_0(v_x)\hat{C}(v_x). \end{aligned} \tag{3.48}$$

Here and henceforth the product of the matrices will imply convolution with respect to the internal variable. Using (3.48) and (3.49) after some simple manipulation we obtain

$$\check{G}_0(v_x) = (\check{1} + \operatorname{sign} v_x \cdot \check{Q}_-(v_x))\, \check{Q}_+^{-1}(v_x), \tag{3.49}$$

$$\check{Q}_\pm(v_x) = \frac{1}{2}\{\hat{C}^+(v_x)\, \check{A}_1 \hat{C}(v_x) \pm \hat{C}(v_x)\, \check{A}_2 \hat{C}^+(v_x)\}. \tag{3.50}$$

We find from (3.49) and (3.50) for the matrix $\check{G}$ [78]:

$$\begin{aligned}
&\hat{G}_0(v_x) = \hat{G}^r(v_x) + \hat{G}^a(v_x),\\
&\hat{G}^r(v_x) = \hat{P}^R \hat{n}_+ - \hat{n}_+ \hat{P}^A,\\
&\quad \hat{P}^{R(A)} = (\hat{1} + \operatorname{sign} v_x\, \hat{Q}_-^{R(A)})\, (\hat{Q}_+^{R(A)})^{-1},\\
&\hat{G}^a(v_x) = -(\hat{Q}_+^R)^{-1} (\hat{Q}_-^R \hat{n}_- - \hat{n}_- \hat{Q}_-^A)\, (\hat{Q}_+^A)^{-1} + \operatorname{sign} v_x [(\hat{Q}_+^R)^{-1} \hat{n}_- (\check{Q}_+^A)^{-1} +\\
&\qquad + \hat{Q}_-^R (\hat{Q}_+^R)^{-1} \hat{n}_- \hat{Q}_-^A (\hat{Q}_+^A)^{-1} - \hat{n}_-],\\
&\quad \hat{n}_\pm = \frac{1}{2}(\hat{C} \hat{n}_1 \hat{C}^+ + \hat{C}^+ \hat{n}_2 \hat{C}).
\end{aligned} \tag{3.51}$$

The current density in the system is determined by formula (3.6). Multiplying (3.6) by the area of the hole S, we obtain

$$I = \frac{1}{4eR_0} \operatorname{Sp} \hat{\tau}_3 \int_{-1}^{1} \alpha\, d\alpha\, \hat{G}_0(v_F \alpha), \tag{3.52}$$

where $R_0 = \pi^2/p_F^2 S e^2$ is the resistance of the junction in the normal state. Formula (3.52) determines the current at the point $x = 0$. It is clear that the current may be calculated at any other point of the N-metal and by virtue of the continuity equation the result is independent of the selection of this point.[3]

Thus, relations (3.8), (3.10), (3.47), (3.50)-(3.52) provide a complete solution of the problem of determining $I(t)$ in pure SNS-bridges with arbitrary values of $V(t)$. When $d = 0$ our expressions for the Green's functions of the systems, as would be expected, become the corresponding formulae from study [39] obtained for pure SNS-junctions.

In addition we note that we may represent the matrix $\hat{G}_0^a$ in the following form:

$$\hat{G}_0^a(v_x) = \hat{P}^R \hat{F}^a - \hat{F}^a P^A. \tag{3.53}$$

The function $\hat{F}^a$ in fact describes the deviation of the distribution function from equilibrium. It may be found from the equations

---

[3] Strictly speaking this statement refers only to the time-independent current component. We will discuss this issue in greater detail below.

$$\hat{Q}^R_{\pm}\hat{F}^a - \hat{F}^a\hat{Q}^A_{\pm} = \hat{Q}^R_{\mp}\hat{n}_- - \hat{n}_-\hat{Q}^A_{\mp}. \tag{3.54}$$

Relations (3.53), (3.54) are also a direct generalization of the corresponding formulae from [38, 39] ($d = 0$).

We will now proceed to an investigation of the derived general expressions.

2. Linear response. Assume a d.c. current I_0 flows through the junction; this current is parametrized by the time-independent phase difference φ_0. When a low a.c. voltage appears across the junction the phase difference may take the form $\varphi(t) = \varphi_0 + \varphi_1(t)$, $\varphi_1 \ll 1$. The existence of the a.c. component $\varphi_1(t)$ produces additional current through the junction $I_1(t)$. We will obtain the expression for this current in a linear approximation in φ_1.

We will employ formulae (3.51). We should note that the φ_1 linear terms will be found in the expansions of the matrices $\hat{P}^{R,A}$ and $\hat{n}_-$. Using this fact after some simple calculations we will have

$$I_{1\omega} = \frac{i\varphi_{1\omega}}{8R_0}\int_{-1}^{1}\alpha\,d\alpha\cos\frac{\omega d}{2v_F\alpha}\int d\varepsilon\left\{\operatorname{th}\frac{\varepsilon_+}{2T}Z^{AA} - \operatorname{th}\frac{\varepsilon_-}{2T}Z^{RR} + \right.$$
$$\left. + \left(\operatorname{th}\frac{\varepsilon_-}{2T} - \operatorname{th}\frac{\varepsilon_+}{2T}\right)Z^{RA}\right\}, \qquad Z^{ij} + \operatorname{sign}\alpha =$$
$$= \{\operatorname{sign}\alpha\,[g^i(\varepsilon_+)g^j(\varepsilon_-) + f^i(\varepsilon_+)f^j(\varepsilon_-)\cos\chi_+\cos\chi_-][1 -$$
$$- f^i(\varepsilon_+)f^j(\varepsilon_-)\sin\chi_+\sin\chi_-] - i\,[f^i(\varepsilon_+)\sin\chi_+ +$$
$$+ f^j\sin\chi_-][g^i(\varepsilon_+)f^j(\varepsilon_-)\cos\chi_- + f^i(\varepsilon_+)g^j(\varepsilon_-)\cos\chi_+]\}\times$$
$$\times[1 + (f^i(\varepsilon_+)\sin\chi_+)^2]^{-1}[1 + (f^j(\varepsilon_-)\sin\chi_-)^2]^{-1},$$
$$\chi_\pm = \varphi_0/2 + \varepsilon_\pm d/v_F\alpha, \qquad \varepsilon_\pm = \varepsilon \pm \omega/2. \tag{3.55}$$

Here $I_{1\omega}$, $\varphi_{1\omega}$ are the Fourier-transforms of the quantities $I(t)$, $\varphi_1(t)$. Here we will not deal in great detail with the issue of investigating expression (3.55) in various critical cases. We note only that this expression is significantly simplified in the conditions $T \sim T_c$, $\omega \ll \Delta_0(T) < v_F/d$ and becomes the equality

$$I_{1\omega} = \frac{i\omega\varphi_{1\omega}}{2R_0} + \varphi_{1\omega}\frac{\pi\Delta_0^2\sin^2\frac{\varphi_0}{2}}{4eR_0T}[1 - 1/i\tau_\varepsilon\omega]^{-1} + I_c\varphi_{1\omega}\cos\varphi_0. \tag{3.56}$$

Here I_c is the critical current of the bridge, $\tau_\varepsilon = 1/2\,\gamma_1 \propto T^3 (T \sim T_c)$ is the electron-phonon relaxation time at the borders.

When $d \to 0$ (3.56) becomes Zaytsev's result [39]. In the case $d \gg \xi_0$ $I_c \propto \exp\{-d/\xi_0\}$ since the last term in (3.56) may be ignored. then with low frequencies $\omega \ll 1/\tau_\varepsilon$ we have

$$I_1 = \frac{V(t)}{\tilde{R}_0(\varphi_0)}, \quad V(t) = \frac{1}{2e}\frac{\partial\varphi_1}{\partial t}, \quad \tilde{R}_0(\varphi_0) = R_0\left[1 - \sin^2\frac{\varphi_0}{2}\,\frac{\pi\Delta_0^2\tau_\varepsilon}{2T}\right]^{-1}. \tag{3.57}$$

Thus, there is an increase in the conductivity in *SNS*-bridges (as well as other types of weak links with direct conductivity); this is due to the deviation of the distribution function from an equilibrium function. We will also focus on the fact that when $\omega \ll v_F/d$ the expression for current (3.56) is in fact independent of d.

In the low temperature range the behavior of the bridge is also rather interesting. When $T = 0$ we have

$$I_{1\omega} = I^r_{1\omega} + I^a_{1\omega};$$
$$I^r_{1\omega} = \begin{cases} \varphi_{1\omega} \dfrac{\Delta_0 \cos(\varphi_0/2)}{eR_0}, & d \ll \xi_0, \\ \dfrac{\varphi_{1\omega} v_F}{6eR_0 d}, & d \gg \xi_0; \end{cases}$$
$$I^a_{1\omega} = \frac{i\omega\varphi_{1\omega}}{2R_0 \cos^2(\varphi_0/2)}. \qquad (3.58)$$

Relations (3.58) are valid for sufficiently low frequencies $\omega \ll \min\{\Delta_0, v_F/d\}$ and values of φ_0 that are not too close to π. At high frequencies $\omega \gg v_F/d$ and $d \gg \xi_0$:

$$I_{1\omega} = \frac{2v_F}{eR_0 d} \varphi_{1\omega} \left[\ln\left(2 \operatorname{tg} \frac{\omega d}{4v_F}\right) + \cos \frac{\omega d}{2v_F} \right]. \qquad (3.59)$$

The logarithmic features in the expression for the linear response of the *SNS*-junctions (3.59) when $\omega = 2\pi v_F/d$ are similar to analogous features in the E-I characteristic of *SNINS*-junctions. These are related to the electron "jumps" between the Andreev levels under the action of an external variable field.

Thus, even with low voltages the behavior of the *SNS*-junctions is sufficiently nontrivial and, as we see, it may be differentiated significantly at high and low temperatures. Below we will examine in greater detail the physical causes behind this differential.

3. Critical current enhancement in a microwave field. In Section 7 we have already demonstrated that microwave irradiation may be used to drive the distribution function of an *SNS*-junctions out of equilibrium thereby causing the critical current to increase significantly.

In order to investigate the enhancement of supercurrent from pure *SNS*-bridges we will use general formulae (3.51). Assume we have an a.c. voltage $V(t) = V_1 \cos \omega_1 t$ across the junction. Then

$$\varphi_{1\omega} = eV_1/\omega_1 \left(\delta(\omega + \omega_1) - \delta(\omega - \omega_1)\right). \qquad (3.60)$$

We will calculate the critical current of the *SNS*-junctions assuming $eV_1 \ll \omega_1$ and $d \gg \xi_0$. For this case, as in the case of an *SNINS*-junction, it is necessary to expand the general expression for current up through terms proportional to the second power of $\varphi_{1\omega}$.

Carrying out the corresponding calculations and averaging over time, we obtain the final result

$$I^a = \frac{eV_1^2}{32\omega_1^2 R_0}\int d\varepsilon \int_{-1}^{1} \alpha\, d\alpha \cos\frac{\omega_1 d}{2v_F\alpha}\sum_{k=1}^{2} T_k Y^{RA}\{\operatorname{sign}\alpha \times$$
$$\times [iL_1^R N_+^{RA} - iL_1^A N_+^{AR} + L_2^{AR} N_-^{RA} - L_2^{RA} N_-^{AR}] -$$
$$- [N_1 + L_3^{RA} N_2^{RA} - L_3^{AR} N_2^{AR}]\}; \tag{3.61}$$

$$L_1^j = [f^j(\varepsilon)\sin\chi(\varepsilon) + f^j(\varepsilon+\omega_k)\sin\chi(\varepsilon+\omega_k)]\,[1 + (f^j(\varepsilon+\omega_k) \times$$
$$\times \sin\chi(\varepsilon+\omega_k))^2]^{-1}, \qquad L_2^{ij} = if^i(\varepsilon)\sin\chi(\varepsilon)\,[1 - f^j(\varepsilon) f^j(\varepsilon+\omega_k) \times$$
$$\times \sin\chi(\varepsilon)\sin\chi(\varepsilon+\omega_k)]\,[1 + (f^j(\varepsilon+\omega_k)\sin\chi(\varepsilon+\omega_k))^2]^{-1},$$
$$L_3^{ij} = L_1^i\,[f^j(\varepsilon)\sin\chi(\varepsilon) + f^i(\varepsilon+\omega_k)\sin\chi(\varepsilon+\omega_k)],$$
$$Y^{RA} = [1 + (f^R(\varepsilon)\sin\chi(\varepsilon))^2]^{-1}\,[1 + (f^A(\varepsilon)\sin\chi(\varepsilon))^2]^{-1};$$
$$N_\pm^{ij} = H_1^i g^j(\varepsilon) \pm H_2^i f^j(\varepsilon)\cos\chi(\varepsilon), \qquad N_2^{ij} = H_1^i f^j(\varepsilon)\cos\chi(\varepsilon) + H_2^i g^j(\varepsilon); \tag{3.62}$$
$$H_1^i = g^i(\varepsilon) f^i(\varepsilon+\omega_k)\cos\chi(\varepsilon+\omega_k) + g^i(\varepsilon+\omega_k) f^i(\varepsilon)\cos\chi(\varepsilon),$$
$$H_2^i = g^i(\varepsilon) g^i(\varepsilon+\omega_k) + f^i(\varepsilon) f^i(\varepsilon+\omega_k)\cos\chi(\varepsilon)\cos\chi(\varepsilon+\omega_k);$$
$$N_1 = (g^R(\varepsilon) g^A(\varepsilon) + f^R(\varepsilon) f^A(\varepsilon)\cos^2\chi(\varepsilon))\,(g^R(\varepsilon) - g^R(\varepsilon+\omega_k) - g^A(\varepsilon) +$$
$$+ g^A(\varepsilon+\omega_k)) - (g^R(\varepsilon) f^A(\varepsilon) + g^A(\varepsilon) f^R(\varepsilon))\cos\chi(\varepsilon)\,[(f^R(\varepsilon) -$$
$$- f^A(\varepsilon))\cos\chi(\varepsilon) + (f^R(\varepsilon+\omega_k) - f^A(\varepsilon+\omega_k))\cos\chi(\varepsilon+\omega_k)];$$

$$T_k = \operatorname{th}\frac{\varepsilon+\omega_k}{2T} - \operatorname{th}\frac{\varepsilon}{2T}, \quad \chi(\varepsilon) = \frac{\varphi_0}{2} + \frac{\varepsilon d}{v_F\alpha}, \quad \omega_k = (-1)^k\omega_1. \tag{3.63}$$

Formulae (3.61)-(3.63) are used to provide the general expression for the "nonequilibrium" part of the current flowing through the junction Here we have not been interested in the expression for I^r since when $\gg v_F/d$ (i.e., at temperatures where the enhancement effect is manifest) $I^r \propto \exp\{-d/\xi_T\}$. The derived general expression is significantly simplified in the case $T \sim T_c$, $\omega_1 \ll \Delta_0(T) < v_F/d$. In these conditions it becomes ($\varphi_0 = \pi/2$):

$$I_c = \frac{\pi}{8}\frac{\Delta_0^2}{R_0 T}\frac{eV_1^2}{\omega_1^2 + 1/\tau_\varepsilon^2}\,. \tag{3.64}$$

Thus, the expression for the critical current in a microwave field for *SNS*-bridges in certain conditions ($\omega_1 \ll v_F/d$) may generally be independent of the *N*-layer thickness. It is clear that when $\omega_1 \ll 1/\tau_e$ $I_c \sim \tau_\varepsilon^2$, while when $\omega \gg 1/\tau_\varepsilon$ the critical current becomes in the order of I_c for narrow bridges in equilibrium conditions. In other words, the nonequilibrium nature of the distribution function serves to "constrict" the bridge.

Clearly when $\omega_1 \gtrsim v_F/d$ the characteristic energies at which the distribution function goes significantly out of equilibrium is in the order of (or greater than) the reverse transit time of the electrons through the junction, so that in this case I_c is already dependent on

d. However, this relation (in certain conditions) will be a power relation rather than an exponential relation as in the equilibrium case. Here the "nonequilibrium" critical current may exceed the equilibrium value I_c by several orders of magnitude, which has been observed in experiment [86].

The derived results are valid ignoring the contribution of electron-phonon scattering in the N-layer. The relaxation of the distribution to an equilibrium state in our case occurs at the superconducting borders. Another physical situation was considered in study [87] whose authors proposed that the energy relaxation of nonequilibrium quasi particles "trapped" in the N-layer occurs as a result of the electron-phonon interaction in this layer. The nonequilibrium correction to the distribution function was calculated by perturbation theory based directly on the kinetic equation for the N-metal with regular "equilibrium" boundary conditions for the Green's functions at the NS-boundaries. Here it is also clear that only a slight deviation from equilibrium may cause a significant increase in the critical current of the junction.

4. The nonstationary Josephson effect. Low and high voltages. As noted previously the derived general expressions for the Green's functions make it possible in principle to determine the current through the junction with an arbitrary voltage. However in a number of cases (for example, when $eV > \Delta_0$) it is quite difficult to carry out the calculations "to the number" and at the same time to completely determine the E-I characteristic of the junction. Hence below we will limit our investigation to the nonstationary Josephson effect with low ($eV \ll \Delta_0$) and high ($eV \gg \Delta_0$) voltages.

a) Low voltages. We will first consider the temperature range $T \sim T_c$. In this case $\Delta_0(T)$ is small, so in any actual experimental situation the condition $d < v_F/\Delta_0(T)$ will be satisfied. Here this examination does not differ appreciably from that carried out in studies [38, 39] for the case of short bridges. Finally we obtain

$$I = \frac{1}{2eR_0}\frac{\partial\varphi}{\partial t} + \frac{\pi\Delta_0^2}{4eR_0T}P\{\varphi\} + I_c \sin\varphi. \tag{3.65}$$

Thus, the form of the "current versus phase" relation for SNS-bridges in the resistive state near T_c is the same as in the case of short SCS-junctions [38, 39] (the form of the functional $P\{\varphi\}$ is found in study [39]). However if in the short junctions the multipliers with $P\{\varphi\}$ and $\sin\varphi$ are equal and coincide with the critical current of the junction, they differ significantly in SNS-junctions. Thus, when $d \gg \xi_0$, I_c is exponentially small, while the first two terms, as we see, are independent of d, i.e., the relative contribution of the "nonequilibrium" components in the case of rather long junctions is greater than for short junctions. The functional $P\{\varphi\}$ was investigated in study [39] and hence it is not necessary here to write out the expressions for the current in various limiting cases. We note only that with small voltages $eV \ll 1/\tau_\varepsilon$ the E-I characteristic of the junction will take the form

$$V = I\tilde{R}_0, \qquad \tilde{R}_0 = r_0\left(1 - \frac{\pi\Delta_0^2}{4T}\tau_\varepsilon\right), \tag{3.66}$$

i.e., as in the case of a small a.c. level, deviation from equilibrium produces an effective increase in the conductivity of the junction.

When reducing the temperature the situation changes significantly. We will consider the case $T \to 0$ and $eV \ll \min\{\Delta 0,\ v_F/d\}$. Moreover, we will assume that V is constant or varies slowly over time. In order to find the current it is convenient to use formula (3.53). The matrices $\hat{Q}_+^{B,A}$, $(\hat{Q}_+^{B,A})^{-1}$ for the range of energies of interest to us in the adiabatic approximation are obtained by substituting φ_0 with $\varphi(t)$ in the corresponding equilibrium expressions. We find the $\hat{F}^1$ function from (3.59). As a result we have

$$I(\varphi) = I^r(\varphi) + I^a(\varphi), \tag{3.67}$$

where $I^r(\varphi)$ is a generalization of the results from studies [19, 40] to the nonstationary case ($T \to 0$):

$$I^r(\varphi) = \begin{cases} \dfrac{2v_F}{3eR_c d}\varphi, & -\pi < \varphi < \pi, \quad d \gg \xi_0, \\ \dfrac{\Delta_0}{eR_0}\sin\dfrac{\varphi}{2}\,\mathrm{sign}\cos\dfrac{\varphi}{2}. \end{cases} \tag{2.68}$$

The expressions for $I^a(\varphi)$ differ significantly depending on the range of variation of the parameters of the problem. Thus, I^a when $\cos\frac{\varphi(H)}{2} \gg \max\left\{\frac{eV}{\Delta_0}, \frac{eVd}{v_F}, \frac{T}{\Delta_0}, \frac{Td}{v_F}\right\}$ is independent of d:

$$I^a = \frac{V}{R_0\cos^2\frac{\varphi}{2}}. \tag{3.69}$$

With arbitrary values of $\varphi(t)$ yet when $T \ll eV \ll v_F/d < \Delta$ we have

$$I^a = \frac{V}{R_0\cos^2\frac{\varphi}{2}}\left(1 + \frac{b^2}{2}\left(1 + \sin^2\frac{\varphi}{2}\right)\mathrm{tg}^2\frac{\varphi}{2}\ln\left|1 - \frac{1}{b^2}\mathrm{ctg}^2\frac{\varphi}{2}\right|\right), \tag{3.70}$$

where $b = eVd/2v2F$.

In the case $\cos\varphi/2 \gg eVd/v_F$ (3.70) coincides with (3.69). The resonant natures in expression (3.70), as in the other cases, result from the existence of a discrete structure of the Andreev levels in the junction.

Formula (3.70) is valid when we may ignore inelastic electron scattering in the system. When accounting for such scattering we have

$$I^a = \frac{V}{R_0\left(\cos^2\frac{\varphi}{2} + \gamma^2\right)}. \tag{3.71}$$

here $\varkappa = \gamma/\Delta_0$, $eV \ll \gamma$. The quantity γ describes inelastic relaxation of the electron states in the system which may result from various scattering mechanisms.

Thus, the $I(\varphi)$ relation at low temperatures for low voltages differs significantly from the equilibrium relation (3.68) (and moreover, from the sinusoidal relation). This differential is directly related to the additional $I^a(\varphi)$ component in (3.67) which describes the deviation of the distribution function from equilibrium. We call attention to the fact that the time-average value of this component is nonzero.

Thus, by averaging expression (3.70) over the period of $\varphi(t)$ and assuming here that V varies only slightly over this period, we obtain

$$\bar{I} = \frac{e^2V^3d}{R_0v_F^2}\ln\frac{2v_F}{eVd}. \tag{3.72}$$

An analogous averaging technique applied to formula (3.71) yields

$$\bar{I} = V/R_0\varkappa, \tag{3.73}$$

i.e., the E-I characteristic of the *SNS*-bridge in these conditions is significantly different from Ohm's Law. This differential is due to the fact that when $T \to 0$ there are no quasi-particle excitations in the system, and the existence of the current is related to the flow of the condensate. From this viewpoint the derived result may appear to be strange. Indeed, the fact that I is nonzero indicates the existence of energy dissipation in the system. However what form of dissipation do we have if all the current is transferred by the superconducting condensate ($T = 0$, $eV \ll \Delta_0$)? We remember in this regard that in tunnel junctions when $T = 0$ and $eV < 2\Delta_0$, $\bar{I} = 0$, i.e., there is no dissipation. Later in this discussion we will attempt to explain the physical nature of this result. For now we will consider the range of high voltages within which it is possible to obtain an expression for the current through the junction.

b) High voltages. In the case $eV \gg \Delta_0$ the normal current significantly exceeds the Josephson current and the interferential component. Here $\varphi = \varphi_0 + 2Vt$ while for $(\hat{Q}_+^{R,A})^{-1}$ completely analogous to study [39], we have

$$[\hat{Q}_+^{R(A)}(\varepsilon, \varepsilon')]^{-1} = \frac{2}{g^{R(A)}\left(\varepsilon + \frac{eV}{2}\right)g^{R(A)}\left(\varepsilon - \frac{eV}{2}\right) + 1}\hat{Q}_+^{R(A)}(\varepsilon, \varepsilon'). \tag{3.74}$$

The current through the junction in the principal approximation takes the form

$$I = I_0 + I_1 \sin\varphi + I_2 \cos\varphi. \tag{3.75}$$

Calculating I_0 by means of (3.71) we obtain

$$I_0=\frac{V}{R_0}+I_{ex}, \qquad I_{ex}=\frac{8}{3}\frac{\Delta_0}{eR_0}\,\mathrm{th}\,\frac{eV}{2T}\,. \tag{3.76}$$

We note that result (3.73) is independent of the thickness of the *N*-metal and is valid with various thicknesses (evidently for pure systems only). Thus, the expressions for the excess current for pure *SNS*-bridges and short *SCS*-junctions [39] coincide.

The expressions for I_1 and I_2 will be provided for the case of short *SNS*-bridges $d \ll v_F/eV$. In the principal approximation in Δ_0/eV the Josephson and interferential current components appear as:

$$\begin{aligned}
I_1=&\frac{i}{4eR_0}\int d\varepsilon\left\{\frac{1}{2}\left[T_+\left(\frac{3V}{2}\right)+T_+\left(\frac{V}{2}\right)\right]\left[f^R\left(\varepsilon+\frac{eV}{2}\right)f^R\left(\varepsilon-\frac{eV}{2}\right)Z^R(\varepsilon)-\right.\right.\\
&-f^A\left(\varepsilon+\frac{eV}{2}\right)f^A\left(\varepsilon-\frac{eV}{2}\right)Z^A(\varepsilon)\Big]+T_-\left(\frac{V}{2}\right)\left[f^R\left(\varepsilon+\frac{eV}{2}\right)\times\right.\\
&\left.\left.\times f^A\left(\varepsilon-\frac{eV}{2}\right)M_++f^R\left(\varepsilon-\frac{eV}{2}\right)f^A\left(\varepsilon+\frac{eV}{2}\right)M_-\right]\right\},
\end{aligned}\tag{3.77}$$

$$\begin{aligned}
I_2=&\frac{1}{4eR_0}\int d\varepsilon\left\{\frac{1}{2}\left[T_-\left(\frac{3V}{2}\right)+T_-\left(\frac{V}{2}\right)\right]\left[f^R\left(\varepsilon+\frac{eV}{2}\right)f^R\left(\varepsilon-\frac{eV}{2}\right)\times\right.\right.\\
&\times Z^R(\varepsilon)+f^A\left(\varepsilon+\frac{eV}{2}\right)f^A\left(\varepsilon-\frac{eV}{2}\right)Z^A(\varepsilon)\Big]-\\
&-T_-\left(\frac{V}{2}\right)\left[f^R\left(\varepsilon+\frac{eV}{2}\right)f^A\left(\varepsilon-\frac{eV}{2}\right)M_++\right.\\
&\left.\left.+f^R\left(\varepsilon-\frac{eV}{2}\right)f^A\left(\varepsilon+\frac{eV}{2}\right)M_-\right]\right\},
\end{aligned}\tag{3.78}$$

$$\begin{aligned}
M_\pm=&Z^R(\varepsilon\pm eV)Z^A(\varepsilon)+\frac{1}{4}\left\{\left[\pm g^R\left(\varepsilon\mp\frac{3eV}{2}\right)\mp\right.\right.\\
&\mp g^R\left(\varepsilon\mp\frac{eV}{2}\right)\Big]Z^R(\varepsilon\mp eV)\pm\left[g^R\left(\varepsilon+\frac{eV}{2}\right)+\right.\\
&\left.+g^R\left(\varepsilon-\frac{eV}{2}\right)\right]Z^R(\varepsilon)\Big\}\left\{\left[g^A\left(\varepsilon-\frac{eV}{2}\right)-g^A\left(\varepsilon+\frac{eV}{2}\right)\right]Z^A(\varepsilon)\pm\right.\\
&\left.\pm\left[g^A\left(\varepsilon\pm\frac{eV}{2}\right)+g^A\left(\varepsilon\pm\frac{3eV}{2}\right)\right]Z^A(\varepsilon\pm eV)\right\},\\
Z^{R(A)}(\varepsilon)=&\left[1+g^{R(A)}\left(\varepsilon+\frac{eV}{2}\right)g^{R(A)}\left(\varepsilon-\frac{eV}{2}\right)\right]^{-1},\\
&T_\pm(x)=\mathrm{th}\,\frac{\varepsilon+ex}{2T}-\mathrm{th}\,\frac{\varepsilon-ex}{2T}\,.
\end{aligned}\tag{3.79}$$

When $T \ll \Delta_0$ formulae (3.77)-(3.79) are significantly simplified. In this case we have

$$I_1=\frac{\pi\Delta_0^2}{2e^2R_0V}, \qquad I_2=\frac{\Delta_0^2}{2e^2R_0V}\left(\frac{3}{4}-\ln\frac{4eV}{\Delta_0}\right). \tag{3.80}$$

For comparison purposes we will provide expressions for the currents I_1 and I_2 in the *SIS*-junctions [27, 28] for $T \ll \Delta_0 \ll eV$:

$$I_1=\frac{\pi\Delta_0^2}{e^2RV}, \qquad I_2=-\frac{2\Delta_0^2}{e^2RV}\ln\frac{2eV}{\Delta_0}\,. \tag{3.81}$$

It is clear that I_1 and $|I_2|$ in our case are less than the corresponding expressions in (3.18). We note in addition that result (3.80) is qualitatively consistent with experimental data from study [88]. Indeed the quantity $\gamma(V) = I_1/I_c$ in our case is equal to

$$\gamma(V) = \Delta_0/2eV, \quad eV \gg \Delta_0 \gg T.$$

An analogous quantity for *SIS*-junctions in this temperature and voltage range is equal to $\gamma_{SIS} = 2\Delta_0/eV$. A drop in $\gamma(V)$ in point junctions with direct conductivity compared to the case of tunnel junctions was observed in study [88]. However a quantitative comparison of our results to results from these experiments is rather complex, since study [88] employed superconductors with a high quantity of impurities and, moreover, rather moderate voltages of $eV \leqslant 4\Delta_0$. In the case of wide *SNS*-junctions the expressions for the currents I_1 and I_2 will, as in the case of *SNINS*-systems, contain an additional (compared to expressions (3.80)) smallness with respect to the parameter v_F/eVd.

Thus the examination has revealed that many of the nonstationary properties of *SNS*-bridges are significantly different from the properties of tunnel junctions examined previously [27, 28] and short superconducting bridges [38, 39]. At high temperatures $T \sim T_c$ the nonequilibrium nature of the distribution function plays a determinant role, since the Josephson current at these values of T and $d > \xi_0$ is exponentially small, while the "nonequilibrium" components do not contain this smallness (at sufficiently low voltages and frequencies there is generally no dependence on d). The critical current enhancement in *SNS*-junctions by a microwave field is in effect related to this. We remember that in short bridges there is a suppression of supercurrent under irradiation by low intensity microwave emission [38], since the reduction in pair current I^r in these conditions predominates over the anomalous component I^a (which is absent in the equilibrium situation). In sufficiently long *SNS*-bridges I^a may exceed I^r by several orders of magnitude, which would serve to "constrict" the bridge under irradiation.

At low temperatures $T \ll \min\{\Delta 0, v_F/d\}$ there are in fact no quasi-particle excitations in the system. In the case of low voltages the current flowing through the junction is nonzero only by virtue of condensate flow. One significant difference between systems with tunnel and direct conductivity arises here. In tunnel junctions, as we know, when $T = 0$ the quasi-particle and interferential current components are equal to zero (in the first order with respect to transparency) when $eV < 2\Delta_0$ (*SIS*) or $eV < \Delta_0$ (*SINS, SNINS*). In junctions with direct conductivity I^a is nonzero even when $T = 0$ and at any low voltages, while the mechanism "responsible" for current flow in this case is Andreev reflection with $E < \Delta_0$ from the *NS*-boundaries (for *SNS*-junctions) or the direct transition of electrons from condensate to condensate (for superconducting bridges). In tunnel junctions these phenomena make a contribution to the current in the second order with respect to transparency only and hence do not play a significant role in current transport. In systems with direct conductivity I^a generated by electron transport between the superconducting condensates of two metals, obviously, will have a significant dependence on

$\varphi(t)$ (see (3.70)-(3.71)). The E-I characteristic of the junction here differs from Ohm's Law.

As noted above the fact that the current I^a is nonzero indicates that dissipation exists in the system. But where is the energy going? We know the current is being transported solely by the Cooper pairs. The answer is simple. The chemical potentials of the two superconductors differ (when calculated for a single Cooper pair) by $2eV$. Consequently in order to transport a pair from one Cooper condensate to another (i.e., from one superconductor to another) it is necessary to expend energy of $2eV$. We emphasize that here we refer specifically to the direct transport of a pair from condensate to condensate. In tunnel junctions in the first order with respect to transparency tunneling is a single-particle process. Consequently, as predicted by theory [27, 28], when $T = 0$ and $eV < 2\Delta_0$ there is no dissipative current in such junctions (and it appears only when accounting for second order terms with respect to transparency). In direct conductivity junctions the Cooper pair need not be broken for transport of electrons from one superconductor to another. In this regard such junctions assume a resistive state even when $T = 0$ and with low voltages.

One additional significant difference of direct conductivity junctions is the existence of excess current in the E-I characteristic; this current is absent in tunnel junctions. This is due to the fact that in tunnel systems quasi particles with $E < \Delta_0$ do not make a contribution to the current in the first order with respect to transparency, at the same time that all states make a contribution to current in *SNS*- and *SCS*-bridges.

Thus, the specific nature of the nonstationary behavior of the *SNS*-bridges is largely related to the contribution to the conductivity of quasi particles with $E < \Delta_0$ "trapped" in the *N*-layer. Discreteness of the spectrum of such oscillations causes a number of characteristic logarithmic features to appear on the E-I characteristic. Clearly these logarithmic peaks on the E-I characteristic "spread out" when we account for the finiteness of the width of the Andreev levels which is the result of quasi particle scattering (elastic and inelastic) processes.

A discrete quasi-particle spectrum when $E < \Delta_0$ is characteristic of pure *SNS*-systems, since these features may be realized only in such systems and do not exist in *SNS*-junctions with a large quantity of impurities. In the remaining aspects the physical situation in pure and dirty *SNS*-bridges is qualitatively the same. The role of the quasi-particle transit time in the *N*-layer $\tau_e \sim d/v_F$ in dirty junctions is played by the characteristic diffusion time $\tau_d \sim d^2/D$ $\left(D = \frac{v_F l}{3}\right)$. The complexity in calculating nonstationary formulae in dirty *SNS*-junctions derives from the impossibility of finding an analytical expression for the Green's functions of the junction at energies $\varepsilon \sim D/d^2$ (as opposed to the case of pure *SNS*-systems whose Green's functions are known for all energies). Nonetheless in certain limiting cases the current in dirty *SNS*-junctions may also be calculated exactly. We will illustrate this using a sample calculation of the critical current of a dirty *SNS*-bridge in a microwave field.

We will consider the case $T \gg D/d^2$ and will assume that we have an a.c. voltage $V(t) = V_1 \cos \omega_1 t$ across the junction, where $eV_1 \ll \omega_1 \ll D/d^2$. In order to calculate the current we must know the Green's functions for low energies only $\varepsilon \ll D/d^2$ whose expressions are known [38]. Consequently we may directly employ the result from study([38], formula (54)). Neglecting the Josephson current and carrying out time averaging we obtain

$$I_c = \frac{\pi}{4}\left(1 - \frac{\pi}{4}\right)\frac{\Delta_0^2}{RT}\,\frac{eV_1^2}{\omega_1^2 + 1/\tau_\varepsilon^2}, \tag{3.82}$$

i.e., the critical current enhancement effect by microwave emission exists: the current does not contain exponential smallness ~ exp $\{-d/\xi_N\}$, $\xi_N = (D/2\pi T)^{1/2}$ (the bridge is "constricted").

We note in addition that at low voltages $V \ll \Delta_0(T) < D/d^2$ and $T \sim T_c$ the expression for current will take the form of (3.65) and the functional $P\{\varphi\}$ is calculated in study [38], while the expression for I_c for dirty *SNS*-bridges is given by formula (2.15).

As already noted all results for current obtained in this section refer to the point $x = 0$, i.e., to the center of the bridge. At the same time the matrix $\check{G}$ at other points in the *N*-layer, generally speaking, differs from G_0:

$$\begin{aligned}
&\check{G}(v_x, x; t, t') = \hat{W}(v_x, t)\check{G}^{(0)}(v_x, x; t, t')\hat{W}^*(v_x, t'),\\
&\hat{W} = \begin{pmatrix} \exp\{i\chi(v_x, t)\} & 0 \\ 0 & \exp\{i\chi(-v_x, t)\}\end{pmatrix},\\
&\chi = -e\int^{t} dt_1\, \Phi(x - v_x(t - t_1)t_1),
\end{aligned} \tag{3.83}$$

where $\check{G}^{(0)}$ is the solution of equations (3.1) in the *N*-layer with $\Phi(x, t) = 0$. We focus on the fact that when $t \to t'\check{G} \equiv \check{G}^{(0)}$, i.e., in order to calculate the current at any point in the *N*-layer it is sufficient for us to know the matrix $\check{G}^{(0)}(\varepsilon)$ which is expressed by the relation

$$\check{G}^{(0)}(v_x, x; \varepsilon, \varepsilon') = \hat{C}^+(v_x, x)\check{G}_0(v_x)\hat{C}(v_x, x), \tag{3.84}$$

where $\check{C}(v_x, x)$ is determined by formula (3.47) in which we replace $d/2$ by x. We emphasize that the result for the current in this case will be valid for any $V(x, t)$, $|x| \geqslant d/2$. In the case of low voltages $eV \ll v_F/d$ the matrix G in the principal approximation in this parameter is independent of x. However, at high voltages this relation becomes significant. For example, in the case $eV \gg \Delta_0 \gg v_F/d$ we have

$$\begin{aligned}
&I = I_0 + I_+ + I_-,\ I_\pm = {}^1\!/_2\,[I_1 \sin\varphi_\pm(x, t) + I_2 \cos\varphi_\pm(x, t)],\\
&\varphi_\pm(x, t) = \varphi_0 + 2eV(t \pm x/v_F).
\end{aligned} \tag{3.85}$$

The electrical field in the N-layer (not too close to the NS-boundaries) takes the form

$$E = E_+ + E_-, \quad E_\pm(x) = (\pi/VS)\ [I_1 \cos\varphi_\pm(x, t) - I_2 \cos\varphi_\pm(x, t)]. \tag{3.86}$$

Thus, in the N-layer in this situation there exist unique current and electrical field waves. Clearly we refer to an electrical field within the N-filament only. Near the NS-boundaries in superconductors there are rather sharp field jumps that occur at distances in the order of $a \ll \xi_0$, since the potential difference between the superconducting borders is equal to V. We emphasize that the existence of such oscillations with the Josephson frequency in the N-metal does not contradict the result that the characteristic frequency of the electron component in the metal is equal to the plasma frequency (which, obviously, is much greater than the Josephson frequency). The problem is that in our case we are discussing stimulated oscillations. After the voltage has been applied to the SNS-junction we then fix the rate of change in the phase difference, i.e., the current frequency. If this frequency is comparable to the reverse quasi particle transit time through the N-metal (or is greater than this value), the electrical field and current oscillations outlined above arise; their frequency coincides with the Josephson frequency.

Finally we will make one additional comment. In this section we have discussed a bridge-type SNS-structure, and the weak nature of the coupling was due to the small transverse size of the N-filament. However all derived results are valid (in the stationary case as well) for sandwich type SNS-structures when $v_{Fn}0 \ll v_{FS}$. In this case the weak coupling will result from the low Fermi velocity of the N-metal, while the formula for R_0 will take the same form: $R_0 = \pi^2/p_{FN}^2 Se^2$.

Chapter 4

VORTEX STATES IN JOSEPHSON JUNCTIONS

In the preceding chapters we examined the quasione-dimensional situation, i.e., we presumed that the problem was homogeneous on y and z (i.e., in the plane perpendicular to the current flowing through the junction). The current flow itself generates a magnetic field that in turn influences the current. At sufficiently high transverse dimensions of the junction this effect becomes significant, so the current density will no longer be homogeneous transverse to the junction. As demonstrated in a number of studies [48-51, 92, 93] the magnetic field may penetrate these extended junctions in the form of vortices (Josephson vortices) that are significantly different from the vortices in type-II bulk superconductors (Abrikossov vortices). In a Josephson vortex there is no suppression of the order parameter and in this regard for a Josephson junction there is no field analogous to

H_{c2}. However the topologic charge of such a vortex is equal to the charge of the Abrikossov vortex (i.e., both vortices have an identical magnetic flux equal to a single quantum $\Phi_0 = \pi\hbar/e = 2\cdot10^{-7}$ G/cm^2). Vortex states of this type may be achieved in junctions having macroscopic dimensions (*SIS, SNS*-junctions as well as superconductor-semiconductor-superconductor structures).

The magnetic field and current distribution in the Josephson junction is described by the equation [49-51]

$$\left(\frac{\partial^2}{\partial t^2} - \frac{\partial^2}{\partial x^2} - \frac{\partial^2}{\partial y^2} + \varkappa_0 \frac{\partial}{\partial t}\right)\varphi + \sin\varphi = 0. \tag{4.1}$$

Here φ is the gradient-invariant phase difference that we have already encountered; the term $\varkappa_0 \partial\varphi/\partial t$ accounts for dissipation associated with the normal current component ($\varkappa_0$ is proportional to the conductivity of the junction).

We will use dimensionless variables. The unit of length will be the Josephson length $\lambda_J = (\Phi_0/2\pi L j_c)^{1/2} \sim 0.1$ mm while the unit of time will be the inverse Josephson plasma frequency $\omega_J^{-1} = (2\pi j_c/C\Phi_0)^{1/2} \equiv c_0/\delta_J \sim 10^{-12}$ s; c_0 is the Swihart wave velocity; the magnetic field is measured in units of $H_J = \Phi^0/2\pi\lambda_J\Lambda \sim 0.1$-$1$ G; Φ_0 is the effective layer thickness at the junction in which a magnetic field exists, L and C represent the inductance and capacitance per unit of the length of the junction.

We note that the equation in (4.1) is widely used to investigate a variety of processes in steady-state physics, phase transition theory, and quantum field theory. In this regard many problems that arise in investigating the vortex states of Josephson junctions (such as the problem of the stability of stationary field and current distributions) may be of additional interest.

9 Vortex free energy in the presence of current

A number of studies have carried out a thermodynamic examination of the various properties of Josephson tunnel barriers [47-51, 93]. An expression was derived for the field H_{c1} that corresponds to the beginning of vortex penetration into the tunnel junction, and magnetization curves are found together with the thermodynamic potentials of weak superconductors of finite size. However the examination was carried out for the case of zero transport current through the barrier and consequently the expression for the free energy of the weak superconductor was drafted for this case only [47].

We will consider the single-dimensional case, i.e., we will assume that φ is dependent only on a single spatial coordinate z and that the magnetic field lies on the axis y while the electrical field lies on the axis x (i.e. in the direction of current flow). The amplitude of the electrical and magnetic fields in the junction is expressed through the derivatives of φ:

$$E(z, t) = \partial\varphi/\partial t, \quad H(z, t) = \partial\varphi/\partial z. \tag{4.2}$$

The stationary distribution of $\varphi(z)$ is given by the solution of equation

$$d^2\varphi/dz^2 = \sin\varphi, \tag{4.3}$$

that follows directly from (4.1). The boundary conditions on this equation describing the various physical situations on the boundaries of the junction may be expressed as

$$\left(\alpha_{0(L)}\varphi + \beta_{0(L)}\frac{d\varphi}{dz}\right)\Big|_{z=0(L)} + \gamma_{0(L)} = 0, \tag{4.4}$$

where L is the dimensions of the junction ($0 \leqslant z \leqslant L$). With fixed values of the magnetic fields at the boundaries we have

$$\frac{d\varphi}{\partial z}\Big|_{z=0(L)} = H_{0(L)} = -\frac{\gamma_{0(L)}}{\beta_{0(L)}}. \tag{4.5}$$

One case is also possible where the Josephson junction "jumps" to the cavity in the superconductor (Fig. 8). Here the boundary condition at the point $z = 0$ is obtained by means of the familiar integration of the expression for current with respect to the contour Γ:

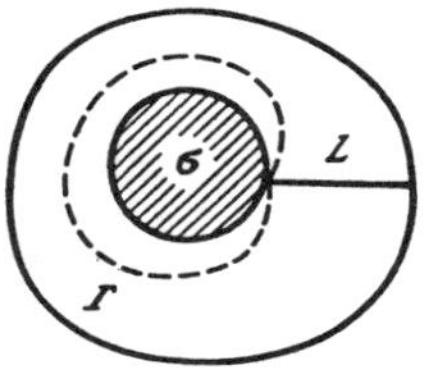

$$\frac{d\varphi}{dz}\Big|_{z=0} = \frac{\varphi(0)}{\sigma} = -\frac{\alpha_0}{\beta_0}\varphi(0), \quad \frac{d\varphi}{dz}\Big|_{z=L} = H_L. \tag{4.6}$$

Fig. 8. Superconducting ring containing the Josephson junction

Here σ is the dimensionless area of the cavity ($\sigma = S/\lambda_J\Lambda$).

First we will consider the case of (4.5). We will assume that H_0 and H_L are time-independent. Multiplying equation (4.1) by $\partial\varphi/\partial t$ and integrating it with respect to z within the limits from 0 to L and with respect to time within the limits from t' to t, we obtain subject to (4.2)

$$E(t) - E(t') = (\varphi_L(t) - \varphi_L(t'))H_L - (\varphi_0(t) - \varphi_0(t')H_0 - \\ - \varkappa_0 \int_{t'}^{t}\int_0^L \left(\frac{\partial\varphi}{\partial\tau}\right)^2 dz\,d\tau. \tag{4.7}$$

Here

$$E(t) = \int_0^L \left\{\frac{1}{2}\left(\frac{\partial\varphi}{\partial t}\right)^2 + \frac{1}{2}\left(\frac{\partial\varphi}{\partial z}\right)^2 + 1 - \cos\varphi\right\} dz \tag{4.8}$$

is the energy of the Josephson barrier (the first two terms represent the electrical and magnetic field energies at the barrier, while the third term represents the coupling energy of the two superconductors). Energy E is expressed in units of $\hbar j_c/2e$.

After differentiating (4.8) with respect to t we will have

$$\frac{dE}{dt} = E_L H_L - E_0 H_0 - \varkappa_0 \int_0^L \left(\frac{\partial \varphi}{\partial t}\right)^2 dz. \tag{4.9}$$

here E_L and E_0 are the values of the electrical field at the edges of the barrier (time-dependent). The physical meaning of relation (4.9) is obvious: the first two terms in the right half of the equality are equal to the values of the Poynting vector at the barrier points $z = 0$ and $z = L$, while the third term represents the resistive losses at the barrier at time t.

Thus, relation (4.9) determines the law of variation of energy of a weak superconductor.

We will now return to formula (4.7). Introducing the function

$$G(t) = E(t) - (\varphi_L(t) H_L - \varphi_0(t) H_0), \tag{4.10}$$

we obtain

$$G(t) - G(t') = -\varkappa_0 \int_{t'}^{t} \int_0^L \left(\frac{\partial \varphi}{\partial \tau}\right)^2 d\tau\, dz. \tag{4.11}$$

We are interested in the case when a static field and current distribution is possible in the barrier ($I < I_c$). Here the nonstationary processes related to the switching on of the field and the current are damped over time due to dissipation, and a stationary distribution of $\varphi(z)$ is established in the barrier as described by equation (4.3). Since the right half of (4.11) is negative, it is clear that when $t \to \infty$ the function $G(t)$ diminishes, i.e., it achieves a minimum value in the steady state. Consequently, the quantity

$$G = \lim_{t\to\infty} G(t) \tag{4.12}$$

is nothing other than the free energy of a weak superconductor with current flow in a magnetic field.

The validity of this statement may be determined in the following manner as well. From the mathematical viewpoint the solution of equation (4.3) will be the extremal of the free energy functional G , i.e., the variation in G will be equal to zero in the solutions of equation (4.3). We may easily observe that this condition is satisfied with certain boundary conditions for the functional

$$G[\varphi] = E[\varphi] - \int_0^L \frac{d}{dz}\left(\varphi \frac{d\varphi}{dz}\right) dz = E[\varphi] - (\varphi_L H_L - \varphi_0 H_0),$$

$$E[\varphi] = \int_0^L \left\{\frac{1}{2}\left(\frac{d\varphi}{dz}\right)^2 + 1 - \cos\varphi\right\} dz. \tag{4.13}$$

This expression is also obtained from (4.10)-(4.12). Expression (4.13) may be rewritten as

$$G[\varphi] = E(\varphi) H_e \overline{H} L - I\overline{\varphi} L,$$

$$\overline{H} = \frac{1}{L}\int_0^L \frac{d\varphi}{dz}\, dz = [\varphi(L) - \varphi(0)]/L, \quad \overline{\varphi} = [\varphi(L) + \varphi(0)]/2L. \tag{4.13a}$$

Thus, the functional $G[\varphi]$ (4.13) is the free energy of a weak superconductor with current I in external magnetic field H_e. The physical meaning of the last two terms in the right half of (4.13a) is easily understood: the quantity H plays the role of the induction of the weak superconductor B [47], while the last term is the direct generalization of the **jA** term.

If boundary conditions (4.6) are valid, $G[\varphi]$ will appear in a somewhat different form:

$$G[\varphi] = E[\varphi] - \varphi_L H_L + \frac{\sigma}{2} H_0^2. \tag{4.14}$$

The last term in (4.14) is simply the magnetic field energy in the cavity (in dimensional units of $SH_0^2/8\pi$). Thus, (4.14) represents the free energy functional for the "junction plus cavity" system. We note that we did not introduce the last term in brackets in (4.14); this is automatically obtained by using conditions (4.6) corresponding to the existence of the cavity.

When necessary it is not difficult to find the expression for g [I] with arbitrary values of the quantities α, β, γ.

We should note that with given boundary conditions several different configurations of $\varphi_n(t)$ are possible, each of which has its own value of G_n. Such values of $G[\varphi_n]$ $(n > 1)$ contain both the minima of this functional (absolute and local) as well as its maxima. Here, the issue of the stability of a given configuration of φ_n remains open and should be investigated by incorporating temporal equation (4.1).

The general technique that allows analytic investigation of the stability of any stationary distributions of $\varphi(z)$ (including those that are not represented analytically) with random boundary conditions was developed in study [90]; here we will not investigate this issue in any greater depth.

10. The spectrum of oscillations in a Josephson junction system

A fairly large number of studies [47, 50, 51, 94, 95] have been devoted to the issue of the dispersion of small oscillations in a Josephson junction. Here we will consider such oscillations in a somewhat different system. We will assume a large number of Josephson junctions separated by holes in a bulk superconductor (Fig. 9). We will assume that the system is in an external magnetic field and in a

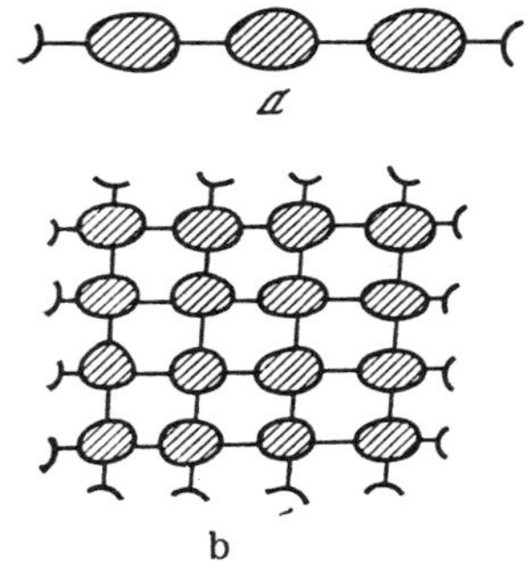

Fig. 9. One-dimensional (a) and two-dimensional (b) Josephson junction systems

stable steady state and that a certain distribution of the phase difference of the order parameters $\varphi_0(z)$ exists in the junction. The function $\varphi_0(z)$ satisfies equation (4.3) and the boundary conditions

$$\sigma \frac{d\varphi_{0n}}{dz}\bigg|_0 = \varphi_{cn}(0) - \varphi_{0n-1}(L),$$

$$\frac{d\varphi_{0n}}{dz}\bigg|_0 = \frac{d\varphi_{0n-1}}{dz}\bigg|_L, \qquad (4.15)$$

where $d\varphi_{0n}/dz = H_n$ is the magnetic field in the hole with number n.

We will now assume that the magnetic field distribution and the urrent distribution go out of equilibrium in the system:

$$\varphi_n(y, z, t) = \varphi_{0n}(z) + \psi_n(y, z, t), \quad \psi_n \ll 1. \qquad (4.16)$$

f we ignore dissipation, the function φ_n will satisfy equation (4.1) ith $\varkappa_0 = 0$. Using (4.1) and (4.16) and representing ψ_n as

$$\Psi_n(y, z, t) = \Psi_n(z)\, e^{i\omega t - iky},$$

e directly obtain

$$d^2\Psi/dt^2 + (\omega^2 - k^2)\Psi_n = \Psi_n \cos\omega_{0n}. \qquad (4.17)$$

quation (4.17) is reduced to Lamé's equation whose solution may be ritten through elliptical Jacobi functions. Here the problem in rinciple may be solved analytically (which is done in a number of ases), although the corresponding calculations are rather cumbersome. t the same time the physical picture of the propagation of small osillations in this system is rather simple and may be described without incorporating such calculations.

Let $L \ll 1$. Then we may ignore the dependence of the phase diference φ_{0n} on z. Using the boundary conditions in this case we may btain

$$\sigma H_n = \varphi_{n+1} - \varphi_n, \; \varphi_{n+1} - \varphi_{n-1} - 2\varphi_n = L\sin\varphi_n. \qquad (4.18)$$

 will establish the dispersion law of small oscillations in the sysm if the condition $\cos\varphi_{0n} = \cos\varphi_0$, where φ_0 is a constant is tisfied for all n. It is easily established that this condition may satisfied if the flow through the hole is quantized, i.e., if the antity σH_n is an even multiple of 2π (since $L \ll 1$ we will ignore the low through the tunnel barriers). Moreover, this condition is also lid in another case. Specifically, let $\varphi_{0n} = (-1)^n\varphi_0$; it is clear from (4.18) that this distribution of the phase difference among th junctions does exist if

$$4\varphi_0 + L\sin\varphi_0 = 0.$$

The stability requirement of these field and current distribu tions in the system is provided by $\cos\varphi_0 > 0$. We may represent th function $\psi_n(z)$ as

$$\Psi_n(z) = \Psi(z)\, e^{iqLn}. \qquad (4.19$$

It is clear that the φ_n function will satisfy the same boundary con ditions (4.15) as φ_{0n}. From here if we account for (4.19) we directl find the boundary conditions for the function $\psi(z)$

$$\sigma \frac{d\Psi}{dz}\Big|_0 = \Psi(0) - \Psi(L)\, e^{-iqL}, \quad \frac{d\Psi}{dz}\Big|_0 = \frac{d\Psi}{dz}\Big|_L e^{-iqL}. \qquad (4.20$$

As is normally the case the solution of equation (4.17) is sought as

$$\Psi(z) = Ae^{i\alpha z} + Be^{-i\alpha z}, \quad \alpha^2 = \omega^2 - k^2 - \cos\varphi_0. \qquad (4.21$$

Substituting (4.21) into boundary conditions (4.20) and equating th determinant of the derived system of equations for A and B to zero, w find

$$\tfrac{1}{2}\sigma\alpha\sin\alpha L - \cos\alpha L + \cos qL = 0,$$

and from here with moderate values of α for the case of $L \ll 1$ we ar rive at the dispersion equation

$$\omega_{qk}^2 = \omega_J^2\cos\varphi_0 + c_0^2k^2 + c_0^2\,(4\Lambda/SL)\sin^2(qL/2). \qquad (4.22$$

In relation (4.22) we converted to dimensional variables.

It is clear that waves may propagate in the system in the direc tion of the magnetic field at a velocity equal to the velocity of th Swihart waves c_0. This result is completely consistent with th result from Kulik [51] who demonstrated that elastic waves whos velocity is also equal to c_0 propagate along the vortex filaments in Josephson junction. Moreover, it follows from (4.22), that the wave propagate in a direction perpendicular to the magnetic field. Suc oscillations may be compared to acoustic oscillations of an atomi chain. The role of the "atomic mass" in this case is played by th inductance of the "elementary cell" of the system under examinatior The "Brillouin zone", as is normally the case, is determined by th condition $-\pi \leqslant qL < \pi$. For longwave oscillations $qL \ll 1$ the spectru has a "plasma" form

$$\omega_{qk}^2 = \omega_J^2\cos\varphi_0 - c_0^2k^2 + c_0^2\frac{L\Lambda}{S}q^2. \qquad (4.23$$

The velocity of wave propagation in the z direction is indeper dent of the magnetic field. We note that the velocity of waves propa gating transverse to the vortex lattice in a long Josephson junctic is highly dependent on the magnetic field and is generally determined by a rather complex expression [51, 94]. We also note that spectrum (4.22) has a threshold frequency which is determined by the level of currents in the system. In the case $\varphi_0 = 0$ (zero field and currents) this frequency is equal to ω_J and coincides with the threshold frequency of the spectrum of small oscillations in a long junction that is also in the Meissner state [50]. When there is a vortex lattice in the junction, there is no such frequency [51] since the average value of the quantity cos φ in this case is equal to zero. In this regard we may also say that the spectrum of oscillations in the system will always have a threshold frequency, since the distribution of the phase difference φ_{0_n} is stable only when $\varphi_{0_n} > 0$ which will be satisfied for all n.

We note that the small oscillations in a two-dimensional system of Josephson junctions may be considered in an entirely analogous manner. Such a system models two-dimensional superconducting granular films that today are under intensive theoretical and experimental investigation.

Thus, small oscillations of the type

$$\omega^2 = \omega_J^2 + \sum_{j=1}^{2} \frac{4c_0^2 \lambda}{SL_j} \sin^2 \frac{\chi_j}{2}, \qquad \chi_j = q_j L_j. \tag{4.24}$$

may exist in two-dimensional granular superconductors. Here we assume that there is no magnetic field in the superconductor (i.e., $\varphi_0 = 0$). According to (4.24) granular superconducting films may contain propagating small oscillations that are either oscillations in current (or in the phase difference, which is the same thing) in Josephson junctions interconnecting superconducting granules. Such oscillations have a threshold frequency ω_J.

This type of oscillation, obviously, is not accompanied by any deviation from equilibrium of the quasi-particle distribution function. The existence of such oscillations is the one significant difference between granular superconductors and "bulk" superconductors in which, as a rule, collective modes exist only with a nonequilibrium distribution of quasi particles (see, for example, [37]). In addition we note that we have not considered oscillations related to charge fluctuations in the granules (i.e., we have not incorporated the capacitance of the granule). Such oscillations are examined in study [96]. Our result (4.22) has included only the capacitance and inductance of the Josephson junctions as well as the inductance of the "ring" $l(=\sigma L)$.

CONCLUSION

We will enumerate the primary results from this study.

1. We have formulated a microscopic theory of the proximity effect in pure NS-systems at low temperatures $T \ll T_c$. It is demonstrated that at such temperatures the order parameter of the superconductor changes at distances in the order of $v_F/2\Delta_0 \sim \xi_0$ from the NS-boundary, amounting (near the boundary in the case of similar values of the Fermi velocities of the metals) to approximately 1/2 its value Δ_0 in the homogeneous situation. It is also demonstrated that the f-function of the superconductor at low Mössbauer frequencies $\omega \ll \Delta_0$ is not suppressed in practice up through the NS-boundary in spite of the significant reduction in the order parameter.

We also investigate the proximity effect when $T \sim T_c$ and it is demonstrated that in this temperature range there exist two scales of variation in $\Delta(x)$ near the NS-boundary ($\sim\xi_0$ and $\sim\xi_T$).

2. We formulate a microscopic theory of the Meissner effect in a normal metal layer bounding a superconductor. Expressions are derived for the screening current for both high and low concentrations of non-magnetic impurities in the system. It is demonstrated that in the pure limit the magnetic field causes a shift of the Andreev levels which produces a homogeneous current in the N-layer. The dependence of the current on the vector potential in this case is nonlocal. At low temperatures $T \lesssim v_F/d$ there may exist a "rescreening" effect i.e., the magnetic field that penetrates the system may change sign here the Fourier transform of the magnetic permeability of an NS-sandwich $\mu(k)$ is negative at certain values of k. In the dirty limit significant screening currents in the N-layer flow at distances of $\lesssim\xi_N = \sqrt{D/2\pi T}$ from the NS-boundary. This result is completely consistent with existing experimental data.

3. It is demonstrated that in wide SNS-sandwiches when $T \ll T_c$ the exact expression for the stationary Josephson current is nearly identical to the calculated value ignoring the suppression of Δ at the borders, which is directly related to the absence of the suppression of f_ω when $\omega \ll \Delta_0$. A theory of the stationary Josephson effect in broad SNS-junctions containing impurities ($l \ll \xi_0$) is formulated. The derived expressions are valid for bridge structures as well as sandwiches with $\sigma_N \ll \sigma_S$ for virtually any T as well as for sandwiches with $\sigma_N \sim \sigma_S$ and $T \ll T_c$. These expressions are found consistent with experimental results. The phase dependence of the current $j(\varphi_0)$ differs significantly from a sinusoidal curve when $T \lesssim D/d^2$.

The boundary conditions are derived for quasi-classical Eilenberger and Usadel equations that make it possible to account for the existence of a dielectric interlayer of arbitrary transparency in the superconducting system. These conditions are used to formulate a microscopic theory of the stationary Josephson effect in an $SNINS$-system. The influence of the position of the dielectric in the N-layer on the discrete spectrum and the critical current is investigated. Also investigated is the influence of impurities on this current. This influence is qualitatively different in the cases $l \gg \xi_0$ and $l \ll \xi_0$. It is demonstrated that a shift in the Andreev levels of an $SNINS$-sys tem in a magnetic field has an influence on the stationary Josephson effect when $T \lesssim v_F/d$.

4. A microscopic theory of the nonstationary Josephson effect is formulated for pure *SNS*-junctions. Two cases are considered: *SNINS*-sandwiches and *SNS*-bridges. In the first case a general expression is derived for the current that is valid for random values of $V(t)$. Expressions for the normal, the Josephson and the interferential current components are investigated. The existence of a discrete spectrum in the *SINS*-junctions results in characteristic oscillations in the E-I characteristic (which are manifest in experiments), while in *SNINS*-junctions, moreover, these may produce logarithmic divergences of the Josephson current at certain values of V. There is significant enhancement of the critical current in a microwave field caused by the nonequilibrium nature of the quasi-particle distribution; this process may have a power (though not exponential) dependence on d. In *SNINS*-systems the "superposition" of this effect and the resonant tunneling effect produces "double" supercurrent enhancement.

In the case of *SNS*-bridges a general expression is derived for the current. The linear response to a low a.c. voltage is investigated together with the critical current enhancement effect and the nonstationary Josephson effect at low and high voltages. The phase dependence of the current is highly nonsinusoidal for all values of T. At high temperatures a significant role is played by the nonequilibrium nature of the distribution function that produces supercurrent enhancement in a microwave field and also increases the conductivity of the bridge; the corresponding expressions for the current (in certain conditions) are independent of d. At low temperatures ($T \to 0$) the current in the system is related to the flow of the superconducting condensate, which produces a complex dependence of the current flowing through the junction on the phase difference $\varphi(t)$. The expression for the excess current appearing in the E-I characteristic when $V \gg \Delta_0$ is independent of d when there are no impurities.

5. The vortex states in Josephson junctions are investigated. The expression is generalized for the free energy of the vortices to the case of current flowing through the junction.

An expression is derived for the spectrum of small oscillations in a Josephson junction system. A two-dimensional system of such junctions simulates a granular superconducting film. Thus, in granular superconductors (unlike "bulk" superconductors) there are collective oscillations without deviations from equilibrium of the distribution function.

APPENDIX 1

We will rewrite equation (1.12) as

$$2\pi T \sum_{\omega>0} \int_0^1 d\alpha \frac{1}{\Omega_0} \frac{\lambda_0^2}{\omega^2 - \lambda_0^2} = 0, \qquad \lambda_0^2 = \left(\frac{v_F p\alpha}{2}\right)^2 - \Delta_0^2. \qquad \text{(P.1}$$

We first note that the solution of this equation p_0 will satisfy th double inequality

$$\frac{4\Delta_0^2}{v_F^2} < p_0^2 < \frac{4(\Delta_0^2 + \pi^2 T^2)}{v_F^2}. \qquad \text{(P.2}$$

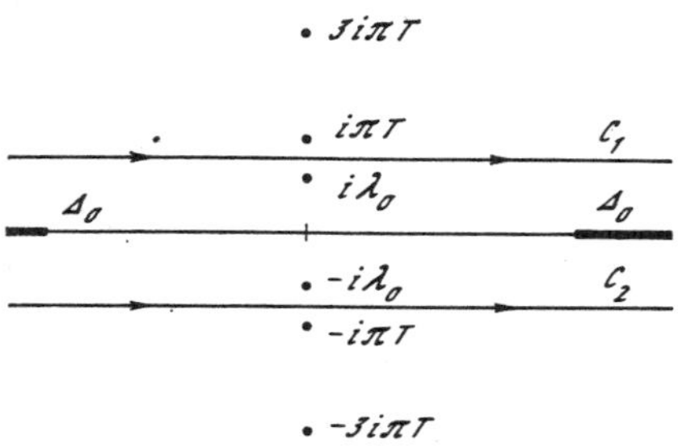

Fig. 10

The left inequality is an obviou consequence of (P.1) while the righ inequality follows from formul (1.11) when accounting for the posi tive value of δh^+_ω. We will decompos the integral with respect to α i (P.1) into two integrals:

$$\int_0^1 d\alpha(\ldots) = \int_0^{2\Delta_0/v_F p} d\alpha(\ldots) + \int_{2\Delta_0/v_F p}^1 d\alpha(\ldots).$$

The first integral after summation with respect to ω and some simpl transformations accounting for the condition $T \ll T_c$ and inequalit (P.2) is represented as a series

$$A = \sum_{n=2}^{\infty} \frac{2^{n-1}(n-1)!}{(4n^2-1)(2n-1)!!} \simeq 0{,}08. \qquad \text{(P.3}$$

We will transform the second integral to

$$B = \int_{C_1} Z(\varepsilon)\, d\varepsilon,$$

$$Z(\varepsilon) = \frac{i}{2} \int_{2\Delta_0/v_F p}^1 d\alpha\, \operatorname{th}\frac{\varepsilon}{2T} (\Delta_0^2 - \varepsilon^2)^{-1/2} \frac{\lambda_0^2}{\varepsilon^2 + \lambda_0^2}, \qquad \text{(P.4}$$

where C_1 is the contour in the plane of the complex variable ε (Fig 10). Obviously integral (P.4), ignoring terms of the order T^2/Δ_0^2, ma also be written in another form:

$$B = \int_{C_2} d\varepsilon\, Z(\varepsilon) - 2\pi i\left(\operatorname{Res} Z(i\lambda_0) + \operatorname{Res} Z(-i\lambda_0)\right). \qquad \text{(P.5}$$

We may easily determine that the integral with respect to the contou C_2 (Fig. 10) is equal to

$$\int_{C_2} d\varepsilon\, Z(\varepsilon) = -B. \qquad \text{(P.6}$$

We obtain directly from (P.5) and (P.6):

$$B = \frac{\pi}{2\Delta_0} \int_{2\Delta_0/v_F p}^{1} d\alpha \sqrt{\left(\frac{v_F p \alpha}{2}\right)^2 - \Delta_0^2} \operatorname{tg} \left[\frac{\sqrt{\left(\frac{v_F p \alpha}{2}\right)^2 - \Delta_0^2}}{2T} \right]. \tag{P.7}$$

We will represent the solution of (P.1) p_0 as

$$p_0^2 = \frac{4(\Delta_0^2 + \gamma^2)}{v_F^2}, \quad 0 < \gamma < \pi T. \tag{P.8}$$

when $T \ll T_c$ we will have

$$B = \frac{\pi}{2} \int_0^{\gamma/\Delta_0} y^2 \, dy \operatorname{tg}\left(\frac{\Delta_0}{2T} y\right). \tag{P.9}$$

By virtue of (P.1) $A + B = 0$. It is clear from here that the quantity γ will differ only insignificantly from $\pi T (\pi T - \gamma \ll T)$, since otherwise B will be of the order T^3/Δ_0^3 and (P.1) will have no solutions. Accounting for this the primary contribution to integral (P.9) comes from the region y near $\pi T/\Delta_0$. Finally we obtain

$$B = \left(\frac{\pi T}{\Delta_0}\right)^3 \ln \cos \frac{\gamma}{2T}, \tag{P.10}$$

from which we directly obtain formula (1.14).

APPENDIX 2

In the case of few impurities $l \gg \xi_0$ equation (2.17) for the matrix Green-Gorkov functions may be written in the following manner:

$$\hat{G}_\omega(x, x') = \begin{pmatrix} G_\omega(x, x') & F_\omega(x, x') \\ F_\omega^+(x, x') & -G_{-\omega}(x', x) \end{pmatrix} \tag{P.11}$$

where

$$\hat{G}_\omega(x, x') = \hat{G}_\omega^{(0)}(x, x') + \hat{G}_\omega^{(0)}(x, x_0) V_0 \hat{\tau}_3 [1 - \hat{G}_\omega(x_0, x_0) V_0 \hat{\tau}_3]^{-1} \hat{G}_\omega(x_0, x'),$$

is the desired matrix for the *SNINS*-system, while $\bar{G}_\omega^{(0)}$ is the analogous matrix for an *SNS*-junction whose expression is known [20]. Equation (P.11) is an algebraic equation and is easily solved. Finally we have

$$G_\omega(x, x') = G_\omega^{(0)}(x, x') + \frac{V_0^2}{1+\Pi} \Big\{ G_\omega^{(0)}(x, x_0) \Big[G_\omega^{(0)}(x_0, x') \frac{1}{V_0} -$$
$$- G_{-\omega}^{(0)}(x_0, x_0) G_\omega^{(0)}(x_0, x') - F_\omega^{(0)}(x_0, x_0) F_\omega^{+(0)}(x_0, x') \Big] -$$
$$- F_\omega^{(0)}(x, x_0) \Big[F_\omega^{+(0)}(x_0, x') \frac{1}{V_0} - G_\omega^{(0)}(x_0, x_0) F_\omega^{+(0)}(x_0, x') +$$
$$+ F_\omega^{+(0)}(x_0, x_0) G_\omega^{(0)}(x_0, x') \Big] \Big\},$$
$$\Pi = - V_0^2 \det \hat{G}_\omega^{(0)}(x_0, x_0) - V_0 \operatorname{Sp} \hat{\tau}_3 \hat{G}_\omega^{(0)}(x_0, x_0); \tag{P.12}$$

$$F^+_\omega(x,\ x') = F^{+(0)}_\omega(x,\ x') + \frac{V_0^2}{1+\Pi}\Big\{F^{+(0)}_\omega(x,\ x_0)\Big[G^{(0)}_\omega(x_0,\ x')\frac{1}{V_0} -$$
$$- G^{(0)}_\omega(x_0,\ x_0)\, G^{(0)}_\omega(x_0,\ x') - F^{(0)}_\omega(x_0,\ x_0)\, F^{+(0)}_\omega(x_0,\ x')\Big] +$$
$$+ G^{(0)}_{-\omega}(x_0,\ x)\Big[F^{+(0)}_\omega(x_0,\ x')\frac{1}{V_0} - G^{(0)}_\omega(x_0,\ x_0)\, F^{+(0)}_\omega(x_0,\ x') +$$
$$+ F^{+(0)}_\omega(x_0,\ x_0)\, G^{(0)}_\omega(x_0,\ x')\Big]\Big\}. \qquad \text{(P.13)}$$

We note that expressions (P.12)-(P.13) are valid for an arbitrary value of V_0. Substituting (P.12) into the known expression for the current in an inhomogeneous system [47]:

$$j = \frac{em^2}{2\pi i}\, T \sum_\omega \int_0^{v_F} v_x\, dv_x \iint_{-\infty}^{\infty} dx\, dx' (\operatorname{sign} x - \operatorname{sign} x') \times$$
$$\times\, \Delta(x)\, \Delta^*(x')\, G^n_\omega(x,\ x')\, G_{-\omega}(x',\ x), \qquad \text{(P.14)}$$

after some rather cumbersome calculations we arrive at expression (2.24) for the current (G^{n}_{ω} is the Green's function of the *NIN*-system whose expression is easily obtained from (P.12) if we set $\Delta = 0$). With large values of V_0 we may represent the function G_ω in the form of the expansion

$$G_\omega(x,\ x') = G_{0\omega}(x,\ x') + \frac{1}{V_0} G_{1\omega}(x,\ x') + \ldots \qquad \text{(P.15)}$$

For simplicity we will represent the results of this expansion in diagram form

$G_{0\omega}(x, x') =$

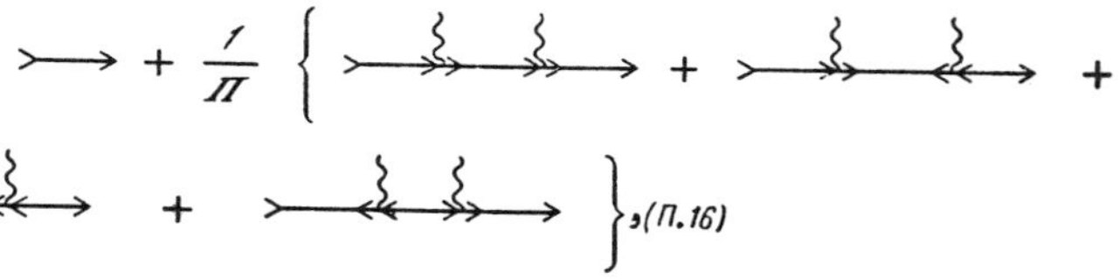

(П.16)

$G_{1\omega}(x, x') = \frac{V_0}{\Pi}$

(П.17)

$\Pi =$

Here we have accounted for the fact that the following identity is satisfied in this problem

$= 0.$

When impurities are present we may use the regular averaging technique [9]. With large values of V_0 we will have

$$\omega^* = \omega + i\Sigma_{\omega*}(x), \qquad \Sigma_{\omega*}(x) = \frac{2\pi}{mp_F}\frac{v_F}{2l}\langle G_{0\omega*}(x,\ x')\rangle, \qquad \text{(P.18)}$$

from which we have formula (2.31) when $l \ll d$ or $\xi_T \ll d$.

In conclusion the author expressions his gratitude to G. F. Zharkov for many useful discussions and extensive assistance in this study.

BIBLIOGRAPHY

1. Eilenberger, G. "Transformation of Gorkov's equation for type II superconductors into transport-like equations" ZTSCHR. PHYS., 1968, V. 214, No. 2, p. 195-213.

2. De Gennes, P. "Boundary effects in superconductors" REV. MOD. PHYS., 1964, V. 36, No. 1, p. 225-237.

3. Zaytsev, R.O. "The solution of superconductivity equations for a normal metal/superconducting metal system" ZhETF, 1965, V. 48, No. 2, p. 644-651.

4. Zaytsev, R.O. "The boundary conditions for superconductivity equations at near-critical temperatures" ZhETF, 1965, V. 48, No. 6, p. 1759-1771.

5. Zaytsev, R.O. "Boundary conditions and surface superconductivity" ZhETF, 1966, v. 50, No. 1, p. 1055-1063.

6. Deutscher, G., De Gennes, P. "Proximity effects" In: Superconductivity, NY, 1969, V. 2, p. 1005-1034.

7. Galayko, V.P., Svidzinskiy, A.V., Slyusarev, V.A. "The theory of the proximity effect in superconductors" ZhETF, 1969, V. 56, No. 3, p. 835-840.

8. Savchenko, S.M., Svidzinskiy, A.V. "A theory of Josephson current in a junction containing impurities" TMF, 1978, V. 35, No. 2, p. 277-281.

9. Abrikosov, A.A., Gor'kov, L.P., Dzyaloshinskiy, I.Ye. "Metody kvantovoy teorii polya v statisticheskoy fizike" [Methods of quantum field theory in statistical physics] Moscow: Fizmatgiz, 1962, 443 p.

10. Larkin, A.I., Ovchinnikov, Yu.N. "A quasi-classical method in superconductivity theory" ZhETF, 1968, V. 55, No. 6(12), p. 2262-2272.

11. Usadel, K.D. "Generalized diffusion equation for superconducting alloys" PHYS. REV. LETT., 1970, V. 25, No. 8, p. 507-508.

12. Ivanov, Z.G., Kupriyanov, M.Yu., Likharev, K.K., Snigirev, O.V "Boundary condition for the Usadel equations and properties of dirty SNS sandwiches" J. PHYS., 1978, suppl. No. 8, p. C6-556-C6-557.

13. Barone, A., Ovchinnikov, Yu.N. "Boundary conditions and critica current in *SNS*-junctions" ZhETF, 1979, V. 77, No. 4, p. 1463-1470

14. Ivanov, Z.G., Kupriyanov, M.Yu., Likharev, K.K., et al. "Boundar conditions for the Eilenberger and Usadel equations and the properties of "dirty" *SNS*-sandwiches" FNT, 1981, v. 7, nO. 5, P. 560-574.

15. Josephson, B.D. "Possible new effects in superconducting tunnelling" PHYS. LETT., 1962, V. 1, No. 7, p. 251-253.

16. Kulik, I.O. "Spatial quantization and the proximity effect in *SNS* junctions" ZhETF, 1969, V. 57, No. 5, p. 1745-1759.

17. Andreev, A.F. "Thermal conductivity of the intermediate state o superconductors" ZhETF, 1964, V. 46, No. 5, p. 1823-1828.

18. Andreev, A.F. "The electron spectrum of the intermediate state o superconductors" ZhETF, 1965, V. 49, No. 2(8), p. 655-660.

19. Ishii, C. "Josephson current through junctions with normal meta barriers" PROGR. THEOR. PHYS., 1970, V. 44, No. 6, p. 1525-1547.

20. Svidzinskiy, A.V., Antsygina, T.N., Bratus', Ye.H. "Superconducting current in wide *SNS*-junctions" ZhETF, 1971, V. 61, No. 4, p 1612-1619.

21. Bardeen, J., Johnson, J.L. "Josephson current flow in pure superconducting-normal-superconducting junctions" PHYS. REV., 1972, V 5, No. 1, p. 72-78.

22. Kulik, I.O. "Weak superconductivity" Prepr. INFORMATION AN SSR Sverdlovsk, 1973, 34 p.

23. Kulik, I.O., Mitsay, Yu.N. "The influence of impurities on th Josephson current in *SNS*-junctions" FNT, 1975, V. 1, No. 7, p 906-913.

24. Galayko, V.P. "The quantization of electron excitations in *SNS* film junctions in a magnetic field" ZhETF, 1969, V. 57, No. 3(9) p. 941-948.

25. Gogadze, G.A. Kulik, I.O. "The oscillatory and resonance effect in *SNS*-junctions in a magnetic field" ZhETF, 1971, V. 60, No. 5 p. 1819-1831.

26. Antsygina, T.N., Bratus', Ye.N., Svidzinskiy, A.V. "The Josephson current in an *SNS*-junction in a magnetic field" FNT, 1975, V. 1, No. 1, p. 49-56.

27. Werthamer, N.R. "Nonlinear self-coupling of Josephson radiation in superconducting tunnel junctions" PHYS. REV., 1966, V. 147, No. 1, p. 255-263.

28. Lapkin, A.I., Ovchinnikov, Yu. N. "The tunnel effect between superconductors in an alternating field" ZhETF, 1966, V. 51, No. 5(11), p. 1535-1543.

29. Gor'kov, L.P., Eliashberg, G.M. "Generalization of equations from Ginzburg-Landau theory for nonstationary problems in the case of alloys containing paramagentic impurities" ZhETF, 1968, V. 54, No. 2, p. 612-626.

30. Eliashberg, G.M. "Inelastic electron collisions an nonequilibrium steady states in superconductors" ZhETF, 1971, V. 61, No. 3(9), p. 1254-1271.

31. Gor'kov, L.P., Kponin, N.B. "Viscous flow of vortices in superconducting alloys" ZhETF, 1973, V. 65, No. 1(7), p. 396-410.

32. Ivlev, B.I., Lisitsyn, S.G., Eliashberg, G.M. "Nonequilibrium excitations in superconductors in high frequency fields" J. LOW TEMP. PHYS., 1973, V. 10, No. 3/4, p. 449-468.

33. Schmid, A., Schön, G. "Linearized kinetic equation and relaxation processes of superconductors near T_c" J. LOW TEMP. PHYS., 1975, V. 20, No. 1/2, p. 207-227.

34. Artemenko, S.N., Bolkov, A.F. "A thermoelectric field in superconductors" ZhETF, 1976, V. 70, No. 3, p. 1051-1060.

35. Larkin, A.I., Ovchinnikov, Yu. N. "The nonlinear conductivity of superconductors in a mixed state" ZhETF, 1975, V. 68, No. 5, p. 1915-1923.

36. Larkin, A.I., Ovchinnikov, Yu.N. "Nonlinear effects from vortex motion in superconductors" ZhETF, 1977, V. 1(7), p. 299-312.

37. Artemenko, S.N., Bolkov, A.F. "The electrical field and collective oscillations in superconductors" UFN, 1979, V. 128, No. 1, p. 3-30.

38. Artemenko, S.N., Bokkov, A.F., Zaytsev, A.V. "The theory of the nonstationary Josephson effect in short superconducting junctions" ZhETF, 1979, V. 76, No. 5, p. 1816-1832.

39. Zaytsev, A.V. "The theory of pure short *SCS* and *SCN* microbridges"

ZhETF, 1980, V. 78, No. 1, p. 221-233. Poprabka, V. 79, No. 5 (11), p. 2015.

40. Kilik, I.O., Omel'yanchuk, A.N. "The Josephson effect in superconducting bridges: microscopic theory" FNT, 1978, V. 4, No. 3, p 296-311.

41. Aslamazov, L.G., Lapkin, A.I. "The Josephson effect in point superconducting junctions" PIS'MA V ZhETF, 1969, V. 9, No. 7, p 150-154.

42. Likharev, K.K., Yakobson, L.A. "The dynamical properties of superconducting filaments of finite length" ZhETF, 1975, V. 68, No. 3 p. 1150-1159.

43. Tinkham, M. "Tunneling generation relaxation and tunneling detection of hole-electron imbalance in superconductors" PHYS. REV. B 1972, V. 6, No. 5, p. 1747-1756.

44. Artemenko, S.N., Volkov, A.F., Zaitsev, A.V. "On the contribution of the superconductor to the resistance of a superconductor-normal metal system" J. LOW TEMP. PHYS., 1978, V. 30, No. 3/4, p. 487 502.

45. Zaitsev, A.V. "Voprosy teorii neravnovesnykh protsessor v sverkhprovodyashchikh strukturakh" [Issues in the theory of non-equilibrium processes in superconducting structures] Dissertation for the degree of Doctor of Physics and Mathematics, Moscow: MFTI 1978, p. 46-63.

46. Aslamazov, L.G., Lapkin, A.I., Ovchinnikov, Yu.N. "The Josephson effect in superconductors separated by a normal metal" ZhETF 1968, V. 55, No. 1(7), p. 323-335.

47. Kulik, I.O., Yunson, I.K. "Effekt Dzhozefsona v sverkhprovodyashchikh tunnel'nykh strukturakh" [The Josephson effect in superconducting tunnel structures] Moscow, Nauka, 1970, 272 p.

48. Likharev, K.K., Ul'rikh, B.T. "Sistema s dzhozefsonovskimi kontaktami" [A system with Josephson junctions] Moscow: Izd-vo MGU 1978, 446 p.

49. Ferrell, R.A., Prange, R.E. "Self-field limiting of Josephson tunneling of superconducting electron pairs" PHYS. REV. LETT., 1963 V. 10, No. 11, p. 479-491.

50. Josephson, B.D. "Supercurrents through barriers" ADV. PHYS., 1965 V. 14, No. 5/6, p. 419-450.

51. Kulik, I.O. "Wave propagation in a tunnel Josephson junction with vortices and the electrodynamics of weak superconductivity" ZhETF 1966, V. 51, No. 6(12), p. 1952-1969.

52. Zaikin, A.D., Zharkov, G.F. "A theory of weak superconductivity in *SNINS*-systems" ZhETF, 1980, V. 78, No. 2, p. 721-732.

53. Zaikin, A.D. "Meissner effect in superconductor-normal metal proximity sandwiches" SOLID STATE COMMUN., 1982, V. 41, No. 7, p. 533-536.

54. Zaikin, A.D., Zharkov, G.F. "On the microscopic theory of proximity effect in normal metal-superconductor sandwiches" PHYS. LETT. A, 1983, V. 95, p. 331-334.

55. Kupriyanov, M.Yu., Lukichev, V.F. "The influence of the proximity effect in electrodes on the steady-state properties of *SNS*-Josephson structures" FNT, 1982, V. 8, No. 10, p. 1045-1052.

56. McMillan, W.L. "Theory of superconductor-normal metal interfaces" PHYS. REV., 1968, V. 175, No. 2, p. 559-568.

57. Arnold, G.B. "Theory of thin proximity-effect sandwiches" PHYS. REV. B, 1978, V. 18, No. 3, p. 1076-1100.

58. Haeschlonck, C. van, Dries, L. van den, Bruynseraede, Y., Gilabert, A. "On the spatial dependence of the order parameter in superconducting proximity sandwiches" J. PHYS., 1981, V. F11, p. 2381-2388.

59. Pippard, A.B. "An experimental and theoretical study of the relation between magnetic field and current in a superconductor" PROC. ROY. SOC. A, 1953, V. 216, p. 547-568.

60. Drangeid, K.E., Sommerhalder, R. "Observed sign reversal of a magnetic field penetrating a superconductor" PHYS. REV. LETT., 1962, V. 8, No. 12, p. 467-469.

61. Gradshtein, N.S., Ryzhik, N.M. "Tablitsy integralov, summ ryadov i proizvedeniy" [Tables of integrals, sums, series and products] Moscow: Fizmatgiz, 1963, p. 918-919.

62. Oda, Y., Nagano, H. "Meissner effect in Cu of thick Cu clad Nb" SOLID STATE COMMUN. 1980, V. 35, p. 631-634.

63. Kupriyanov, M.Yu. "The steady-state properties of pure *SNS*-sandwiches" FNT, 1981, V. 7, No. 6, p. 700-708.

64. Zaikin, A.D., Zarkov, G.F. "A theory of wide "dirty" *SNS*-junctions" FNT, 1981, V. 7, No. 3, p. 375-378.

65. Clarke, J. "Supercurrents in lead-copper-lead sandwiches" PrOC. ROY. SOC. LONDON A, 1969, V. 308, p. 447-471.

66. Likharev, K.K. "The $T_S(\varphi)$ relation for *SNS*-bridges of variable thickness" PIS'MA V ZhETF, 1976, V. 2, No. 1, p. 29-34.

67. Makeev, A.I., Svidginskiy, A.V. "Theory of current states in a critically doped *SNS*-junction in a near zero temperature range" TMF, 1980, V. 44, No. 1, p. 85-92.

68. Niemeyer, J., Minnigerode, G. von. "Effects of nonmagnetic impurities in the normal metal layer of SNS junctions" ZTSCHR. PHYS. B, 1979, V. 36, No. 1, p. 57-66.

69. Hsiang, T.Y., Finnemore, D.K. "Superconducting critical currents for thick, clean superconductor - normal metal - superconductor junctions" PHYS. REV. B, 1980, V. 22, No. 1, p. 154-163.

70. Rowell, J.M., McMillan, W.L. "Electron interference in a normal metal induced by superconducting contacts" PHYS. REV. LETT., 1966, V. 16, No. 11, p. 453-456.

71. Rowell, J.M. "Tunneling observation of bound states in a normal metal-superconductor sandwich" PHYS. REV. LETT., 1973, V. 30, No. 5, p. 167-170.

72. Bellanger, D., Klein, S., Leger, A. et al. "Tunneling measurements of electron interference effects on Cu-Pb sandwiches" PHYS. LETT. A, 1973, V. 42, No. 6, p. 459-460.

73. Zaikin, A.D., Zharkov, G.F. "The influence of external fields and impurities on the Josephson current in *SNINS*-junctions" ZhETF, 1981, V. 81, No. 5(11), p. 1781-1802.

74. Ambegaokar, V., Baratoff, A. "Tunneling between superconductors" PHYS. REV. LETT., 1963, V. 10, No. 11, p. 486-489; V. 11, No. 2, p. 104.

75. Haberkorn, W., Knauer, H., Richter, J. "A theoretical study of the current phase relation in Josephson contacts" PHYS. STATUS SOLIDI (A), 1978, V. 47, p. K161-K164.

76. Bezuglyy, A.I., Kulik, I.O., Mitsay, Yu.N. "A theory of superconducting junctions containing a normal metal interlayer" FNT, 1975, V. 1, No. 1, p. 57-67.

77. Zaikin, A.D., Zharkov, G.F. "The "double" critical current enhancement in pure *SNINS*-systems" PIS'MA V ZhETF, 1982, V. 35, No. 12, p. 514-516.

78. Zaikin, A.D. "The nonstationary Josephson effect and nonequilibrium properties of *SNS*-junctions" ZhETF, 1983, V. 84, No. 4, p. 1560-1574.

79. Dikharev, K.K. "Superconducting weak links: stationary processes" UFN, 1979, V. 127, No. 2, p. 185-220.

80. Aslamazov, L.G., Lapkin, A.I. "The influence of a microwave field on the critical current of superconducting junctions" ZhETF, 1978, V. 74, No. 6, p. 2184-2195

81. Schmid, A., Schön, G., Tinkham, M. "Dynamic properties of superconducting weak links" PHYS. REV. B, 1980, V. 21, No. 11, p. 5076-5086.

82. Keldysh, L.V. "The diagram technique for nonequilibrium processes" ZhETF, 1964, V. 47, No. 4(10), p. 1515-1527.

83. Bezuglyy, A.I., Mitsay, Yu.N. "The theory of the Tomash effect in *SNINS*-junctions" FNT, 1975, V. 1, No. 4, p. 451-457.

84. Mitsay, Yu.N. "The nonstationary Josephson effect in a junction containing a normal metal interlayer" FTT, 1980, V. 22, No. 6, p. 1761-1765.

85. Eliashberg, G.M. "The superconductivity of high frequency field-enhanced films" PIS'MA V ZhETF, 1970, V. 11, No. 3, p. 186-188.

86. Warlaumont, J.M., Brown, J.C., Foxe, T., Buhrman, R.A. "Microwave enhanced proximity effect in superconductor-normal metal-superconductor microjunctions" PHYS. REV. LETT., 1979, V. 43, No. 2, p. 169-172.

87. Asmamazov, L.G., Lempitskiy, S.V. "Microwave field-enhancment of superconductivity in superconductor-normal metal-superconductor junctions" ZhETF, 1982, V. 5, p. 1671-1678.

88. Weitz, F.A., Skocpol, W.J., Tinkham, M. "Fer-infrared frequency dependence of the ac Josephson effect in niobium point contacts" PHYS. REV. B, 1978, V. 18, No. 7, p. 3282-3292.

89. Zharkov, G.F., Zaikin, A.D. "The free energy of a weak superconductor with current in an external magnetic field" KRAT. SOOBSHCH PO FIZIKE FIAN, 1978, No. 7, p. 21-26.

90. Bulanov, S.V., Zharkov, G.F., Zaikin, A.D. "An investigation of the stability of a system of magnetic field vortices in a Josephson junction of finite size" TR. FIAN, 1983, V. 148, p. 182-191.

91. Zaikin, A.D., Zharkov, G.F. "The spectrum of oscillations in a system of Josephson junctions" KRAT. SOOBSHCH PO FIZIKE FIAN, 1980, No. 2, p. 28-32.

92. Owen, C.S., Scalapino, D.J. "Vortex structure and critical currents in Josephson junctions" PHYS. REV., 1967, V. 164, No. 2, p. 538-544.

93. Zharkov, G.F. "The penetration of vortices into a Josephson junction of finite width" ZhETF, 1976, V. 71, No. 5(11), p. 1951-1959.

94. Lebwohl, P., Stephen, M.J. "Properties of vortex lines in superconducting barriers" PHYS. REV., 1967, V. 63, No. 2, p. 376-379.

95. Gorbonosov, A.Ye., Kulik, I.O. "Plasma oscillations and metastable states in Josephson tunnel junctions" ZhETF, 1971, V. 60, No. 2, p. 688-694.

96. Akopov, S.G., Lozovik, Yu.Ye. "A phase diagram of a granular superconductor" FNT, 1981, V. 7, No. 4, p. 524-527.

97. Golub, A.A., Bezzub, O.P. "The critical current of pure *SNS*-junctions." FNT, 1982, V. 8, No. 1, p. 54-58.

A Nonquasiclassical Description of Inhomogeneous Superconductors

A.D. Zaikin, S.V. Panyukov

Abstract: Exact microscopic equations are derived that make it possible to describe inhomogeneous superconductors when the quasi-classical approach is not suitable. These equations are simplier than the Gorkov equations. We generalize the derived equations for describing the nonequilibrium states of inhomogeneous superconductors. It is demonstrated that the derived equations (including the case of a nonequilibrium quasi particle distribution function) may be written in the form of linear differential equations for the simulateneous wave functions u, ν. The quasi-classical limit of such equations is examined. Effective boundary conditions are derived for the u, ν functions that allow description of superconductors with a sharp change in parameters within the scope of the quasi-classical approach.

INTRODUCTION

As we know in all existing superconductors motion of the Cooper pairs is quasi-classical which is due to the high value of the superconducting correlation length $\xi_0 \gg a$, where a is the atomic scale. Accounting for this fact allows us to significantly simplify equations of superconductivity theory. If external fields acting on the superconductor vary slowly on atomic scales, the superconducting system may be described by Eilenberger equations [1] that determine the quasi-classical Green's functions. For the case of a spatially-inhomogeneous superconductor whose parameters jump on atomic scales (the superconductor-normal metal boundary, *SN*; the superconductor-dielectric boundary, *SI*; the superconductor-superconductor and Josephson systems, *SIS*, *SCS*, *SNS*, etc.) these equations do not provide a correct description of the superconducting subsystem in direct proximity to the inhomogeneity boundary. Here the influence of the boundary may be accounted for by imposing boundary conditions on the quasi-classical Green's functions [2-5]. For their derivation we must use an exact system of Gorkov equations [6] that is not constrained by the quasi-classical limits. A situation is also possible in which the various parameters of the superconductor (Fermi velocity, order parameter, effective electron mass) change significantly at distances much greater than the interatomic distance, although much less (or in the order of) the coherence length ξ_0. In this case the quasi-classical equations of superconductivity theory at not applicable. It is also necessary to rely on exact Gorkov equations. Moreover, the quasi-classical condition is violated in describing antiferromagnetic superconductors in which the exchange fields of the magnetic subsystem vary on the atomic scales. In the majority of such cases we may assume that the fields acting on the electrons are dependent on only a single spatial variable which we will designate as x.

The successful utilization of the Eilenberger equations is related primarily to the fact that the quasi-classical Eilenberger functions are dependent only on the single spatial variable **r** at the same time that the single-electron Green's functions are functions of the two variables **r** and **r**$'$. The physical reason for this differential is based on the quasi-classical condition that allows us to use an abbreviated description in the language of the coordinate **R** of the center of inertia of the Cooper pair. This fact means that it is significantly easier to solve the Eilenberger equations for spatially-inhomogeneous superconductors compared to the initial system of Gorkov equations. In addition we note that the solution of Gorkov equations contains excess information on the spectrum of quasi particle excitations, etc. At the same time one advantage of the Eilenberger equations is that they contain the minimum necessary information since in reality only the Green's function with coincident arguments are required.

Hence, the process of finding exact equations that are not limited by quasi-classical considerations for the Green's functions with coincident arguments **r** = **r**p' is of significant interest. The derivation of these equations is discussed in Section 1 of this chapter. As was demonstrated in study [7] the Eilenberger equations may be written as linear differential equations for the wave functions u, υ through which the Green-Eilenberger functions are expressed algebraically. In Section 2 it is demonstrated that such a factorization is possible in the case of exact Green's functions not limited by quasi-classical aspects and equations are found that determine such u- and υ-functions. The quasi-classical limit of the derived equations is examined in Section 3.

If external time-dependent fields act on the superconductor, the electron Green's functions depend not only on the single difference temporal variable, $t - t'$, but also on each of them separately (t, t'). The existence of this relation significantly complicates the solution of equations in superconductivity theory. In order to describe the nonequilibrium states when external fields exist that vary slowly on atomic scales, quasi-classical Eliashberg equations are used for two time Green's functions [8]. As demonstrated in study [7] these equations may be written in the form of equations for simultaneous wave functions through which the quasi-classical Green's functions are expressed algebraically. It is demonstrated in Section 4 that such factorization is also possible for exact equations determining the Green's functions. Here in addition to the functions u, υ it is also necessary to incorporate the retarded and advanced wave functions [7]. Finally in Chapter 5 we derive simple boundary conditions for the u- and v-functions that allow us to describe superconducting systems whose properties jump (at distances in the order of the interatomic distances) by means of quasi-classical equations for such functions.

1. Derivation of equations for the Green's functions

We will assume further that the external fields and, consequently, the superconducting order parameter are dependent only on the coordinate x. In this case the momentum is conserved in perpendicular directions $\mathbf{p}_\perp = (0, p_y, p_z)$ and the temperature Green's functions [6] $G_\omega(\mathbf{r}_1, \mathbf{r}_2)$ are in fact dependent only on the variables x_1, x_2, $\mathbf{r}_{2\perp} - \mathbf{r}_{1\perp}$. In the Gorkov equations we will carry out a Fourier transformation over a difference variables $\mathbf{r}_{2\perp} - \mathbf{r}_{1\perp}$ and for the Fourier components of the Green's functions we find the matrix equation

$$\frac{1}{2m}\frac{\partial^2}{\partial x_1^2}\hat{G}_\omega(x_1, x_2, \mathbf{p}_\perp) + [-\xi_\perp(x_1) + \hat{H}(x_1)]G(x_1, x_2, \mathbf{p}_\perp) = \hat{1}\delta(x_1 - x_2),$$

$$\xi_\perp(x) = \begin{pmatrix} \xi_\perp(x), & 0 \\ 0, & \xi_\perp^*(x) \end{pmatrix}, \quad \hat{H}(x) = \begin{pmatrix} i\omega, & \Delta(x) \\ -\Delta^*(x), & -i\omega \end{pmatrix} - U(x)\hat{1}, \tag{1}$$

where we have used the gauge of the vector potential $A_x(x) = 0$; 1 is the identity matrix; $U(x)$ is the external potential. The function $\xi_\perp(x)$ is determined by the expression

$$\xi_\perp(x) = \frac{1}{2m}[(p_y - ieA_y(x))^2 + (p_z - ieA_z(x))^2] - \mu, \tag{2}$$

where μ is the chemical potential.

We will now find the equation determining the function $\hat{G}_\omega(x, x, \mathbf{p}_\perp)$. For this we will first transform equation (1) to a first order matrix system

$$\left[\frac{\partial}{\partial x_1} + \check{H}(x, \mathbf{p}_\perp)\right]\check{G}_\omega(x_1, x_2, \mathbf{p}_\perp) = \check{1}\delta(x_1 - x_2),$$

$$\check{G}_\omega = \begin{pmatrix} \hat{G}_{1\omega}, & \hat{G}_\omega \\ \hat{G}_\omega, & -\hat{G}_{2\omega} \end{pmatrix}, \quad \check{H}(x, \mathbf{p}_\perp) = \begin{pmatrix} 0, & -\xi_\perp(x) + \hat{H}(x) \\ -2m, & 0 \end{pmatrix}. \tag{3}$$

We note that unlike (1) equation (3) also describes the case of mass m dependent on the coordinate $m = m(x)$. The corresponding term in the Hamiltonian of the electron system in the secondary quantization representation takes the form

$$\frac{1}{2}\int dx \sum_\sigma m^{-1}(x) \colon \frac{\partial\hat{\psi}_\sigma^+(x, \tau)}{\partial x}\frac{\partial\hat{\psi}_\sigma(x, \tau)}{\partial x}\colon,$$

where the dots designate normal ordering. Consequently we should replace the second derivative in equation (1) with

$$\frac{1}{2}\frac{\partial}{\partial x_1}\left[m^{-1}(x)\frac{\partial}{\partial x_1}\cdots\right],$$

and all subsequent results are also applicable to the case of a coordinate- (position-) dependent mass. Second order equation (1) for the Green's function $\hat{G}_\omega$ may be obtained from (3) by eliminating the function $G_{2\omega}$ from equation (3). The matrix functions $G\omega$ and $G_{1\omega}$ are introduced to expand $\hat{G}_\omega$ to a square matrix. We note that equation (3) was derived by differentiating the Green's function $\hat{G}(\mathbf{r}_1, \mathbf{r}_2, \tau_1 - \tau_2)$ with respect to τ_1 [9]. In addition to this equation we will consider one additional equation found by differentiation of the Green's function with respect to the variable τ_2. Repeating the calculations which led us to formula (3) we obtain

$$-\frac{\partial}{\partial x_2}\check{G}_\omega(x_1, x_2, \mathbf{p}_\perp)+\check{G}_\omega(x_1, x_2, \mathbf{p}_\perp)\check{H}(x_2, \mathbf{p}_\perp)=\delta(x_1-x_2). \tag{4}$$

We now find the equation that directly determines the Green's function $G_\omega(x, x, \mathbf{p}_\perp)$ with coincident arguments. For this it is convenient to introduce the coordinates of the center of inertia of the Cooper pair $x = (x_1 + x_2)/2$ and the coordinate with respect to motion $r = x_2 - x_1$. In the new coordinates equations (3), (4) are written as

$$\begin{gathered}\left(\frac{1}{2}\frac{\partial}{\partial x}-\frac{\partial}{\partial r}\right)\check{G}_\omega+\check{H}\left(x-\frac{r}{2}, \mathbf{p}_\perp\right)\check{G}_\omega=\delta(r),\\ \left(-\frac{1}{2}\frac{\partial}{\partial x}-\frac{\partial}{\partial r}\right)\check{G}_\omega+\check{G}_\omega\check{H}\left(x+\frac{r}{2}, \mathbf{p}_\perp\right)=\delta(r),\\ \check{G}_\omega\equiv\check{G}_\omega\left(x-\frac{r}{2}, x+\frac{r}{2}; \mathbf{p}_\perp\right).\end{gathered} \tag{5}$$

We will compare the difference of these equations. In subtraction, the terms with the r derivatives and the δ-functions drop out. In the equation derived we may set $r = 0$. Thus, the equation for the function $\check{g}_\omega(x, \mathbf{p}_\perp) \equiv 2G_\omega(x, x, \mathbf{p}_\perp)$ takes the form

$$\frac{\partial\check{g}_\omega(x, \mathbf{p}_\perp)}{\partial x}+\check{H}(x, \mathbf{p}_\perp)\check{g}_\omega(x, \mathbf{p}_\perp)-\check{g}_\omega(x, \mathbf{p}_\perp)\check{H}(x, \mathbf{p}_\perp)=0. \tag{6}$$

In this form this equation is similar to the system of Eilenberger equations for the quasi-classical Green's functions and in fact is a generalization of the latter to the case of random external fields that may change rapidly in space.

We will first find the function $\check{g}_\omega(x, \mathbf{p}_\perp)$ for a spatially-homogeneous superconductor in the absence of external fields. In this case the function $g_\omega(\mathbf{r}_1 = \mathbf{r}_2)$ is dependent only on the difference variables and we find from (7) for its Fourier-component

$$[ip_x+\check{H}(\mathbf{p}_\perp)]\,\check{G}_\omega(p_x, \mathbf{p}_\perp)=\check{1}. \tag{7}$$

We will multiply equation (7) from the left by the matrix $(-ip_x + H(\mathbf{p}_\perp))$ and will use relation $H_2 = 2m(\xi_\perp - H(\mathbf{p}_\perp))$ that follows directly from definition (3). Here and henceforth for simplicity we will not provide the identity matrices. As a result we arrive at the equation

$$[\xi-\check{H}]\,\check{G}_\omega(p_x, \mathbf{p}_\perp)=(\check{H}-ip_x)/2m, \tag{8}$$

where $\xi = \xi_\perp + p_x^2/2m$. We find the Green's function $G_\omega(p_x, \mathbf{p}_\perp)$ from (8) by multiplying this equation by $\xi + \check{H}$. Integrating the function with respect to momentum p_x, we find the final expression for the function $\hat{g}_\omega(x, \mathbf{p}_\perp)$ of a homogeneous superconductor:

$$\check{g}_\omega(\mathbf{p}_\perp) = \left[\alpha_+ + \alpha_- \frac{\hat{H}}{(\omega^2 + |\Delta|^2)^{1/2}}\right] \frac{\check{H}(\mathbf{p}_\perp)}{\sqrt{2m(\xi_\perp^2 + \omega^2 + |\Delta|^2)}},$$

$$\alpha_\pm^2 = \left[\sqrt{\xi_\perp^2 + \omega^2 + |\Delta|^2} \pm \xi_\perp\right]/2. \tag{9}$$

We will now establish certain relations for derived function (9). Taking the spur of matrix (9) we find

$$\mathrm{Sp}\,\hat{g}_\omega(\mathbf{p}_\perp) = 0. \tag{10}$$

We will then square Green's function (9). For this it is most convenient to use the commutation capability of the matrices $\check{H}\hat{H} = \hat{H}\check{H}$ which may be directly tested, together with the relations

$$\alpha_+\alpha_- = (\sqrt{\omega^2 + |\Delta|^2})/2, \qquad \alpha_+^2 - \alpha_-^2 = \xi_\perp, \tag{11}$$

deriving from (9). As a result we find a simple normalization relation for the function $\hat{g}_\omega(x, \mathbf{p}_\perp)$ in a spatially-homogeneous superconductor

$$\check{g}_\omega^2(\mathbf{p}_\perp) = \check{1}. \tag{12}$$

We will now proceed to an examination of equation (6) in the general spatially-inhomogeneous case and will establish certain of its general properties. We will take the spur of equation (6). The commutator of $\check{g}$ and $\check{H}$ is dropped here and we obtain

$$\frac{\partial}{\partial x}\,\mathrm{Sp}\,\check{g}_\omega(x, \mathbf{p}_\perp) = 0. \tag{13}$$

It follows from here that $\mathrm{Sp}\,\check{g}_\omega(x, \mathbf{p}_\perp)$ is independent of the coordinate x. We will assume that when $x \to -\infty$ a spatially-homogeneous solution described by the matrix Green's function (9) is achieved. It follows from relation (13) and (10) that in the entire superconductor volume

$$\mathrm{Sp}\,\check{g}_\omega(x, \mathbf{p}_\perp) = 0. \tag{14}$$

We will now multiply equation (6) by $\check{g}_\omega$ from the right and then from the left and then add the derived equations. As a result we find

$$\frac{\partial \check{g}_\omega^2(x, \mathbf{p}_\perp)}{\partial x} + \check{H}(x, \mathbf{p}_\perp)\check{g}_\omega^2(x, \mathbf{p}_\perp) - \check{g}_\omega^2(x, \mathbf{p}_\perp)\check{H}(x, \mathbf{p}_\perp) = 0. \tag{15}$$

It follows from condition (12) when $x \to -\infty$ and from equation (15) that the normalization relation

$$\check{g}_\omega^2(x, \mathbf{p}_\perp) = \check{1} \tag{16}$$

must also be satisfied in the entire superconductor volume. Writing out equations (16) and (14) in component form we derive the following relations

$$\mathrm{Sp}\,\hat{g}_2 = \mathrm{Sp}\,\hat{g}_1, \qquad \hat{g}\hat{g}_1 = \hat{g}_2\hat{g}, \qquad \hat{g}_2^2 + \hat{g}\hat{\tilde{g}} = \hat{1}, \tag{17}$$

as well as the following relations that may be derived from (17):

$$\hat{g}_1\hat{\tilde{g}} = \hat{\tilde{g}}g_2, \qquad \hat{g}_1^2 + \hat{\tilde{g}}\hat{g} = \hat{1}, \qquad \check{g}_\omega(x,\ \mathbf{p}_\perp) = \begin{pmatrix} \hat{g}_1, & \hat{\tilde{g}} \\ \hat{g}, & -\hat{g}_2 \end{pmatrix}. \tag{18}$$

We will also provide the self-consistency equation for the superconducting order parameter

$$\Delta(x) = g_{\text{э}\Phi}\frac{\pi T}{2}\sum_\omega \int (d\mathbf{p}_\perp) f_\omega(x,\ \mathbf{p}_\perp), \qquad \hat{g}_\omega = \begin{pmatrix} g_\omega, & f_\omega \\ f_\omega^+, & \tilde{g}_\omega \end{pmatrix},$$

$$\int (d\mathbf{p}_\perp) \equiv \int_0^{\sqrt{2m\mu}} \frac{p_\perp\, dp_\perp}{2\pi}\int_0^{2\pi}\frac{d\varphi}{2\pi}, \qquad \mathbf{p}_\perp = (0,\ \mathbf{p}_\perp\cos\varphi,\ p_\perp\sin\varphi) \tag{19}$$

and the expression for the current density in the gauge $A_x = 0$:

$$\mathbf{j}_\perp(x) = \frac{eT}{m}\sum_\omega \int (d\mathbf{p}_\perp)\,\mathbf{p}_\perp g_\omega(x,\ \mathbf{p}_\perp) - \frac{e^2}{mc}\mathbf{A}(x)\,n, \tag{20}$$

where n is the electron density. The x-current component requires special consideration since it is expressed through the derivative of the Green's function with respect to the difference variable $r = x_2 - x_1$:

$$j_x(x) = \frac{ieT}{m}\sum_\omega \int (d\mathbf{p}_\perp)\, g'_\omega(x,\ \mathbf{p}_\perp),$$

$$\hat{g}'_\omega(x,\ \mathbf{p}_\perp) = \begin{pmatrix} g'_\omega, & f'_\omega \\ f_\omega^{+\prime}, & \tilde{g}'_\omega \end{pmatrix} \equiv \lim_{r\to 0}\frac{d}{dr}\,\hat{g}\left(x - \frac{r}{2},\ x + \frac{r}{2},\ \mathbf{p}_\perp\right). \tag{21}$$

We will show that the function $\hat{g}'_\omega$ is expressed through the functions $\hat{g}_1$ and $\hat{g}_2$ in (18). For this we will sum equations (5) and here terms with x derivatives drop out. We will consider the derived equation for the function $G_\omega(x - r/2,\ x + r/2,\ \mathbf{p}_\perp)$. We may easily see that the δ-function also drops and we may set $r = 0$ in the derived equation. As a result we find

$$\hat{g}'_\omega(x,\ \mathbf{p}_\perp) = m\,(\hat{g}_2 - \hat{g}_1). \tag{22}$$

Thus, the matrix function $\hat{g}_\omega(x,\ \mathbf{p}_\perp)$ is the solution of equations (6) completely describing the thermodynamic properties of a superconductor with random fields $A(x)$, $U(x)$ which may have, specifically, δ-functional features.

In the conclusion of this section we will provide a generalization of the derived results to the case of chaotically-distributed nonmagnetic impurities. Henceforth we will consider only the impurities with a δ-functional interaction potential $V(\mathbf{r}-\mathbf{r}')=V\delta(\mathbf{r}-\mathbf{r}')$. The equation for the Green's function averaged over the impurity position [6] takes the form of (1) in which we replace $\hat{H}(x)$ by

$$\hat{H}(x) - cV^2\hat{\sigma}_z\hat{G}_\omega(\mathbf{r},\ \mathbf{r}')\hat{\sigma}_z, \tag{23}$$

where $\hat{\sigma}_z$ is the Pauli matrix and c is the concentration of impurity atoms. Repeating the calculations that led us to equation (6), we find the equation that determines the averaged Green's function:

$$\frac{\partial \check{g}_\omega(x,\ \mathbf{p}_\perp)}{\partial x} + \check{H}(x,\ \mathbf{p}_\perp)\check{g}_\omega(x,\ \mathbf{p}_\perp) - \check{g}_\omega(x,\ \mathbf{p}_\perp)\check{H}(x,\ \mathbf{p}_\perp) = 0,$$

$$\check{H}(x) = \begin{pmatrix} 0, & -\xi_\perp(x) + \hat{H}(x) - \frac{cV^2}{2}\int (d\mathbf{p}_\perp)\,\hat{\sigma}_z\hat{g}_\omega(x,\ \mathbf{p}_\perp)\,\hat{\sigma}_z \\ -2m, & 0 \end{pmatrix}. \tag{24}$$

In dirty superconductors with a free path length of $l \gg \xi_0$ we may obtain from (24) more simple equations that are a generalization of the Usadel equation to the case of fields that vary rapidly in space.

2. Factorization of the Green's functions of a superconductor

It is noted in a study by one of the authors [7] that the quasiclassical Eilenberger equations have symmetry that makes it possible to write these equations as a system of linear differential equations for the wave functions. These functions are a temperature analog of the coefficient functions of the Bogolubov transform [11] that diagonalizes the superconducting Hamiltonian. In this section we will show that such factorization is also possible in the case of exact equations (6) which makes it possible to significantly simplify their solution. We will then largely follow the primary techniques from study [7].

We will reduce equation (6) to a more convenient form. We will introduce the functions u_α, u^+_α which are solutions of the equations

$$\frac{\partial u_\alpha(x,\ \mathbf{p}_\perp)}{\partial x} - \sum_\beta u_\beta(x,\ \mathbf{p}_\perp)\check{H}_{\beta\alpha}(x,\ \mathbf{p}_\perp) = 0,$$

$$\frac{\partial u^+_\alpha(x,\ \mathbf{p}_\perp)}{\partial x} + \sum_\beta \check{H}_{\alpha\beta}(x,\ \mathbf{p}_\perp)u^+_\beta(x,\ \mathbf{p}_\perp) = 0, \tag{25}$$

where $\alpha,\ \beta = 1,\ \ldots,\ 4$. Let $u_{\alpha s}$, $u^+_{\alpha s}$ be the various solutions of these equations (25) where the index s labels these solutions. We will then multiply the first equation in (25) for $u_{\alpha s}$ by $u^+_{\alpha s'}$, and will sum the equation over α. Analogously we will multiply the second equation in (25) for $u^+_{\alpha s'}$ by $u_{\alpha s}$. Summing over α and adding to the first equation derived here we find the following first integrals of equations (25):

$$\sum_{\alpha=1}^{4} u^+_{\alpha s}(x,\ \mathbf{p}_\perp)u_{\alpha s'}(x,\ \mathbf{p}_\perp) = \text{const}. \tag{26}$$

We will show that the Green's function $\check{g}_{\omega\alpha\beta}$ is expressed through two linearly-independent solutions of equations (25):

$$\check{g}_{\omega\alpha\beta}(x, \mathbf{p}_\perp) = 2\sum_{s=1}^{2} u^+_{\alpha s}(x, \mathbf{p}_\perp) u_{\beta s}(x, \mathbf{p}_\perp) - \sigma\delta_{\alpha\beta}, \tag{27}$$

where $\sigma = \pm 1$. We note that as demonstrated in study [7] a formula of the type shown in (27) also describes the quasi-classical Eilenberger function with spin-nondiagonal interactions. It is assumed in (27) that the functions $u_{\alpha s}$ and $u^+_{\alpha s'}$ are normalized by the condition

$$\sum_{\alpha=1}^{4} u^+_{\alpha s}(x, \mathbf{p}_\perp) u_{\alpha s'}(x, \mathbf{p}_\perp) = \sigma\delta_{ss'}, \tag{28}$$

and according to (26) this condition is the first integral of equations (25).

We will substitute the function $\check{g}_\omega$ (27) into equation (6) and will use equations (25) for the functions u_α, u^+_β. We may easily see that equation (6) becomes an identity and in order for representation (27) to be valid it is sufficient to test for satisfaction of normalization conditions (13), (16) for the function $\check{g}_\omega$. We will set $s = s'$ and sum over s in (28). We may easily see that the derived relation is equivalent to (13). Normalization condition (18) may also be tested by using (28). Thus, we have shown that matrix function (27) is a solution of equations (6) and (13), (18).

The Green's function of interest to use $\hat{g}_{\alpha\beta}$ is expressed only through the 1,2-components u_α and the 3,4-components u^+_α. We will designate these as $u_1 = u$, $u_2 = v$, $u^+_3 = u^+$, $u^+_4 = v^+$. Eliminating the remaining u_α, u^+_β components from (25), we find the equations for the indicated wave functions only:

$$\left[\frac{1}{2m}\frac{\partial^2}{\partial x^2} - \xi_\perp(x) + i\omega - U(x)\right] u(x, \mathbf{p}_\perp) - \Delta^*(x) v(x, \mathbf{p}_\perp) = 0,$$
$$\left[\frac{1}{2m}\frac{\partial^2}{\partial x^2} - \xi^*_\perp(x) - i\omega - U(x)\right] v(x, \mathbf{p}_\perp) + \Delta(x) u(x, \mathbf{p}_\perp) = 0, \tag{29}$$

and analogous equations for the functions u^+ and v^+:

$$\left[\frac{1}{2m}\frac{\partial^2}{\partial x^2} - \xi_\perp(x) + i\omega - U(x)\right] u^+(x, \mathbf{p}_\perp) + \Delta(x) v^+(x, \mathbf{p}_\perp) = 0,$$
$$\left[\frac{1}{2m}\frac{\partial^2}{\partial x^2} - \xi^*_\perp(x) - i\omega - U(x)\right] v^+(x, \mathbf{p}_\perp) - \Delta^*(x) u^+(x, \mathbf{p}_\perp) = 0. \tag{30}$$

Orthogonality condition (28) in terms of these functions takes the form

$$u_s\frac{\partial u^+_{s'}}{\partial x} - \frac{\partial u_s}{\partial x} u^+_{s'} + v_s\frac{\partial v^+_{s'}}{\partial x} - \frac{\partial v_s}{\partial x} v^+_{s'} = 2m\sigma\delta_{ss'}, \tag{31}$$

and we may easily determine its validity directly from equations (29), (30). These equations may be considered the temperature analog of the Bogolubov equations [11]. We note that unlike (28), (29) the Bogolubov equations are equations for the eigenvalues of the quasi-particle energy spectrum. According to this analogy the spectrum of the quasi particles and their eigenfunctions in pure superconductors may be obtained by analytic continuation of the solution of (29), (30) in the range of imaginary frequency values $\omega = -i\varepsilon$. However the solution of equations (29), (30) for pure superconductors may be simpler since it does not require solution of the dispersion equation for the eigenvalues of the energy ε_ν. In addition we note that unlike the Bogolubov equations, equation system (29), (30) makes it possible to describe dirty superconductors. According to formulae (24) with chaotically-distributed nonmagnetic impurities we must replace the quantities ω, $\Delta(x)$ and $\Delta^*(x)$ in equations (29), (30) by $\tilde{\omega}(x)$, $\tilde{\Delta}(x)$ and $\tilde{\Delta}^+(x)$, respectively; these latter are determined by the expressions

$$\tilde{\omega}(x)=\omega+icV^2\int(d\mathbf{p}_\perp)\sum_{s=1}^{2}u_s^+(x,\mathbf{p}_\perp)u_s(x,\mathbf{p}_\perp),$$

$$\tilde{\Delta}(x)=\Delta(x)+cV^2\int(d\mathbf{p}_\perp)\sum_{s=1}^{2}u_s^+(x,\mathbf{p}_\perp)v_s(x,\mathbf{p}_\perp),$$

$$\tilde{\Delta}^+(x)=\Delta^*(x)+cV^2\int(d\mathbf{p}_\perp)\sum_{s=1}^{2}v_s^+(x,\mathbf{p}_\perp)u_s(x,\mathbf{p}_\perp). \tag{32}$$

The case of the existence of spin-nondiagonal interactions (for example, the exchange field in antiferromagnetic superconductors) has been investigated in detail in the quasi-classical limit in study [7]. Using the method from this section we may easily generalize the results obtained in study [7] to the case of fields rapidly varying in space.

Self-consistency equation (19) takes the following form in terms of u-, and v-functions

$$\Delta(x)=g_{\text{эф}}\pi T\sum_{\omega}\int(d\mathbf{p}_\perp)\sum_{s=1}^{2}u_s^+(x,\mathbf{p}_\perp)v_s(x,\mathbf{p}_\perp). \tag{33}$$

Using formulae (20), (21) we also find the current density in the superconductor

$$\mathbf{j}_\perp(x)=\frac{2eT}{m}\sum_{\omega}\int(d\mathbf{p}_\perp)\sum_{s=1}^{2}u_s^+(x,\mathbf{p}_\perp)u_s(x,\mathbf{p}_\perp)-\frac{e^2}{mc}\mathbf{A}(x)n,$$

$$\mathrm{j}_x(x)=\frac{ieT}{m}\sum_{\omega}\int(d\mathbf{p}_\perp)\sum_{s=1}^{2}\left[\frac{\partial u_s(x,\mathbf{p}_\perp)}{\partial x}u_s^+(x,\mathbf{p}_\perp)-u_s(x,\mathbf{p}_\perp)\frac{\partial u_s^+(x,\mathbf{p}_\perp)}{\partial x}\right]; \tag{34}$$

It is expressed only through the functions u, u^+.

3. The quasi-classical limit

In this section we will consider the case where the external fields slowly vary on the atomic scales. As demonstrated in study [7] in this case the superconducting subsystem may be described by the quasi-classical wave functions $u(x, \mathbf{v}_F)$ and $v(x, \mathbf{v}_F)$, where v_F is the Fermi velocity. Hence we must find the relation between these functions and the quasi-classical asymptotics of the functions u_s, v_s introduced in the preceding section. Establishing this relation allows us to write the boundary conditions for the quasi-classical Green's functions directly in terms of the functions $u(x, \mathbf{v}_F)$, $v(x, \mathbf{v}_F)$. It is demonstrated in study [7] that these functions are algebraically related to the quasi-classical Green-Eilenberger functions.

In the quasi-classical limit we may isolate the component of the functions $u_s(x, \mathbf{p}_\perp)$ $v_s(x, \mathbf{p}_\perp)$ that varies rapidly on the atomic scales. We will search the solution of equations (29), (30) in the form

$$\begin{pmatrix} u_s(x, \mathbf{p}_\perp) \\ v_s(x, \mathbf{p}_\perp) \end{pmatrix} = \begin{pmatrix} u_s(x, \mathbf{v}_F) \\ v_s(x, \mathbf{v}_F) \end{pmatrix} \exp(isp_x x),$$
$$\begin{pmatrix} u_s^+(x, \mathbf{p}_\perp) \\ v_s^+(x, \mathbf{p}_\perp) \end{pmatrix} = \begin{pmatrix} u_s^+(x, \mathbf{v}_F) \\ v_s^+(x, \mathbf{v}_F) \end{pmatrix} \frac{i}{v_x} \exp(-isp_x x), \tag{35}$$

where $s = \pm 1$ and it is assumed that the functions $u(x, \mathbf{v}_F)$, and $v(x, \mathbf{v}_F)$ vary slowly on the atomic scales. The vector $\mathbf{v}_F$ is determined by the expression

$$\mathbf{v}_F = \mathbf{p}_F/m, \qquad \mathbf{p}_F = (p_x, \mathbf{p}_\perp), \qquad p_x^2 + \mathbf{p}_\perp^2 = 2m\mu. \tag{36}$$

Functions (35) describe waves propagating at velocity sv_x on axis x. By substituting (35) into equations (29), (30) we may ignore the second derivatives of the functions $u(x, \mathbf{v}_F)$, $v(x, \mathbf{v}_F)$, since their contribution is small in terms of the parameter $a/\xi_0 \ll 1$, and we may also ignore terms of the order $\mathbf{A}^2$. Accounting for this after carrying out the gauge transformation $\Delta(x) \to i\Delta(x)$ we find the equations

$$\left[\omega + (\mathbf{v}_F\mathbf{A}(x)) - sv_x\frac{d}{dx}\right] u_s(x, \mathbf{v}_F) + \Delta^*(x) v(x, \mathbf{v}_F) = 0,$$
$$\left[\omega + (\mathbf{v}_F\mathbf{A}(x)) + sv_x\frac{d}{dx}\right] v_s(x, \mathbf{v}_F) - \Delta(x) u(x, \mathbf{v}_F) = 0 \tag{37}$$

and analogous equations for the functions u^+, v^+:

$$\left[\omega + (\mathbf{v}_F\mathbf{A}(x)) + sv_x\frac{d}{dx}\right] u_s^+(x, \mathbf{v}_F) + \Delta(x) v^+(x, \mathbf{v}_F) = 0,$$
$$\left[\omega + (\mathbf{v}_F\mathbf{A}(x)) - sv_x\frac{d}{dx}\right] v_s^+(x, \mathbf{v}_F) - \Delta^*(x) u^+(x, \mathbf{v}_F) = 0. \tag{38}$$

We will also find the normalization relations for the u- and v-functions (35) introduced previously. By substituting (35) into (31) in the quasi-classical limit we must differentiate only the rapidly oscillating exponential multiplier, and neglect the contribution of the derivatives of the u- and v-functions. As a result the normalization relations ($s = s'$) for these functions take the form

$$u_s^+(x, \mathbf{v}_F) u_s(x, \mathbf{v}_F) + v_s^+(x, \mathbf{v}_F) v_s(x, \mathbf{v}_F) = s\sigma. \tag{39}$$

For the nondiagonal elements ($s \neq s'$) condition (31) is satisfied automatically.

Equations (37), (38) and normalization conditions (39) coincide with the equations found in study [7] for the quasi-classical functions $u(x, \mathbf{v}_F)$, $\upsilon(x, \mathbf{v}_F)$ through which the Eilenberger functions are expressed [1, 7]. In the general case in the quasi-classical limit the functions $u_s(x, \mathbf{p}_\perp)$ and $\upsilon_s(x, \mathbf{p}_\perp)$ have the form of linear combinations of functions (35) with $s = \pm 1$ and describe both the wave passing through the inhomogeneity ($s = +1$) and the wave reflected off of the inhomogeneity boundary ($s = -1$). The normalization conditions for the amplitudes of the corresponding waves are found analogous to the derivation of relation (38). The amplitude of the wave with a given velocity direction $\mathbf{v}_f$ determines the quasi-classical functions $u(x, \mathbf{v}_F)$, $\upsilon(x, \mathbf{v}_F)$ through which the Eilenberger functions are expressed [7].

4. Generalization of the equations to the nonstationary case

We will now consider the case of time-dependent external fields. As in preceding sections we will limit our examination to their dependence on the single spatial variable x only. Equations describing the Green's function $\hat{g}$ as well as the retarded Green's function $\hat{g}^R$ and the advanced Green's function $\hat{G}^A$ (the Gorkov equations in the Keldysh technique) are used in many studies (see, for example, [4, 5, 12-14]) and are not provided here. Carrying out a Fourier transformation over the variables $\mathbf{r}_\perp$ in these equations we find by using the method from Section 1 the following equations:

$$\frac{\partial \check{g}}{\partial x} + i\check{\sigma}_z \frac{\partial \check{g}}{\partial t} + i \frac{\partial \check{g}}{\partial t'} \check{\sigma}_z + \check{H}(t)\check{g} - \check{g}\check{H}(t') + \check{\Sigma}\check{g} - \check{g}\check{\Sigma} = 0, \tag{40}$$

that determine the Green's function with coincident arguments

$$\check{g}(x, \mathbf{p}_\perp, t, t') \equiv 2\begin{pmatrix} \check{G}^R(x, x, \mathbf{p}_\perp, t, t'), & \check{G}(x, x, \mathbf{p}_\perp, t, t') \\ \check{0}, & \check{G}^A(x, x, \mathbf{p}_\perp, t, t') \end{pmatrix}. \tag{41}$$

Here each of the functions $\check{G}$, $\check{G}^R$, $\check{G}^A$ has the matrix structure of (3). The matrix H in (40) is determined by expression (3) when $\omega = 0$, tensor-multiplied by the identity matrix, and the self-energy part of $\check{\Sigma}$ describes the electron-phonon and -impurity atom interaction:

$$\check{\Sigma}(t, t') = \begin{pmatrix} \hat{\Sigma}^R(t, t'), & \hat{\Sigma}(t, t') \\ \hat{0}, & \hat{\Sigma}^A(t, t') \end{pmatrix}. \tag{42}$$

We will assume that when $t,\ t' \to -\infty$ the external fields are time-independent and normalization condition (16) is asymptotically valid. Then analogous to the derivation of the normalization condition for energy-integrated Green's functions [12, 13] we find

$$\int dt_1 \check{g}(x, \mathbf{p}_\perp, t, t_1)\check{g}(x, \mathbf{p}_\perp, t_1, t') = \check{1}\delta(t - t'). \tag{43}$$

We will show that equation (40) may be significantly simplified if we write the equation in factored form analogous to the process in the stationary case in Section 2. We will introduce the time-dependent classical wave functions (see also [7]):

$$\begin{array}{l}\check{u}=(u^R, u),\\ \check{u}^A=(0, u^A),\end{array}\quad \check{u}^+=\begin{pmatrix}u^+\\ u^{+A}\end{pmatrix},\quad \check{u}^{+R}=\begin{pmatrix}u^{+R}\\ 0\end{pmatrix},\tag{44}$$

where each of the functions introduced in this manner has four components. We will determine these functions as a solution of the equations

$$\begin{gathered}\frac{\partial}{\partial x}u^{+R}+i\hat{\sigma}_z\frac{\partial}{\partial t}u^{+R}+\hat{H}(t)u^{+R}+\hat{\Sigma}^R u^{+R}=0,\\ \frac{\partial}{\partial x}u^A+i\frac{\partial}{\partial t}u^A\hat{\sigma}_z-u^A\hat{H}(t)-u^A\hat{\Sigma}^A=0.\end{gathered}\tag{45}$$

The functions u^{+A} and u^R obey equations analogous to (45) that may be written in the following form in conjunction with equations for u and u^+

$$\begin{gathered}\frac{\partial}{\partial x}\check{u}^+ + i\check{\sigma}_z\frac{\partial}{\partial t}\check{u}^+ + \check{H}(t)\check{u}^+ + \check{\Sigma}\check{u}^+=0,\\ \frac{\partial}{\partial x}\check{u}+i\frac{\partial}{\partial t}\check{u}\check{\sigma}_z-\check{u}\check{H}(t)-\check{u}\check{\Sigma}=0.\end{gathered}\tag{46}$$

We will construct an expression for the Green's functions from solutions (44) of equations (45), (46). For this we will first consider the stationary case describing the $t,\ t' \to -\infty$ asymptotics. Green's function (41) is asymptotically dependent only on the difference argument $t - t'$ over which we may carry out the Fourier transform. The solution of stationary equations (40) will be searched as [7]

$$\check{g}_{\alpha\beta}(t,t')=\int\frac{d\varepsilon}{2\pi}\sum_{s,\sigma}[\check{u}^{+R}_{\alpha,s\sigma}(t)\check{u}_{\beta,s\sigma}(t')+\check{u}^{+}_{\alpha,s\sigma}(t)\check{u}^{A}_{\beta,s\sigma}(t')],\tag{47}$$

where the indices s, σ = ±1 label the four linearly-independent solutions of equations (45), (46):

$$\begin{gathered}(\check{u}(t),\check{u}^A(t))=(\check{u},\check{u}^A)\exp(i\varepsilon t),\\ (\check{u}^+(t),\check{u}^{+R}(t))=(\check{u}^+,\check{u}^{+R})\exp(-i\varepsilon t),\end{gathered}\tag{48}$$

normalized by the condition

$$\begin{gathered}\sum_\alpha u^{+R(A)}_{\alpha,s\sigma}u^{R(A)}_{\alpha,s'\sigma'}=\sigma\delta_{ss'}\delta_{\sigma\sigma'},\\ \sum_{s,\sigma}\sigma u^{+R(A)}_{\alpha,s\sigma}u^{R(A)}_{\beta,s\sigma}=\delta_{\alpha\beta}.\end{gathered}\tag{49}$$

The functions u and u^+ here are determined by the expressions [7]

$$u^{+}_{\alpha,s\sigma}=-\operatorname{th}\frac{\varepsilon}{2T}u^{+A}_{\alpha,s\sigma},\qquad u_{\alpha,s\sigma}=\operatorname{th}\frac{\varepsilon}{2T}u^{R}_{\alpha,s\sigma}.\tag{50}$$

By substituting (47) into (40) stationary equation (40) becomes an identity if the functions $\check{u}$ and $\check{u}^+$ are solutions of equations (45), (46). Normalization and orthogonality conditions (49), (50) make it possible to satisfy condition (43). Thus, formula (47) yields a representation of the solution of equation (40) in the stationary case. We may easily show that relation (47) is also valid in the general case of time-dependent fields. Indeed, substituting (47) into (40) reduces this equation to an identity. Normalization condition (43) is satisfied automatically since it is the first integral of equations (40), while when $t,\ t' \to -\infty$ its validity has already been demonstrated.

Relation (47) allows us to reduce the problem of solving nonstationary Gorkov equations for two time Green's functions in the Keldysh contour to a simpler problem: solving equations (45), (46) for simultaneous classical wave fields. Clearly the "extraneous" components of these fields may be eliminated as was done in Section 2, leaving only the equations for the u- and υ-functions. We note that we require the technique of incorporating the nonstationary interaction only for construction of the solution of (47). In the general case the Green's function may be searched in the form of (47) by selecting the solution of equations (45), (46) so that normalization condition (43) is satisfied.

5. Boundary conditions for the u-, υ-functions

As noted above in many cases it is necessary to deal with superconducting systems whose properties vary significantly over distances in the order of the atomic distances. Here we may describe the properties of such systems by means of quasi-classical equations; the existence of such an inhomogeneity may be accounted for by means of effective boundary conditions. Here we will consider the case where two superconductors are differentiated by a potential of the type $U(x) = V_0\delta(x)$. Such a potential may, for example, be used to simulate the existence of an oxide film between metals making contact. The boundary conditions for quasi-classical Eilenberger functions in the case of identical metals with such a potential are derived in study [4]. Moreover, we will assume that the Fermi velocity of the superconductor jumps at the point $x = 0$ as well:

$$v_F(x) = v_{F_-} + \theta(x)(v_{F_+} - v_{F_-}).$$

This more general case was considered in study [5]. However the boundary conditions obtained in study [5] for the Eilenberger functions have a very complex structure. At the same time the method developed in this study makes it possible to easily obtain the boundary conditions for the functions u_s, υ_s and u_s^+, υ_s^+, which obey equations (37), (38). We may easily obtain from equations (29), (30) ($p_{x\pm}^2 = m^2 v_{x\pm}^2 = p_{F\pm}^2 - p_\perp^2$):

$$\begin{aligned}
&u_{+1}(+0, v_{x+}) + u_{-1}(+0, v_{x+}) = u_{+1}(-0, v_{x-}) + u_{-1}(-0, v_{x-}),\\
&p_{x+}[u_{+1}(+0, v_{x+}) - u_{-1}(+0, v_{x+})] - p_{x-}[u_{+1}(-0, \dot{v}_{x-}) - u_{-1}(-0, v_{x-})] =\\
&\qquad = 2mV_0[u_{+1}(+0, v_{x+}) + u_{-1}(+0, v_{x+})], \quad p_{\perp\pm} \leqslant p_{F-};\\
&u_{+1}(+0, v_{x+}) = u_{-1}(+0, v_{x+}), \qquad p_{\perp+} > p_{F-}.
\end{aligned} \tag{51}$$

Analogous conditions exist for the functions υ, υ^+, u^+. As would be expected the existence of a potential barrier $V_0\delta(x)$ and a jump in υ_F in the system "mix up" the waves $u_{\pm 1}$, $\upsilon_{\pm 1}$. Thus, in order to describe the sharp boundaries between the superconductors it is not necessary to rely on cumbersome boundary conditions [5], but rather it is possible to use the significantly simpler conditions in (51). In addition we note that such conditions will be exactly valid for functions satisfying the nonstationary equations directly generalizing equations (37), (38) [7].

CONCLUSION

In this study we have derived equations (6) for the exact Green-Gorkov functions with coincident arguments. These equations are simpler than the Gorkov equations from study [6] and may be used to describe inhomogeneous superconductors when quasi-classical Eilenberger equations are not directly suitable. It is demonstrated that equations (6) may be written in the form of linear second order differential equations for the wave functions u, υ through which the Green-Gorkov functions with coincident arguments are expressed (formula (27)). The quasi-classical limit of these equations is examined. Moreover, the derived equations are generalized in order to describe the nonstationary and nonequilibrium states of inhomogeneous superconductors and it is demonstrated that the problem of finding the two time Green-Keldysh functions may be reduced to a significantly simpler problem: solving equations (45)-(46) for the one time functions u, υ. The equations derived here are used to find the effective boundary conditions on the functions u, υ for the case of superconductors with a sharp boundary in contact, which allows description of such superconductors within the scope of quasi-classical equations.

BIBLIOGRAPHY

1. Eilenberger, G. "Transformation of Gor'kov equation for type II superconductors into transport-like equations" ZTSCHR. PHYS., 1968, V. 214, No. 2, p. 195-213.

2. Kulik, I.O., Omel'yanchuk, A.N. "The Josephson effect in superconducting bridges: microscopic theory" FNT, 1978, V. 4, No. 3, p. 296-311.

3. Ivanov, Z.G., Kupriyanov, M.Yu., Pikharev, K.K., et al. "Boundary conditions for Eilenberger and Usadel equations and the properties of "dirty" *SNS*-sandwiches" FNT, 1981, V. 7, No. 5, p. 560-574.

4. Zaikin, A.D., Zharkov, G.F. "The influence of external fields and impurities on the Josephson current in *SNINS*-junctions" ZhETF, 1981, V. 81, No. 5(11), p. 1781-1802.

5. Zaitsev, A.V. "Quasi-classical equations of superconductivity theory for contact metals and the properties of constricted microjunctions" ZhETF, 1984, V. 86, No. 5, p. 1742-1757.

6. Abrikosov, A.A., Gor'kov, L.P., Dzyaloshinskiy, I.Ye. "Metody kvantovoy teorii polya v statisticheskoy fizike" [Quantum field theory techniques in statistical optics] Moscow: Fizmatgiz, 1962, 443 p.

7. Panyukov, S.V. "Transformation of the Eilenberger equations to a system of linear differential equations" PREPR. FIAN, No. 245, Moscow, 1983.

8. Eliashberg, G.M. "Inelastic electron collisions and nonequilibrium steady states in superconductors" ZhETF, 1971, V. 61, No. 3(9), p. 21254-1271.

9. Lifshits, Ye.M., Pitaevskiy, L.P. "Statisticheskaya fizika" [Statistical physics] Moscow: Nauka, 1978, Ch. 2, 448 p.

10. Usadel, K.D. "Generalized diffusion equation for superconducting alloys" PHYS. REV. LETT., 1970, V. 25, p. 507-508.

11. Bogolyubov, N.N. "Izbrannye trudy po statisticheskoy fizike" [Selected works on statistical physics] Moscow: Izd-vo MGU, 1979, p. 132-142.

12. Lapkin, A.I., Ovchinnikov, Yu.N. "The nonlinear conductivity of superconductors in a mixed state" ZhETF, 1975, V. 68, No. 5, p. 1915-1923.

13. Lapkin, A.I., Ovchinnikov, Yu.N. "Nonlinear effects from vortex motion in superconductors" ZhETF, 1977, V. 73, No. 1(7), p. 299-312.

14 Zaikin, A.D. "Equilibrium and nonequilibrium phenomena in inhomogeneous and weakly-coupled superconducting structures" This volume.

Quantum Tunneling with Dissipation

A.D. Zaikin, S.V. Panyukov

Abstract: We derive an exact solution to the problem of determining the rate of decay Γ of the metastable state of a quantum system when $T = 0$ with arbitrary dissipation. A potential of the type $V(q) = aq^2 + b(x - q)\Theta(1 - x)$ is considered; this is realized in quantum decay of current states of superconducting bridges with a normal interlayer. The expression for Γ differs significantly from the case $V(q) = q^2 - \beta q^3$ examined previously.

INTRODUCTION

Dissipation may have a significant influence on the probability of quantum tunneling. As demonstrated by Caldeira and Leggett [1,2] the influence of dissipation serves to reduce the decay rate of the metastable state. In these studies dissipation in the system was modeled by introducing interaction between a quasi-classical degree of freedom and a medium consisting of a large number of quantum harmonic oscillators. The Hamiltonian of the "particle plus medium" system will be written as

$$\hat{H} = \frac{m_0 \dot{q}^2}{2} + V_0(q) + \sum_k gqQ_k + \sum_k \hat{H}_k, \tag{1}$$

where q and Q_k are the coordinates of the particle and k-oscillator of the medium, respectively; m_0 and V_0 are the bare mass and potential; g is the interaction constant; H_k is the Hamiltonian of the k-oscillator. If we write a regular quantum mechanical formula for the transition probability as a functional integral and integrating with respect to the coordinates of the medium and assuming the frequency distribution of the oscillators in the medium is continuous, we obtain

$$W = \int \exp\{-S_{\rm eff}[q]\}\, Dq, \tag{2}$$

where the expression for the effective action S_{eff} takes the form [1 2]

$$S_{\rm eff}=\int d\tau\left\{\frac{m\dot{q}^2}{2}+V(q)+\frac{\eta}{4\pi}\int d\tau'\left(\frac{q(\tau)-q(\tau')}{\tau-\tau'}\right)^2\right\}. \qquad (3$$

In expressions (2), (3) we have converted to the Euclidean action b analytically extending the expression for S onto the imaginary tim axis.[1] This technique is convenient for calculating the transitio probability between states separated by a potential barrier (i.e., th tunneling probability). In the expression for action (3) the mass and potential $V(q)$ imply their renormalized values, while η is th coefficient of viscosity. Studies [1, 2] have calculated the deca rate Γ of the metastable state when $T = 0$ for the interesting physica case of $V(q) = \alpha q^2 - \beta q^2$, where the quantity Γ was calculated in bot weak and strong dissipative conditions. The temperature dependence o Γ was investigated by Larkin and Ovchinnikov [3-5].

In this study we will investigate another very important case:

$$V(q)=\frac{m\omega_0^2}{2}q^2+\frac{\Delta}{2}(\chi-q)\theta(q-\chi). \qquad (4$$

The potential $V(q)$ (4) is significantly nonquasi-classical near th point $q = \chi$. Such behavior of $V(q)$ is characteristic of the case o superconducting direct conductivity weak links [6]. As indicated b the expression for the action for superconducting junctions of thi type obtained by means of macroscopic theory [6], the dissipative ter in the general case is nonlinear with respect to q. However, with sufficiently high inelastic relaxation frequency in the superconduc tors and at a very near-critical current, the dissipative term become linear (see study [6] for details). Moreover, in considering tunne junctions [4, 5] the case of linear dissipation may be implemented b connecting a shunting resistor to the junction.

Therefore we will consider dissipation to be linear in our case The role of viscosity η is played by the effective resistance of th system $\eta = R_{eff}/e^2$, while the role of mass is played by the capaci tance of the junction $m = C/e^2$ (generally speaking the capacitance infers its renormalized value). The effective action with potentia $V(q)$ of the type in (4) is used to describe superconducting bridge with a normal metal interlayer (*SNS*-bridges) with a low impurity con centration. Here $\omega_0^2 = 4v_F/3RCd$, $\Delta = 2\pi m\omega_0^2$, d is the width of the N layer, v_F is the Fermi velocity, R is the resistance of the junctio in a normal state.

---

[1] At the final temperatures this technique, generally speaking is not a trivial technique, although when $T = 0$ (and it is this speci fic case that is considered in this study) it does not cause any dif ficulties.

It will be demonstrated that in this case it is possible to exactly calculate the value of Γ (both the exponent and the preexponential factor) with random dissipation. Here we are limiting our examination to the case $T \to 0$ (or more precisely the temperature is assumed to be significantly less than the characteristic instanton frequency, so that integration with respect to τ and τ' in (3) may be carried out within infinite limits).

The quantity Γ is represented as

$$\Gamma = A \exp\{-B\}. \tag{5}$$

We will consider the situation where Γ is sufficiently small ($B \gg 1$).

1. Calculation of the exponent

Functional integral (2) is determined by the trajectories $q(\tau)$ that originate from the point $q = 0$ when $\tau = -\infty$ and return to this point when $\tau \to +\infty$. We will designate as $(-\tau_-)$ and τ_+ the times when the trajectory passes through point χ, i.e., $q(\pm\tau_+) = \chi$. Then by performing a Fourier transform we may write action (3) as

$$S[q_\omega] = \int \frac{d\omega}{2\pi}\left[\frac{\Omega(\omega)}{2} q_\omega q_{-\omega} - \Delta \frac{e^{i\omega\tau_+} - e^{-i\omega\tau_-}}{2i\omega} q_\omega\right] + \\ + \frac{\Delta}{2}\chi(\tau_+ + \tau_-), \qquad \Omega(\omega) = m(\omega^2 + \omega_0^2) + \eta|\omega|, \tag{6}$$

where the quantities $\tau_\pm$ are determined by the relations

$$\int \frac{d\omega}{2\pi} e^{\pm i\omega\tau_\pm} q_\omega = \chi. \tag{7}$$

The quantity B in (5) is determined by the classical trajectory that is found from the minimum action condition (6). Calculating the first variational derivative of (6) we have

$$\delta S[q_\omega] = \int \frac{d\omega}{2\pi} \delta q_\omega \left[\Omega(\omega) q_{-\omega} - \Delta\left(\frac{e^{i\omega\tau_+} - e^{-i\omega\tau_-}}{2i\omega}\right)\right] + \\ + \frac{\Delta}{2}\left[\chi - \int \frac{d\omega}{2\pi} e^{i\omega\tau_+} q_\omega\right]\delta\tau_+ + \frac{\Delta}{2}\left[\chi - \int \frac{d\omega}{2\pi} e^{-i\omega\tau_-} q_\omega\right]\delta\tau_-. \tag{8}$$

Due to condition (7) the last two terms in (8) vanish. Equating (8) to zero we find $\tau_+ = \tau_- = \tau_0$ and

$$\tilde{q}_\omega = \Delta \frac{\sin\omega\tau_0}{\omega\Omega(\omega)}, \tag{9a}$$

$$B = S[\tilde{q}] = \Delta\chi\tau_0 - \Delta^2 \int_0^\infty \frac{d\omega}{2\pi} \frac{\sin^2\omega\tau_0}{\omega^2\Omega(\omega)}, \tag{9b}$$

where the quantity τ_0 is determined from the equation

$$\chi = \int_0^\infty \frac{d\omega}{2\pi} \frac{\Delta \sin 2\omega\tau_0}{\omega\Omega(\omega)}. \tag{10}$$

We note that according to (8) equation (10) may be obtained from maximum condition (9b) with respect to τ_0 if we formally assume that τ_0 is an independent variable. In the case $\varkappa = 2\pi m\omega_0^2\chi/\Delta \ll 1$ we find from (9) and (10):

$$B = \frac{\pi\chi^2 |m^2\omega_0^2 - \eta^2/4|^{1/2}}{2r};$$

$$r = \begin{cases} \frac{\pi}{2} - \operatorname{arctg}\frac{\eta}{\sqrt{4m^2\omega_0^2 - \eta^2}}, & \eta < 2m\omega_0, \\ \frac{1}{2}\ln\frac{\omega_+}{\omega_-}, & 2m\omega_0 < \eta \lesssim m\omega_0/\sqrt{\varkappa}; \end{cases}$$

$$B = \frac{\pi}{2}\chi^2\eta p(1 - p/2), \qquad m\omega_0 \lesssim \eta\sqrt{\varkappa}, \tag{11}$$

where p is determined from the equation $p(1 - C - \ln \varkappa p) = 1$, $C = 0.577$ is Euler's constant, $m\omega_\pm = \eta \pm (\eta^2 - 4m^2\omega_0^2)^{1/2}$.

2. The preexponential factor

The quantity A in (5) is determined by the trajectory bundle $q(\tau) = \tilde{q}(\tau) + \delta q(\tau)$ in the vicinity of classical trajectory (9a) found above. Expanding action (6) accurate to $\delta q(\tau)$-squared terms, we have

$$S[q(\tau)] = S[\tilde{q}(\tau)] + \frac{1}{2}\int \frac{d\omega\, d\omega'}{(2\pi)^2} G^{-1}(\omega, \omega')\,\delta q_\omega \delta q_{\omega'}, \tag{12a}$$

$$G^{-1}(\omega, \omega') = 2\pi\Omega(\omega)\,\delta(\omega - \omega') - \gamma\,[e^{i(\omega-\omega')\tau_0} + e^{-i(\omega-\omega')\tau_0}], \tag{12b}$$

where we used relations (7) in order to express the variation $\delta\tau_\pm$ through δq_ω:

$$\delta\tau_\pm = \int \frac{d\omega}{2\pi}\,\delta q_\omega e^{\pm i\omega\tau_0} \Big/ \int \frac{d\omega}{2\pi}\,\tilde{q}_\omega \omega \sin\omega\tau_0. \tag{13}$$

The quantity γ in (12b) is determined by the expression

$$\gamma = \gamma_0 = [\beta(0) - \beta(2\tau_0)]^{-1}, \qquad \beta(\tau) = \int \frac{d\omega}{2\pi}\,\frac{\cos\omega\tau}{\Omega(\omega)}. \tag{14}$$

In order to calculate the functional integral we will expand $\delta q(\tau)$ in terms of the eigenfunctions of the operator G^{-1} (12b):

$$\delta q(\tau) = c_0 q_0(\tau) + c_1 q_1(\tau) + \sum_{n \geqslant 2} c_n q_n(\tau), \tag{15a}$$

$$S[q(\tau)] = B + \frac{1}{2}\sum_{n=0}^{\infty} c_n^2\lambda_n, \qquad \int d\tau\, q_n(\tau)\, q_{n'}(\tau) = \delta_{nn'}. \tag{15b}$$

λ_n are the eigenvalues that represent solutions of the transcendental equation

$$1 = \gamma \int \frac{d\omega}{2\pi}\,\frac{1 \mp \cos\omega\tau_0}{\Omega(\omega) - \lambda_n}, \tag{16}$$

where the sign $\mp$ is selected for even and odd values of n, respectively. One of the eigenvalues is equal to zero ($\lambda_1 = 0$) and its corresponding eigenfunction

$$q_1(\tau) = \frac{\partial \tilde{q}(\tau)}{\partial \tau}\left[\int d\tau \left(\frac{\partial \tilde{q}}{\partial \tau}\right)^2\right]^{-1/2} \qquad (17)$$

corresponds to the translational mode. Indeed, when the position of the center of the instanton τ_c changes by $\delta\tau_c$ the change in trajectory is

$$\delta q(\tau) = \frac{\partial \tilde{q}}{\partial \tau}\delta\tau_c = q_1(\tau)\left[\int d\tau \left(\frac{\partial \tilde{q}}{\partial \tau}\right)^2\right]^{1/2}\delta\tau_c. \qquad (18)$$

Comparing (15a) and (18) we find (compare to study [7])

$$dc_1 = \left[\int d\tau \left(\frac{\partial \tilde{q}}{\partial \tau}\right)^2\right]^{1/2} d\tau_c. \qquad (19)$$

Proceeding with integration with respect to the coefficients c_n in (2), (15b) we find the formal expression for the single-instanton contribution to the quantity Γ:

$$\left[\prod_{n\neq 1}\lambda_n\right]^{-1/2}\left[\int d\tau\left(\frac{\partial \tilde{q}}{\partial \tau}\right)^2\right]^{1/2} e^{-B}\int_{-1/2 T}^{1/2 T} d\tau_c/\sqrt{2\pi}, T\to 0. \qquad (20)$$

In order to calculate (20) it is not necessary to solve the virtually unachievable problem of finding all eigenvalues of λ_n (16). In (12b) we will set $\gamma = \gamma_0 - \varepsilon$, Re $\varepsilon > 0$, then when $\varepsilon \to 0$ we have the obvious equality

$$A(\gamma)(\lambda_1^{(\varepsilon)})^{1/2} \to \left[\prod_{n\neq 1}\lambda_n\right]^{-1/2}, \qquad (21a)$$

$$A(\gamma) = \int D(\delta q)\exp\left[-\frac{1}{2}\int \frac{d\omega\, d\omega'}{(2\pi)^2} G^{-1}(\omega,\omega')\,\delta q_\omega \delta q_{-\omega}\right], \qquad (21b)$$

where $\lambda_1^{(\varepsilon)}$ is the eigenvalue of the operator G^{-1}, which vanishes when $\varepsilon \to 0$. We find from equation (16) the asymptotics for this value when $\varepsilon \to 0$:

$$\lambda_1^{(\varepsilon)} = \varepsilon\Delta^2[\beta(0) - \beta(2\tau_0)]^2/2\int d\tau\left(\frac{\partial \tilde{q}}{\partial \tau}\right)^2. \qquad (22)$$

In order to calculate the quantity $A(\gamma)$ (21b) we will differentiate $\ln A(\gamma)$ with respect to γ:

$$\frac{\partial \ln A(\gamma)}{\partial \gamma} = \frac{1}{2}\langle(\delta q(\tau_0))^2 + (\delta q(-\tau_0))^2\rangle = G(\tau_0, \tau_0), \qquad (23)$$

where the function $G(\tau, \tau')$ is inverse to (12b). When $\gamma < \gamma_1 = [\beta(0) + \beta(2\tau_1)]^{-1}$ it is found in explicit form:

$$G(\tau,\tau') = \beta(\tau-\tau') + \frac{\gamma}{2}\left\{\frac{[\beta(\tau+\tau_0)-\beta(\tau-\tau_0)][\beta(\tau'+\tau_0)-\beta(\tau'-\tau_0)]}{1-\gamma[\beta(0)-\beta(2\tau_0)]} + \frac{[\beta(\tau+\tau_0)+\beta(\tau-\tau_0)][\beta(\tau'+\tau_0)+\beta(\tau'-\tau_0)]}{1+\gamma[\beta(0)+\beta(2\tau_0)]}\right\}. \qquad (24)$$

By substituting (24) into (23) and integrating with respect to γ we obtain

$$A(\gamma) = A(0)\{[1-\gamma(\beta(0)+\beta(2\tau_0))][1-\gamma(\beta(0)-\beta(2\tau_0))]\}^{-1/2}. \qquad (25)$$

After substituting (25) into (21a) and performing analytic continuation of the function $A(\gamma)$ to $\gamma = \gamma_0$ (14) we have

$$A^{(1)} = A(0)[-i\tilde{\Gamma}/T], \tag{26a}$$

$$\tilde{\Gamma} = \frac{\Delta}{\sqrt{2\pi}} \frac{[\beta(0) - \beta(2\tau_0)]}{2[\beta(2\tau_0)]^{1/2}} e^{-B}. \tag{26b}$$

Summing the contributions of the n-instanton configurations to the quantity A, we find in the approximation of an ideal instanton gas

$$\sum_{n=0}^{\infty} A^{(n)} = e^{-F/T} = A(0) \sum_{n=0}^{\infty} \frac{1}{n!} \left(-\frac{i\tilde{\Gamma}}{T}\right)^n = A(0) \exp\left\{-\frac{i\tilde{\Gamma}}{T}\right\}, \tag{27}$$

where the factor $1/n!$ accounts for the identity of the instantons. As we know (see, for example, [7]) the imaginary part of the quantity F (27) determines the rate of decay of the metastable state of the quantal system when $T \to 0$, i.e., $\Gamma = \tilde{\Gamma}$. We note that in this equality in our case (unlike study [7]) there is no two factor in front of Γ, which is due to our method of analytic continuation of functional integral (2). Calculating the preexponential factor in the case $\varkappa \ll 1$, we find

$$A = \begin{cases} \dfrac{\pi\chi\sqrt{m}\,|\omega_0^2 - \eta^2/4m^2|^{3/4}}{2\sqrt{2}\,r^{3/2}}, & m\omega_0 \gtrsim \eta\sqrt{\varkappa}, \\ \dfrac{\Delta}{2\pi\sqrt{2\eta}} \dfrac{\ln(\eta^2 p\varkappa/m^2\omega_0^2) + C}{[\ln(1/p\varkappa) - C]^{1/2}}, & m\omega_0 \lesssim \eta\sqrt{\varkappa}. \end{cases} \tag{28}$$

We will now find the range of applicability of the derived results. We will first consider the single-instanton configuration. In finding expression (26) we ignored the contribution of trajectories that intersected the boundary $q(\tau_0) = \chi$ more than twice, as well as the change in time τ_0 of its intersection in higher orders in $\delta q(\tau)$. The necessary condition here is smallness of fluctuations of $\delta q(\tau)$ at the boundary $\tau = \tau_0$:

$$\xi \equiv \sqrt{\langle \delta' q(\tau_0))^2 \rangle}/\chi \ll 1. \tag{29}$$

In the remaining region far from $\tau = \tau_0$ the fluctuations are accounted for exactly, since action (3) is $q(\tau)$-quadratic. The function $\delta' q(\tau)$ in (29) is determined by the expression

$$\delta' q(\tau) = \sum_{n \neq 1} c_n q_n(\tau), \tag{30}$$

where we eliminated the contribution of the zeroth mode, since its position may be considered to be fixed in examining fluctuations of a single instanton. We will now find the correlator

$$\langle \delta' q(\tau)\, \delta' q(\tau') \rangle = \sum_{n \neq 1} q_n(\tau) q_n(\tau')/\lambda_n, \tag{31}$$

where we employed (30) and the normalization relation of the eigenfunctions of $q_n(\tau)$ (15b). Correlator (31) may be calculated by the limiting process

$$\langle \delta' q(\tau)\, \delta' q(\tau')\rangle = \lim_{\varepsilon\to 0} [G(\tau, \tau') - q_1(\tau) q_1(\tau')/\lambda_1]. \quad (32)$$

Using expressions (17) and (22) for $q_1(\tau)$ and λ_1 we find

$$\langle \delta' q(\tau)\, \delta' q(\tau')\rangle = \beta(\tau - \tau') - \frac{1}{\beta(2\tau_0)} \frac{\beta(\tau + \tau_0) + \beta(\tau - \tau_0)}{2} \frac{\beta(\tau' + \tau_0) + \beta(\tau' - \tau_0)}{2}. \quad (33)$$

If we set $\tau = \tau' = \tau_0$ in (33) then we finally obtain the smallness condition of fluctuations of $q(\tau)$ as

$$\xi = \frac{\beta(0) - \beta(2\tau_0)}{2\chi [\beta(2\tau_0)]^{1/2}} = \frac{\sqrt{2\pi}}{\Delta\chi} A \ll 1. \quad (34)$$

In fact the quantity ξ for the instantons plays the same role as the Ginzburg number in phase transition theory [8].

We will now consider multi-instanton configurations. In sum (27) all terms with the number $n > n_{xap} \sim \Gamma/T$ are insignificant, and hence the characteristic interval between the instantons is equal to $|\tau_i - \tau_j| \sim 1/Tn_{xap} \sim \Gamma^{-1}$. The approximation of the noninteracting instanton gas is valid when $|\tau_i - \tau_j| \gg \tau_0$ and this condition may be written as

$$\xi B \exp\{-B\} \ll 1, \quad (35)$$

where we used the expression for ξ (34) and the estimate $B \sim \Delta\chi\tau_0$ that follows from (9b). Thus, when $\xi \ll 1$ we may ignore the interaction between the various instantons, and condition (34) determines the range of applicability of our results.

CONCLUSION

Thus we have found the exact solution to the problem of determining the tunneling probability with dissipation with an arbitrary relation between the parameters of the system. With small values of $\varkappa = 2\pi m\omega_0^2\chi/\Delta$ it is expressed through elementary functions. The preexponential factor is significantly nonquasi-classical which is related to the nonquasi-classical nature of the effective potential, and its dependence on the coefficient of viscosity differs from that found in studies [2, 5]. The quantity B is determined by functional integral (12a), which formally diverges due to the existence of a mode with a negative eigenvalue $\lambda_0 < 0$. In order to calculate this value this chapter has proposed using the analytic continuation method based on the interaction constant γ from the region in which the functional integral is well-defined. Within the scope of this approach the problem of calculating the eigenvalues is reduced to the simplest problem: finding the correlation function of fluctuations in the classical trajectory. This problem is solved exactly for the case of the effective action taking the form of (3), (4) and describing the behavior of superconducting bridges with a normal metal interlayer in certain conditions [6].

BIBLIOGRAPHY

1. Caldeira, A.O., Leggett, A.J. "Influence of dissipation on quantum tunneling in macroscopic systems" PHYS. REV. LETT., 1981, V. 46, p. 211-214.

2. Caldeira, A.O., Leggett, A.J. "Quantum tunneling in a dissipative system" ANN. PHYS. (US), 1983, V. 153, p. 374-456; 1984, V. 153, p. 445.

3. Lapkin. A.I., Ovchinnikov, Yu.N. "Quantum tunneling with dissipation" PIS'MA V ZhETF, 1983, V. 37, p. 322-325.

4. Lapkin, A.I., Ovchinnikov, Yu.N. "Superconducting current damping in tunnel junctions" ZhETF, 1983, V. 85, p. 1510-1520.

5. Lapkin, A.I., Ovchinnikov, Yu.N. "Quantum mechanical tunneling with dissipation. The preexponential factor" ZhETF, 1984, V. 86, p. 719-726.

6. Zaikin, A.D., Panyukov, S.V. "The quantum decay of metastable current states in superconducting junctions" ZhETF, 1985, V. 89, No. 1.

7. Callan, C., Coleman, S. "Fate of the false vacuum: First quantum corrections" PHYS. REV. D, 1977, V. 16, p. 1762-1768.

8. Patashinskiy, A.Z., Pokrovskiy, V.L. "Fluktuatsionnaya teoriya fazovykh perekhodov" [Fluctuation theory of phase transitions] Moscow: Nauka, 1982, 382 p.

The Nonequilibrium Properties of Superconductors Under Optical Excitation and Current Tunnel Injection

K.V. Mitsen

Abstract: The properties of nonequilibrium superconductors under excitation by optical pumping and current tunnel injection are investigated experimentally. With high intensities of the excitation source we discover a nonthermal transition of the superconducting films to a spatially-inhomogeneous state. The transition conditions are investigated and it is demonstrated that the experimental results are consistent with the model of "coherent" instability. Techniques are proposed for determining the characteristics of superconductors based on results from optical pumping experiments. Superconductivity enhancement in tin under current injection in quasi particle tunneling related to the increase in the average quasi particle energy is discovered. The causes of a subharmonic gap structure in the E-I characteristics of superconducting tunnel junctions are examined.

INTRODUCTION

Today superconductivity is one of the most important and rapidly developing fields of solid state physics. Due to the extensive progress in theoretical and experimental research the phenomenon of superconductivity is already finding broad applications in various fields of science and technology. Research on equilibrium superconductors has made it possible to obtain a clear picture of the processes of current in these devices and to achieve comparatively high critical parameters.

The further development of superconductivity has focused research on the properties of nonequilibrium superconductors. The importance of this issue derives from the fact that nonequilibrium systems may prove to be the most promising for improving superconducting characteristics [1]. Moreover, the investigation of the behavior of superconductors in a nonequilibrium state largely determines progress in the development and applications of various superconducting quantum devices and instruments. From the theoretical viewpoint the problem of nonequilibrium superconductivity is also interesting in that the superconductor is a very convenient system for testing theoretical ideas and concepts with regard to the behavior of many bodied systems driven from equilibrium. One indicator of the significant interest in nonequilibrium superconductors is the increasing volume of studies devoted to this issue [2-4].

By the time this study was begun experimental investigations of the properties of superconductors in a nonequilibrium state were limited primarily to the investigation of the nonequilibrium spectrum of phonons emitted by a superconductor under excitation [5-7] and investigations of enhancement of superconductivity by microwave irradiation [8-11]. Virtually all these studies were conducted at low intensities of the external excitation source.

At the same time as early as study [12] which investigated the properties of superconducting lead films under optical pumping, it was demonstrated that at high laser intensities a qualitatively new phenomenon could be observed: the nonthermal breakdown of the superconducting state by a high frequency electromagnetic field. However, the nature of this phenomenon and the conditions required for its generation remained unclear. The behavior of superconductors under excitation by current tunnel injection were also little studied.

The purpose of this chapter is to perform an experimental investigation of the properties of superconductors under various excitation conditions using laser pumping and current tunnel injection to create the nonequilibrium state and to explain the nature of the transition of the superconductor to a resistive state that occurs at high pumping levels.

The first chapter provides results from an experimental investigation of the nonequilibrium properties of lead and tin films under low intensity optical pumping. The change in the gap is determined on the basis of measurements of the E-I characteristics of the pumped tunnel junctions as a function of emission power and temperature, and the localization region of the nonequilibrium addition to the distribution function is determined. The derived data are used to determine the quasi particle recombination time in lead.

In the second chapter we give results from an investigation of the properties of superconductors under high intensity optical pumping. Here we focused on the investigation of a new phenomenon discovered during this research: the transition of the superconductor to a spatially-inhomogeneous state when exceeding the critical quasi-particle concentration. This chapter gives a discussion of the derived results and compares them to existing theoretical models. A kinetic picture of the phenomenon is considered.

The third chapter is devoted to an investigation of the nonequilibrium properties of superconductors under excitation by current injection through a tunnel barrier. The measurements were carried out on tin films at voltages across the tunnel junction corresponding to two different states: quasi-particle tunneling and quasi-particle injection. In the first case an increase in the energy gap was observed indicating enhancement of superconductivity by current injection in this state. This phenomenon was first observed in tin. Under pumping in the quasi-particle injection state, when a certain critical value of the injection current is exceeded, characteristic features were observed in the E-I characteristics of the tunnel junctions. We related their appearance to transition of the tin film to a spatially-inhomogeneous state upon reaching the critical quasi-particle concentration analogous to the case of intensive optical pumping. A discussion of the derived results is given together with a comparison to theory.

Chapter 1

THE NONEQUILIBRIUM PROPERTIES OF SUPERCONDUCTORS UNDER LOW INTENSITY OPTICAL PUMPING

1. Experimental methodology

Low intensity optical pumping will infer a dimensionless concentration of excess quasi particles $n \ll 1$. Here $n \equiv (N-N_T)/4\Delta_0 N(0)$; N is the total concentration of quasi particles in a nonequilibrium superconductor; N_T is the equilibrium concentration of quasi particles at temperature T; Δ_0 is the equilibrium value of the gap; $N(0)$ is the state density on the Fermi surface.

In order to measure small deviations of the superconducting parameters from their equilibrium values at a low pumping level, we employed the tunnel technique. The E-I characteristic of a superconducting tunnel junction, as we know, is the integral of the product of the state densities and the corresponding distributions functions of quasi particles in the superconductor. Measuring the E-I characteristics under quasi-particle injection action makes it possible in principle to determine the nonequilibrium addition to the quasi-particle distribution function, as well as to measure the superconducting gap Δ and a number of other quantities determining the kinetics of the relaxation processes.

We will consider the tunnel junction S-I-S (superconductor-insulator-superconductor) biased to a voltage of $V > 2\Delta/e$. In this case the quasi particles will be injected into each of the wafers on the level with energy E lying in the interval $\Delta < E < eV - \Delta$, thereby changing the distribution function $n(E)$ in the injection region. By illuminating one of the wafers (i.e., changing the $n(E)$) and by comparing the E-I characteristics of the illuminated and unilluminated junction, we may determine the change in $n(E)$ in the interval $\Delta < E < eV - \Delta$. By varying the voltage across the junction we may obtain information on the nonequilibrium addition to the distribution function in the energy range of quasi particles of interest to us.

We may determine the change in the superconducting gap $\delta\Delta$ under optical pumping action by measuring the change in voltage δV that occurs in this case across the tunnel junction with a fixed bias corresponding to the position of the operating point at the half amplitude level of the rapidly-increasing part of the E-I characteristic at a voltage of $V \cong 2\Delta/e$. In this case assuming that both films are perturbed, we may set $\Delta < E \ll eV - \Delta$. The quantity $\delta\Delta$ at a fixed temperature T, obviously, will be determined by the optical pumping power as well as the effective recombination time of the quasi particles τ_R^{eff} that determines their excess concentration. We may change the quantity τ_R^{eff} (and, consequently, $\delta\Delta$) by varying the temperature. By analyzing the $\delta\Delta(T)$ relation we may determine τ_R^{eff} as a function of T.

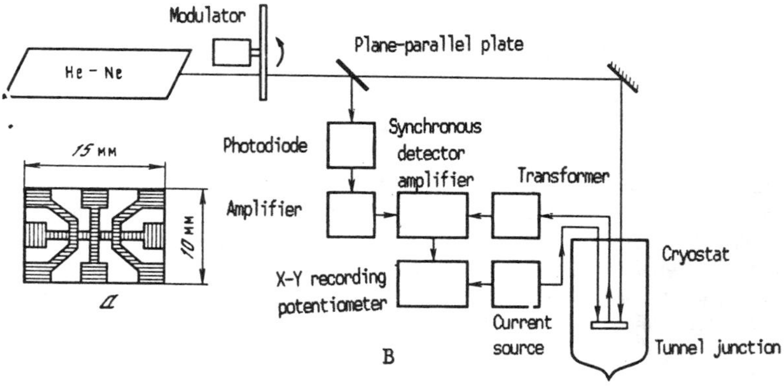

Fig. 1. Geometry of samples (a) and block diagram of the configuration for measuring the conductivity of the tunnel junction under laser irradiation (b)
The shaded area indicates the deposited metallic layers

In the experiment we measured the change δV in voltage across the S-I-S tunnel junction caused by the irradiation of one of the junction wafers. The geometry of the samples is shown in Fig. 1a.

A block diagram of the measurement configuration is shown in Fig. 1b. A LG-75 25 mw He-Ne-laser was used as the pumping source. A series of neutral filters were used to attenuate the emission level the laser beam was modulated by mechanical disk modulator M and was directed through a window to a cryostat on the tunnel junction submerged in liquid helium. A bias current generated by a linear current source was passed through the junction. An a.c. signal with an amplitude equal to the change in voltage across the junction caused by the incident emission was fed through the input transformer to the input of the synchronous detector amplifier. The rectified output signal was taken from the amplifier and injected to the Y input of the X-Y recording potentiometer (PDP-4). A voltage from the current source proportional to the bias current was injected to the X input. A portion of the emission reflected off of the plane-parallel plate impacted photodiode FD-1 and the signal from the photodiode, amplified by broad band amplifier UZ-12, was used as the reference signal for USD-1.

All measurements were carried out at temperatures below λ: the helium point. This made it possible to eliminate effects associated with the laser heating of the sample. Estimates of film heating (Pb) in He II and the contribution of the thermal mechanism to the gap change at an irradiation power of W = 0.2 mw/cm^2 yield $\varepsilon T_H \sim 0.4 \times 10^{-4}$ K and $\delta\Delta_H$ = ~0.2 nv, respectively. The error in determining the absolute value of the maximum power was 10% and was determined primarily by the measurement error of the IMO-2 optical power meter. For the lowest power levels the possible error increased to 15%. When measuring $\delta\Delta$ the error was 30% for a power of 0.22 mw/cm^2 and ~5% for higher power levels. The temperature was measured by the helium vapor pressure accurate to ±0.01 K.

2. Investigation of the quasi-particle distribution function for the case of optical pumping

Fig. 2 (curve 1) gives the change in voltage δV plotted as a function of the bias current I_B obtained for a Pb-PbO$_x$-Pb tunnel junction. The tunnel resistance of the junction was 0.01 ohms with an area of 0.4 × 0.4 mm^2. The E-I characteristic of the junction is also shown (curve 2). The emission power is 15 mw/cm^2. As is clear from Fig. 2 in the current range corresponding to a voltage of $V < 2\Delta/e$ across the junction, the addition to the voltage is $\delta V < 0$ which corresponds to an increase in the tunnel conductivity of the junction under laser emission action at this bias current. In the range $V > 2\Delta/e$ the value of δV becomes positive, i.e., under optical pumping the conductivity of the junction drops in this range. Here (when $V > 2\Delta/e$) two sections should be identified: a deep and comparatively narrow minimum in δV at $V \gtrsim 2\Delta/e$ and then when $V >$ 6 mv a broad range of voltages where $\delta V > 0$. The value of δV at the minimum increases with an increase in pumping intensity.

The results for Sn-SnO$_x$-Sn junctions are shown in Fig. 3. The tunnel resistance is $r_T \cong 0.2$ ohms with a junction area of 0.4 × 0.4 mm^2. The power level of incident emission was 15 mw/cm^2. As we see from the diagram, unlike the results for Pb, for Sn in a voltage range of $V \gtrsim 2\Delta/e$ the voltage variation $\delta V < 0$ although in this case there is a local minimum $|\delta V|$. When $V >$ 5.2 mv, $\delta V > 0$.

The derived results are consistent with the results from study [13]. An increase in junction conductivity when $V < 2\,\Delta/e$ is explained by an increase in the quasi-particle concentration and a reduction in the gap under bombardment. The range of voltages $V \gtrsim 2\Delta/e$ is of special interest; the minimum in junction conductivity under irradiation is observed here. This primarily reveals the nonequilibrium nature of the quasi-particle distribution function under optical pumping, i.e., the emission action does not cause simple heating. In the latter case we would have a monotonic drop in $|\delta V|$ across the entire range of voltages. The observed minimum in the conductivity is due to the reduction in tunneling probability when $V \geqslant 2\Delta/e$ in the illuminated superconductor due to Pauli's principle (the so-called blocking effect [13, 14]). Here the reduction in conductivity due to an increase in the distribution function near the gap edge is not compensated by an increase in conductivity due to the reduction in the gap and the increase in concentration.

The reason for the reduction in conductivity of the junctions at higher voltages corresponding to the range of maxima of the phonon density of states of Pb and Sn is not entirely clear. Study [15] has calculated the nonequilibrium addition to the current in a tunnel junction under optical pumping related to the peaks in the phonon state density. However, the value of the predicted effect is two to three orders of magnitude smaller than the observed level.

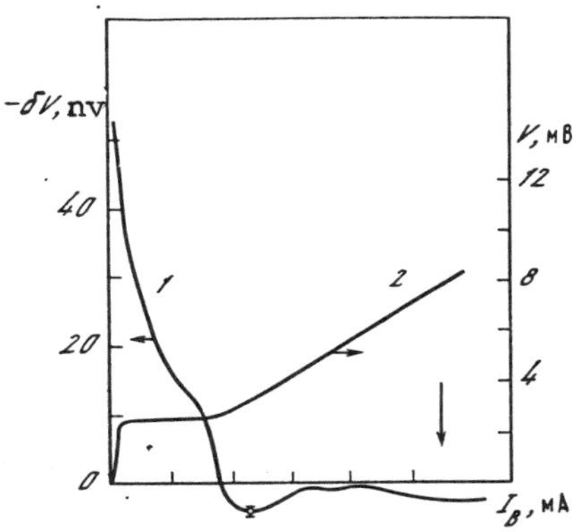

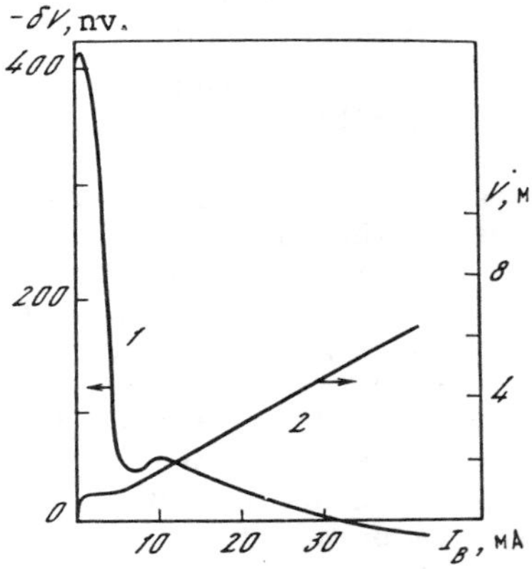

Fig. 2. The change in voltage δV across a Pb-PbO_x-Pb tunnel junction under laser irradiation as a function of bias current I_B (1) and the E-I characteristic of the corresponding tunnel junction (2)
The segment in the lower section of the diagram indicates measurement error. The arrow in the diagram gives the position of the phonon peak of lead at a voltage of $V = \hbar\omega_{ph} + 2\Delta = 7.5$ mv. The incident emission power is 15 mw/cm^2, $T = 2$ K

Fig. 3. The change in voltage δV across a Sn-SnO_x-Sn tunnel junction under laser irradiation plotted as a function of bias current I_B (1) and the E-I characteristic of the corresponding tunnel junction (2)
The phonon frequency ranges of tin correspond to a voltage range of $\hbar\omega_{ph} + 2\Delta = 5$-$20$ mv. The incident emission power is 15 mw/cm^2, $T = 2$ K

The derived results make it possible to determine the range of localization of the nonequilibrium addition δn to the distribution function. The localization interval of the principal part of δn may be considered to be approximately equal to the width of the minimum of $|\delta V|$ when $V \geqslant 2\Delta/e$. For the test Pb and Sn films δn is localized in both cases in an energy range of width $\sim\Delta/e$ over the gap (1.6 mev for

Pb and 0.6 mev for Sn). It was not possible to determine the precise form of δn based on available data. As demonstrated the observed minimum will exist for a variety of nonequilibrium distribution functions [14, 16].

3. Measurements of the dependence of the energy gap on optical pumping intensity

Measurements of the dependence of the change in the superconducting gap under optical pumping for Pb and Sn are given for a temperature range of 1.45 to 2.17 K in a power range of $W = 10^{-4}$–$2 \cdot 10^{-2}$ w/cm^2. Fig. 4 gives an experimental plotting of $\delta\Delta$ against W for Pb obtained for T = 1.55 K. The irradiation power density corresponding to each unit of relative power is 15 mw/cm^2. The bias current I_B was 1.3 ma. The area of the junction was 0.4 × 0.4 mm^2 with a tunnel resistance of R_T = 1.1 ohms. The film thickness is 1200 Å. The value of $\delta\Delta$ plotted in Fig. 4 corresponds to a reduction in the gap. Theory [17, 18] provides a relationship between the change in the gap and the concentration of the excess quasi particles. For small n we have

$$\delta\Delta/\Delta_0 = -2n. \tag{1.1}$$

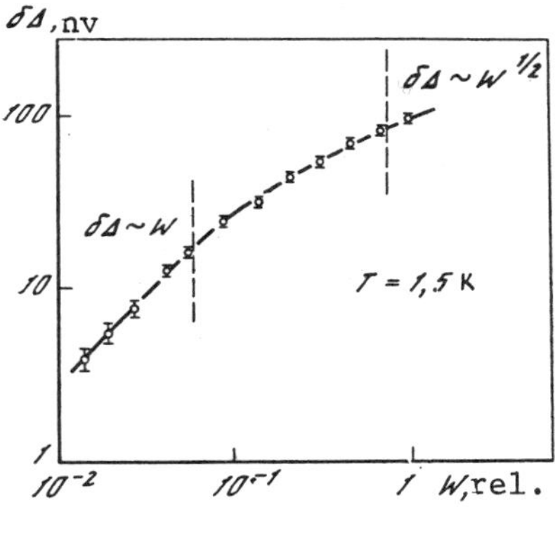

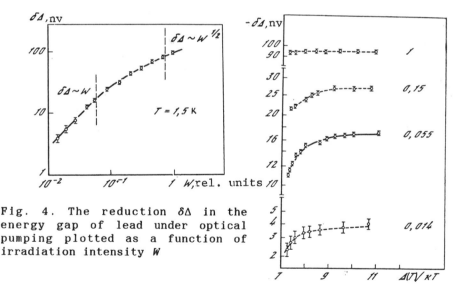

Fig. 4. The reduction $\delta\Delta$ in the energy gap of lead under optical pumping plotted as a function of irradiation intensity W

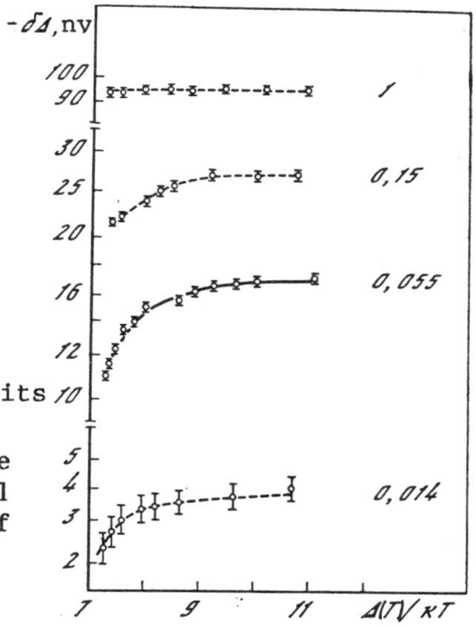

Fig. 5. The reduction $\delta\Delta$ in the energy gap of tin under optical pumping plotted as a function of $\Delta(T)/kT$ for various irradiation intensities

The experimental values are represented by the dots. The solid line curve is calculated by (1.2). The numbers to the right of the curves indicate the transmittances of the optical wedges.

The excess concentration n that occurs in the sample is determined by the irradiation power W and the effective recombination time of the quasi particles τ_R^{eff} which depends on the total concentration. If n is small we would properly anticipate a linear increase in the value of $|\delta\Delta|$ with an increase in W. The deviation in $\delta\Delta(W)$ from a linear law will indicate that the excess concentration is rather high and will correspond to "superinjection." It is clear from the diagram that at low intensities the gap indeed obeys a linear law with pumping intensity which indicates smallness of excess concentration ΔN_L created by optical pumping compared to a stationary concentration of quasi particles N. The latter is determined by the sum of temperature excitations and injection excitations generated from the flow of the tunnel bias current I_B. In this state, recombination of the light-generated excess quasi particles occurs primarily from stationary excitations and the recombination time is determined solely by N. As the power increases there is a gradual conversion to "superinjection" where $\Delta N_L \gg N$. In this case the recombination time is determined only by the value of ΔN_L. In the experimental curve this is manifest as a transformation from a linear relation of $\varepsilon\Delta \propto W$ to $\delta\Delta \propto W^{1/2}$.

We will now consider the temperature dependence of the change in the gap under optical pumping. Fig. 5 gives the $\delta\Delta(\Delta/kT)$ relations for Pb for four different pumping intensities that correspond to irradiation densities of 15, 2.25, 0.892 and 0.21 mw/cm^2. The temperature measurement range of 2.17–1.45 K corresponds to a change in $\Delta(T)/kT$ from 7.2 to 11. It is clear that as the emission intensity increases the temperature dependence of $\delta\Delta$ becomes weaker and for W = 15 mw/cm^2 $\delta\Delta$ is virtually constant. For low intensities an increase in $|\delta\Delta|$ with a reduction in temperature at high T is observed, together with saturation ensuing with a further reduction in temperature. We should note that such behavior cannot be explained by the heating of the film by the laser emission, since in the latter case the value of $\delta\Delta$ would drop with a reduction in temperature due to the fact that the gap would become less dependent on temperature. Consequently, in this case we are dealing with a nonthermal effect.

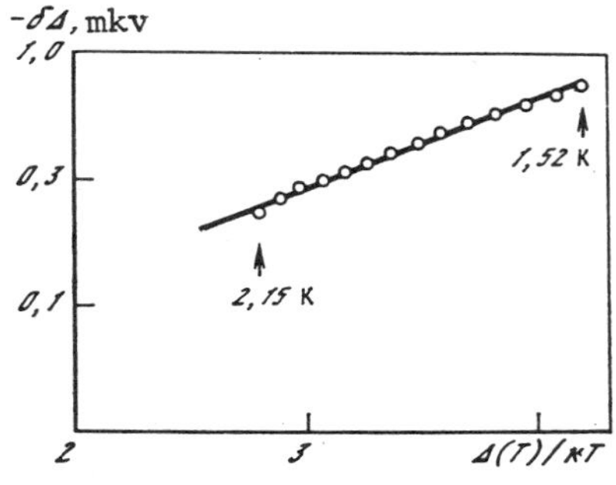

Fig. 6. The change in the gap $\delta\Delta$ in Sn films under laser irradiation plotted as a function of the $\Delta(T)/kT$ ratio
Emission power of 22 mw/cm^2

In order to understand the causes of the observed behavior we will divide the quasi particles in the film into three arbitrary groups: optically excited quasi particles (their concentration is ΔN_L), injected quasi particles (concentration ΔN_B) and thermal quasi particles (their concentration is N_T). We note that the separation process is arbitrary, since any change in the concentration of one group of quasi particles will, generally speaking, change the concentration of the other groups, since the recombination time is determined by the total concentration. We will consider the various cases.

1. $N_T \gg \Delta N_L$ and $N_T \gg \Delta N_B$. Here $N \sim N_T e^{-\Delta/kT}$ and the recombination time is determined only by N_T and will vary with temperature as $\tau_R^{eff} = \tau_0 T^{-1/2} e^{\Delta/kT}$. Here $\delta\Delta \backsim n \backsim \tau_R^{eff}$ and in our logarithmic coordinates the $\delta\Delta$ (Δ/kT) relation will be approximately linear.

2. $\Delta N_L \gg N_T$, $\Delta N_L \gg \Delta N_B$. In this case the recombination time depends only on ΔN_L and is independent of temperature. This corresponds to $\delta\Delta$ = const.

3. $\Delta N_B \gg \Delta N_L$, $\Delta N_B \gg N_T$. Here τ_R^{eff} is determined by ΔN_B and is also independent of temperature. Consequently, $\delta\Delta$ = const.

We will return to Fig. 4. At low power levels the $\delta\Delta(W)$ relation is linear. This means that for these values of W, $\Delta N_L \ll N$ $(=\Delta N_B + N_T)$ at $T = 1.5$ K $(\Delta/kT \simeq 10)$.

Consequently, in the range of saturation of $\delta\Delta(\Delta/kT)$, where $\delta\Delta$ = const (for low power levels!), we have case 3, i.e., recombination of the light-excited quasi particles occurs with the injected excitations. With an increase in temperature N_T increases and when N_T exceeds ΔN_B, recombination begins to occur with temperature excitations (case 1).

For a maximum intensity of 15 mw/cm^2, as indicated from Fig. 5, $\Delta N_L \gg N(=\Delta N_B + N_T)$, and consequently, we have case 2. In this case the light-excited quasi particles recombine primarily with one another. In a power range of 15-0.8 mw/cm^2 the intermediate case is valid.

Fig. 6 gives the $\delta\Delta(\Delta/kT)$ relation for Sn. The area of the tunnel junction is 0.4×0.4 mm^2 with a tunnel resistance of R_T = 0.16 ohms. The emission power is 22 mw/cm^2. In the case of Sn the temperature measurement interval corresponds to a change in Δ/kT from 3 to 5. It is clear from the figure that $\delta\Delta$ increases with a reduction in temperature linearly (in logarithmic coordinates) across the entire temperature range which corresponds to case 1. This is due to the fact that for Sn at these temperatures due to the small value of the gap N_T is large and recombination of the light excitations occurs with the thermal quasi particles.

Analogous measurements were carried out previously for Sn and Pb in studies [14, 19, 20]. However in these studies the experimental data were analyzed ignoring the contribution of injected quasi particles. Although, as is clear from our results, the injected quasi particles in certain conditions (with low pumping and low temperatures) play a determinant role in the recombination process of excess quasi particles.

4. Determining quasi particle recombination time

The experimental results given in the preceding section may be used to determine τ_R^{eff} in Pb. For this we will use the method of processing the experimental relations $\delta\Delta(\Delta/kT)$ developed in studies [21, 22].

We will write the stationary solution of the Rothwarf-Taylor equations [23] for the excess quasi-particle concentration n:

$$n = \frac{N_T}{4N(0)\,\Delta_0}\left[\left(1+\frac{I_0\tau_R^{eff}}{N_T}\right)^{1/2}-1\right].$$

The rate of quasi-particle creation is $I_0 = 2I_B/ev + I_L$ where v is the volume of the tunnel junction, while I_L is the optical excitation volume velocity. The multiplier 2 is due to the fact that in tunneling two excitations with energy Δ are created. If $I_L \ll 2I_B/ev$ and I_L are amplitude-modulated, the corresponding gap modulation will be equal to

$$\delta\Delta = \tau_R^{eff}\left(1+\frac{2I_B\tau_R^{eff}}{evN_T}\right)^{-1/2}\frac{I_L}{4N(0)}\,.$$

Here equation (1.1) is used. At low temperatures

$$\tau_R^{eff} = \tau_0 T^{-1/2} e^{\Delta/kT},$$

where τ_0 is the dimensional numerical factor and

$$N_T = 4N(0)\,(\pi\Delta k/2)^{1/2}T^{1/2}e^{-\Delta/kT},$$

from here we have

$$\delta\Delta = T^{-1/2}e^{\Delta/kT}\,(1+AT^{-1}e^{2\Delta/kT})^{-1/2}\frac{\tau_0 I_L}{4N(0)}\,, \qquad (1.2)$$

where

$$A = \frac{\tau_0 I_B}{2evN(0)\left(\frac{1}{2}\pi\Delta k\right)^{1/2}} = \frac{\tau_0 I_B}{3{,}03v}\,.$$

Here $N(0) = 2.2\cdot10^{22}$ ev^{-1} cm^{-3}, I_B is measured in amperes and v is measured in cubic centimeters.

After selecting A to best match relation (1.2) to the experimental relation $\delta\Delta(\Delta/kT)$ we may determine τ_0.

For processing we may use the relations $\delta\Delta(\Delta/kT)$ for W = 0.82 and W = 0.21 mw/cm^2, since, as demonstrated above, the condition $I_L \ll 2I_B/ev$ is satisfied in these cases only. We will use the first curve since for W = 0.21 mw/cm^2 due to the low signal levels the measurement error is significant. In Fig. 5 the dots represent the experimental values of $\delta\Delta$. The solid line curve for W = 0.82 mw/cm^2 gives the

$\delta\Delta(\Delta/kT)$ relation obtained by (1.2) for $A = 2.5\cdot10^{-5}$ and the adjustment scale factor

$$C = \frac{\tau_0 I_L}{4N(0)} = 1{,}56\cdot10^{-11}.$$

The value determined in this manner is $\tau_0 = 0.85\cdot10^{-10}$ s$\cdot$K$^{1/2}$ (±15%). Thus, the effective recombination time is $\tau_R^{eff}(T) = 0.85\cdot10^{-10}T^{-1/2}e^{\Delta/kT}$ s.

The derived value of τ_0 is consistent with the value $\tau_0 = 1.5\cdot10^{-10}$ s$\cdot$K$^{1/2}$ determined by an analogous technique in study [21]. The agreement should be considered even better since study [21] utilized Pb layers with a total thickness of 3542 Å which is 1.5 times greater than the thickness of the films used in this study (2400 Å) and this means that the corresponding phonon capture factor $(1 + \tau_{es}/\tau_B) \sim \tau_{es}/\tau_B$ is also 1.5 times higher than in our case. Thus, adjusting the results from study [21] for a thickness of 2400 Å we obtain $\tau_0 = 1\cdot10^{-10}$ s$\cdot$K$^{1/2}$ which is in agreement with our result accounting for experimental error.

Chapter 2

THE SPATIALLY INHOMOGENEOUS STATE IN SUPERCONDUCTORS UNDER OPTICAL EXCITATION

1. Experimental technique

In investigating the behavior of superconductors under high intensity optical pumping we employed Pb, *NS*-boundary, Nb_3Sn and V_3Si films. A four-probe pulsed measurement technique was used. A test d.c. current was passed through the sample (a superconducting film). The sample was driven to a resistive state by a laser pulse of specific power; this state was indicated by a drop in voltage across the sample. In this study we measure the parameters of the resulting voltage pulse as a function of the optical pumping power W, temperature T, laser pulse duration τ, and film width d.

For our investigations in the 0.2-10 μs range our emission source was an injection GaAs laser operating at 0.85 μm with an output power of 60 w/pulse (Fig. 7). A power amplifier was used to provide power to the laser; voltage pulses from master oscillator G5-15 were fed to the input of the power amplifier. The level and waveform of the laser pulse were recorded by means of a Ge-Au photocell. The sample, the laser and the photocell were housed in a cryostat. Due to the significant beam divergence (approximately 30°) and the impossibility of placing the laser in immediate proximity to the sample due to spurious emission, a lightguide was used; a quartz rod 80 mm in length and 4 mm in diameter (with a light spot size of 12.5 mm^2) was used as the lightguide. The configuration of elements on the Dewar vessel attachment is shown in Fig. 8.

The photocell signal and the voltage pulse that arises in the sample from laser pulse action are injected to the input of two-channel stroboscopic oscillograph S7-8 with an analog signal output to the PDS-021 X-Y recorder. When necessary the signal from the sample was preamplified by amplifier UZ-7A. This scheme made it possible to record on the recorder both the waveform of the pulses and the dependence of signal level from the sample at any time with respect to the beginning of the laser pulse on the temperature and emission power.

A neodymium glass laser operating at 0.16 μm (Fig. 9) was used for the investigations for a laser pulse duration from 0.01 to 0.1 μs. The emission intensity was altered by means of a set of neutral filters.

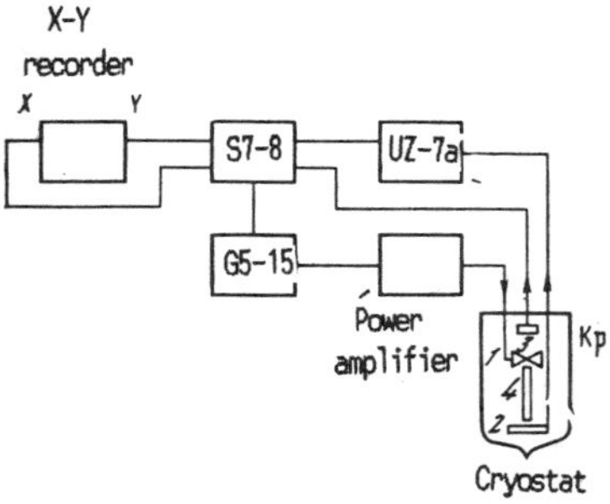

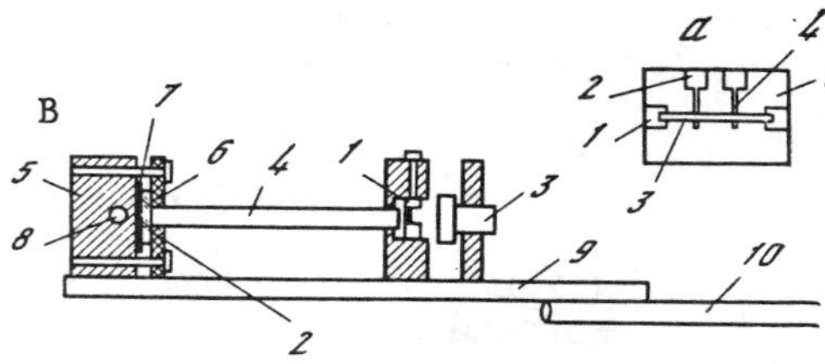

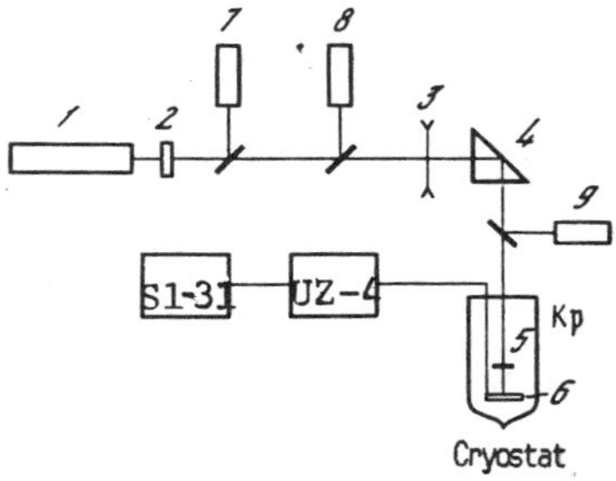

Fig. 7. Block diagram of the assembly for measuring the characteristics of nonequilibrium superconductors under high level optical pumping with a pulse duration from 0.2-10 μs
1 - injection laser; 2 - sample; 3 - photocell; 4 - lightguide

Fig. 8. Geometry of sample (a) and element configuration on the Dewar vessel assembly (b)
a: 1 - current junctions; 2,4- potential junctions; 3 - test metallic strip; 5 - substrate; b: 1- injection laser; 2 - sample; 3 - photocell; 4 - lightguide; 5 - copper unit; 6 - textolite pressure plate; 7 - indium interlayer; 8 - thermometer; 9 - acrylic plastic plate; 10 - piping for lead wiring

Fig. 9. Block diagram of assembly for measuring the characteristics of nonequilibrium superconductors under high level optical pumping with a pulse duration of 0.01-0.1 μs
1 - laser; 2 - filter set; 3 - diverging lens; 4 - rotating prism; 5- diverging plate; 6 - sample; 7 - power meter; 8 - FEK-0.9; 9 - LFD-1

The laser emission was first passed through a scattering lens (f = 150 mm) and a rotatable prism was used to direct the emission through a glass window to the cryostat, and after passing through a frosted scattering plate mounted 40 mm from the support, the emission impacted the sample. The voltage that arises across the sample under laser pulse action was injected to the input of broadband oscillograph S1-31. When necessary the signal was preamplified by means of amplifier UZ-4.

The pulse energy was measured by the OKG-IMO-2 average power and energy meter. In order to measure the pulse duration the signal was taken from a wideband photodiode (bandwidth of 5 GHz and sensitivity of 0.15 amps/w) and was injected through a delay line (~350 ns) to broadband oscillograph S1-31. The oscillograph sweep was triggered by injecting a voltage pulse from the FEK-09 vacuum photocell.

2. Experimental results from an investigation of the properties of superconductors under high level optical pumping

Fig. 10 gives voltage pulse waveforms from a 100 Å Pb sample obtained at various temperatures. The diagram also shows the waveform of GaAs laser emission. The laser pulse duration was 10 μs and the absorbed energy density was equal to 30 w/cm^2. It is clear from Fig. 10 that the signal from the sample related to its jump to a resistive state was injected at temperatures significantly below T_c = 7.2 K and when T_c = 5.2 K its amplitude was 30% of the value of U_N: the voltage across the sample in a normal state when $T = T_c$. (The quantity U_N is represented by the arrow in the left half of the figure.) With an increase in temperature the signal amplitude increases, reaching the value U_N. During the laser pulse the voltage level across the sample is not constant, but rather increases over time, reaching the amplitude value at the end of the laser pulse. Hence the level of the laser-induced voltage across the sample depends on the laser pulse duration. The $U/U_N(\tau)$ relation for Pb at T = 5.75 K in a 0.2-10 μs range is shown in Fig. 11. Here two sections may be identified. In the first section with pulse durations below 1 μs the signal level has only a weak dependence on τ. In the second section when $\tau \geqslant 1$ μs the value of U increases with an increase in the pulse duration.

An important experimental fact is the signal delay $\Delta\tau$ with respect to the beginning of the laser pulse. The value of $\Delta\tau$ drops with an increase in temperature (Fig. 10). When approaching T_c there is a "broadening" of the voltage pulse due to an increase in the signal decay time.

In Fig. 12 the amplitudes of the voltage pulses from lead films 1000 Å in thickness are plotted as a function of the initial temperature of the sample for laser pulse durations of 1 and 10 μs. The emission power density absorbed by the film was 56 and 30 w/cm^2, respectively. It is clear that in both cases the voltage across the

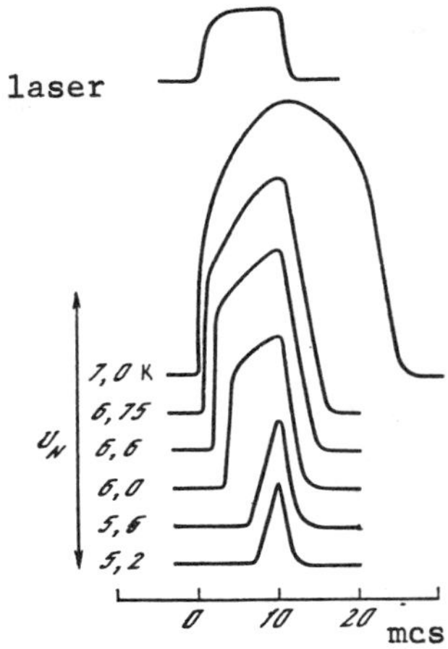

Fig. 10. Voltage pulse waveforms for a lead film under laser irradiation at various temperatures The upper diagram is an oscillogram of the laser pulse. The arrow to the left shows the voltage drop across the sample in a normal state

Fig. 11. The normalized voltage across the sample plotted as a function of laser pulse duration (T = 5.75 K, W = 86 w/cm^2)

Fig. 12. The normalized voltage across the lead sample plotted as a function of the initial temperature for two laser pulse durations τ 1 - 10 μs; 2 - 1 μs

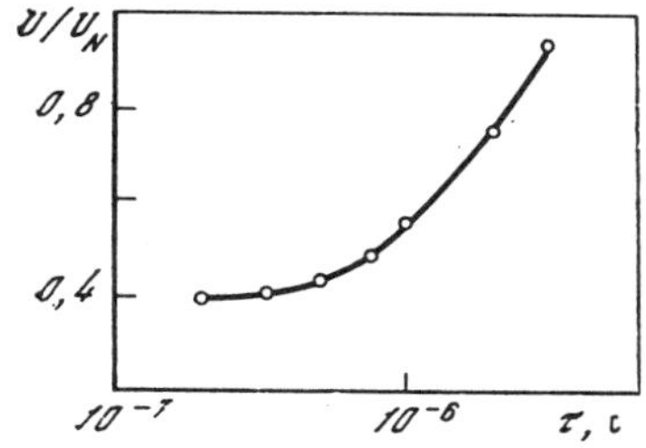

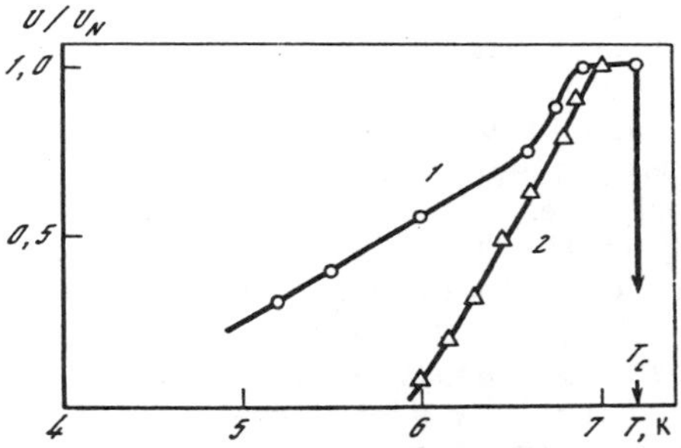

sample under laser irradiation appears at temperatures significantly below T_c and slowly changes with temperature, reaching U_N when $T < T_c$. With a reduction in the pulse duration the voltage appears across the sample, as is evident from the diagram, at a higher temperature. However, in both cases the temperature range of the signal rise significantly exceeds the width of the superconducting junction ΔT_c that amounts to $\leqslant$0.04 K for lead films. We should note that the voltage appears at significantly lower temperatures, as would be anticipated if we assume that the entire laser irradiation effect is reduced to heating only. The heating level ΔH_{H} of the film by a laser emission pulse of this power and duration was measured for the case of Pb independently using a different method: when $T \geqslant T_c$ the change in the resistance of the film under the influence of the laser pulse of this power and duration was measured. The temperature dependence of the resistance was then used to determine the heating of the sample; these values were 0.3 ±0.1 K and 0.45 ±0.1 K, respectively, for 1 and 10 μs pulses (see Fig. 12).

An analogous situation was observed in measurements with Nb_3Sn and V_3Si films in a laser pulse duration range of 0.2–10 μs. The experimental $U/U_N(T)$ relations for Nb, Nb_3Sn and V_3Si are given in Fig.

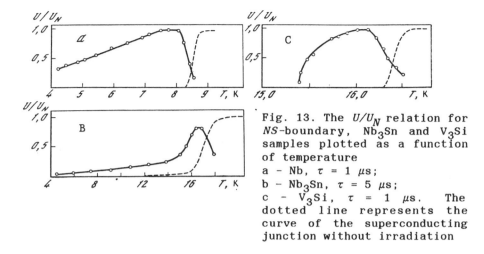

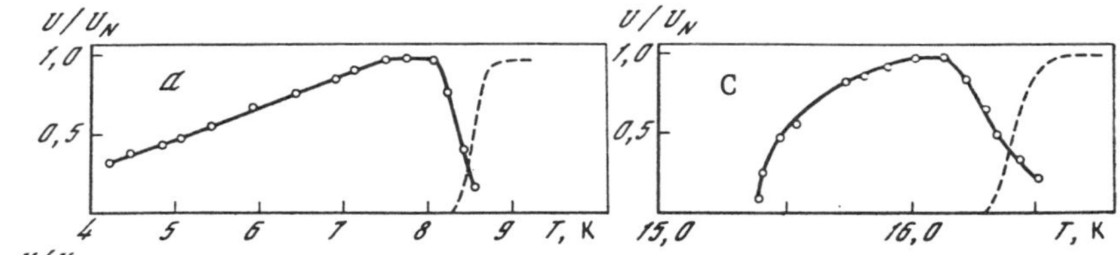

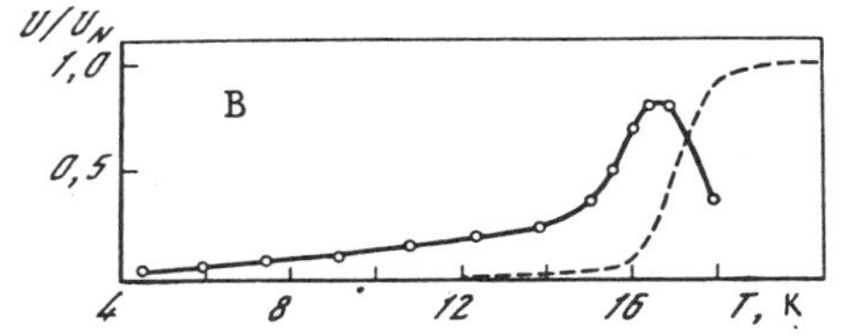

Fig. 13. The U/U_N relation for *NS*-boundary, Nb_3Sn and V_3Si samples plotted as a function of temperature
a - Nb, τ = 1 μs;
b - Nb_3Sn, τ = 5 μs;
c - V_3Si, τ = 1 μs. The dotted line represents the curve of the superconducting junction without irradiation

13. The overall pattern observed from the investigation of the properties of laser-irradiated Nb, Nb_3Sn and V_3Si films is similar to that of Pb. However there were significant differences in the signal waveform and signal level as well as its dependence on power and temperatures. These differences are related primarily to the form of the temperature dependence of the resistance near T_c and the threshold power level. Thus, Nb_3Sn films had a significantly broader transition range to the superconducting state ($\Delta T_c \sim$ 2 K) and a comparatively higher threshold power. For films 200 Å in thickness the threshold power at T = 4.5 K and τ = 5 μs was 30 w (the absorbed power density was 53 w/cm^2). At our power levels these differences did not make it possible to obtain the total signal level for Nb_3Sn and resulted in a smoother signal decay with $T \geqslant T_c$. It is clear from the figures that as in the case of Pb the temperature range of the rise in the signal proportional to the induced resistance exceeds the width of the superconducting junction. Unfortunately, the absence of a noticeable temperature progression of the resistance when $T \geqslant T_c$ made it impossible to determine the value of ΔT_H for these materials.

The measurements on Nb were also conducted by irradiating films with neodymium laser pulses from a laser operating in a single-mode state producing 40 ns pulses. The conversion to another wavelength (from 0.85 to 1.06 μm) was not manifest in the qualitative aspect of the phenomenon, since the ratio $h\omega \gg 2\Delta$ was satisfied as before. Fig. 14 gives the $U/U_N(T)$ relation for this case. It is clear from a comparison of Figs. 14 and 12 that the qualitative aspect of the phenomenon is concerned with a change in laser pulse duration from 40 ns to 1 μs in the transition of the source from single-mode to multi-mode lasing.

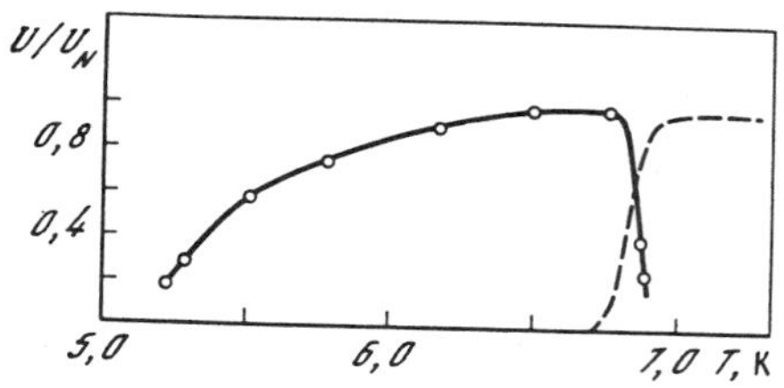

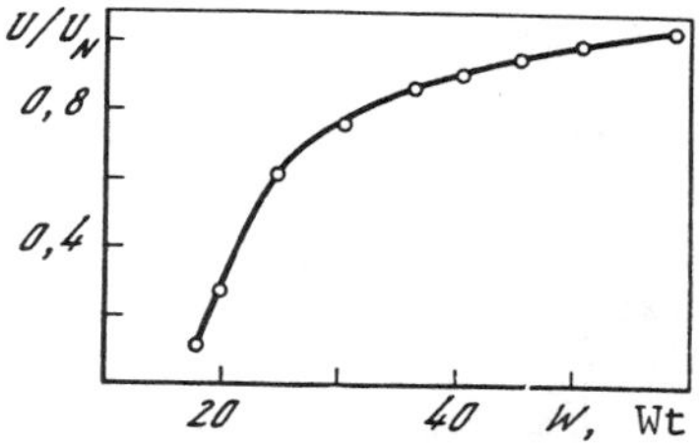

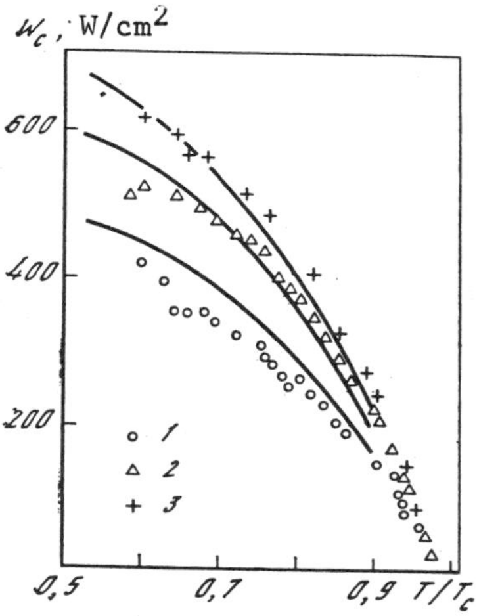

Fig. 14. The $U/U_N(T)$ relation for an Nb sample under neodymium laser irradiation (τ = 40 ns)

Fig. 15. The normalized voltage in a Pb sample under laser irradiation plotted as a function of emission power (τ = 1 μs, T = 6.75 K)

Fig. 16. Critical power plotted as a function of adjusted temperature for Pb films of various thicknesses The solid line curves represent the theoretical relations calculated by (2.1); 1 - 300 Å; 2 - 500 Å; 3- 1000 Å

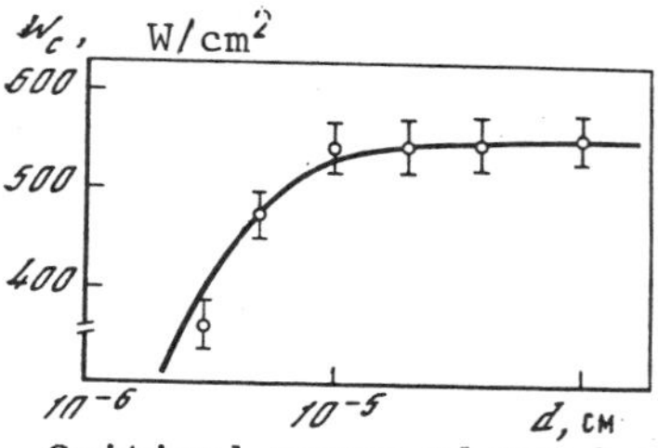

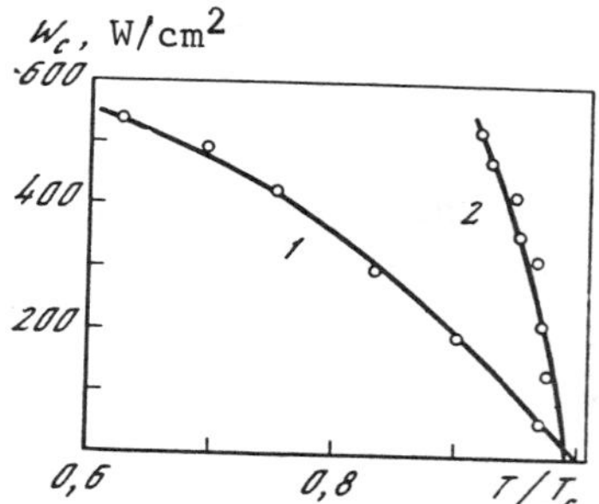

Fig. 17. Critical power plotted as a function of Pb film thickness for T/T_c = 0.7
The solid line curve represents the theoretical value of $W_c(d)$ calculated by (2.1)

Fig. 18. Critical power W_c of the transition to an inhomogeneous state plotted as a function of adjusted temperature T/T_c; τ = 1 μs
1 - Pb; 2 - V_3Si

The dependence of the voltage in the sample on the emission power manifests a threshold nature. Fig. 15 gives the dependence of U/U_N on power W obtained at T = 6.75 K and τ = 1 μs for lead film 1000

Å in thickness. It is clear that the transition of the sample to a resistive state is observed only when a certain critical value W_C of the pumping power is exceeded. The value of W_C may be determined by extrapolating the experimental $U/U_N(W)$ relation to zero voltage. For the case given in Fig. 15, W_C = 17 w, which corresponds to an absorbed power density of 30 W/cm^2. With an increase in power the voltage induced in the sample increases, reaching a value of U_N. This relation is characteristic of all metals tested.

The value of W_C for this material drops with an increase in temperature and increases with a reduction in laser pulse duration and a reduction in film thickness.

Fig. 16 gives the critical power W_C plotted as a function of temperature for Pb films 300, 500 and 1000 Å in thickness. The $W_C(T)$ relations for Pb films 0.2-1 μm in thickness virtually coincide with the $W_C(T)$ curve for a film with d = 1000 Å. Hence they are not shown in Fig. 16. The critical power for Pb plotted as a function of film thickness at T = 5 K (T/T_C) = 0.7 is given in Fig. 17. It is clear from the diagrams that W_C increases with an increase in film thickness from 300 to 1000 Å across the entire range of test temperatures. At higher thicknesses the $W_C(d)$ relation enters saturation. At the same time it was discovered that the heating ΔT_H of the film is independent of its thickness and is determined only by the value of W.

It is also interesting to compare the critical powers for various materials. Fig. 18 gives the temperature dependencies of the critical power for Pb (d = 1000 Å) and V_3Si (d = 2500 Å) films. It is clear from Fig. 18 that the value of W_C for V_3Si significantly exceeds the critical power level for Pb. In both cases the resistive state ensues over a broad (compared to the width of the junction) temperature range up to T_C when the critical power is exceeded.

Studies of the superconducting properties of metals under laser irradiation were first investigated using Pb films in Testardi's study [12]. An argon laser operating at 0.55 μm with pulse durations of 6 and 40 μm were used as the pumping source. The maximum emission power per pulse corresponded to an absorbed power density of 17 w/cm^2. The measurements were carried out on films 275 and 1700 Å in thickness. A transition to the resistive state was discovered in films with d = 275 Å with $T < T_C$ under laser irradiation; this transition process was not related to heating. Here the temperature range in which the resistance remained below its value in the normal state was much broader than the width of the superconducting junction. For films with d = 1700 Å this effect was not observed.

Studies [24, 25] that were published simultaneously with our study [26] investigated the properties of superconducting Sn films under laser irradiation. In study [25] the authors investigated the change in the E-I characteristic of Sn-SnO_x-Sn tunnel junctions under GaAs laser irradiation operating at a power level of 50 w with a pulse duration of 30 ns. The authors discovered a broadening of the current jump in the E-I characteristic at voltages across the junction equal to 2Δ independent of heating.

In study [24] the authors investigated the superconducting properties of Sn films 150-1100 Å in thickness under GaAs laser irradiation with a pulse duration of 60-90 ns. The maximum absorbed power density was 8 w/cm^2. At the same time the change in the reflectance of the microwave emission with laser irradiation was measured together with the resulting resistance. The result of this research was the discovery of the transition of the film to a resistive state at a certain pumping power. In this range of powers corresponding to the resistive state a continuous increase in the microwave reflection coefficient was observed. The authors interpreted these results to be a consequence of the onset of a dynamical intermediate state in which the sample is a superconducting sample at certain times and is a normal sample at other times.

Subsequently a number of studies have appeared [27, 28] investigating the influence of laser irradiation on the superconducting properties of aluminum. Study [28] observed the onset of a resistive state from irradiation of an Al film with d = 500 and 800 Å by dye laser pulses with a pulse duration of 6 ns and a pulse energy of 0.5 μJ. The results are explained based on an assumption of the onset of a spatially-inhomogeneous state or creation of slip-centers.

In study [27] aluminum films with d = 200 Å were irradiated by an argon laser with a maximum output power of 0.5 w. The change in transmittance of microwave emission (100 GHz) under laser irradiation was measured. The derived results are best explained by assuming formation of a certain intermediate state in the film and in the superconducting region the gap and T_c conserve their unperturbed values while the volume of the normal phase increases with an increase in the optical pumping power.

The resistive state that ensues in superconducting films under irradiation is characterized, as we have seen, by incomplete ($U < U_N$) resistance in the film over a broad temperature range. In this sense this state is analogous to the intermediate

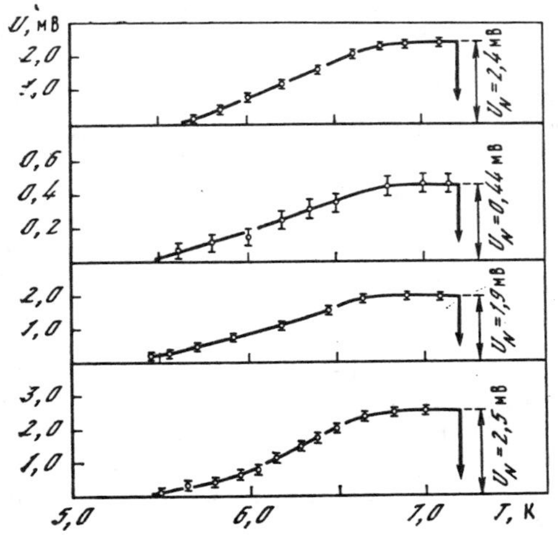

Fig. 19. The voltage across various sections of a Pb film plotted as a function of temperature (τ = 1 μs)
The arrows on the right give the voltage drop across the corresponding section in a normal state

state that occurs in a magnetic field and consists of alternating regions of normal and superconducting phases. In order to estimate the size of the normal and superconducting regions in our case we conducted an experiment in which the potential junctions are separated by different distances. The minimum distance between the junctions was ~0.1 mm. Fig. 19 gives the curves plotting the voltage that arises from laser pulse action in various sections of the Pb film as a function of temperature. The film thickness was 1000 Å with a width of 0.1 mm and a pulse duration of 1 μs. It is clear from the diagram that the qualitative situation remains unchanged, i.e., the interval of signal growth as before somewhat exceeds the width of the superconductor junction even when reducing the distance between the junctions to 0.1 mm. Here we may conclude that the dimensions of the normal and superconducting regions that arise from the transition to the resistive state under irradiation are less than 0.1 mm. Otherwise we would observe an irregular signal rise at a certain temperature when the emission power exceeds the critical value (for a given temperature). This result is in agreement with study [27] which indicates that the dimensions of the regions that occur from the transition to a spatially-inhomogeneous state are significantly smaller than the wavelength of the microwave emission at 100 GHz (the corresponding wavelength was 0.3 cm).

3. Discussion of results. Comparison to theory

As we know, with a homogeneous excitation distribution in the sample with an increase in their concentration N the film resistance remains zero as long as $N < N_0$ (here N_0 is the critical excitation concentration for the transition to a homogeneous normal state). With an increase in N when N reaches the value N_0 the film resistance will jump to its total value in the normal state R_N.

Another situation is observed in experiments: at a certain pumping power a resistance appears that slowly increases to R_N with an increase in W. As we have seen this phenomenon is not related to sample heating (since the heating level was measured experimentally) nor to the possible inhomogeneity of the laser beam (since in the transition from multimode lasing to single mode lasing the basic phenomenon does not change).

In studies [26, 29] we were the first to offer the hypothesis that in certain conditions determined by the temperature and pumping power, a nonequilibrium superconductor under homogeneous excitation assumes a spatially-inhomogeneous state characterized by the simultaneous coexistence of the normal and superconducting phases. An analogous hypothesis was offered in study [25] virtually simultaneously. Here we associated the threshold nature of the power dependence of the resistance with the achievement of a certain critical quasiparticle concentration in the sample. An increase in resistance with an increase in power may be explained by an increase in the volume of the normal phase, while the signal delay was related to the time needed for the normal phase expanding in the volume to intersect all superconducting current lines.

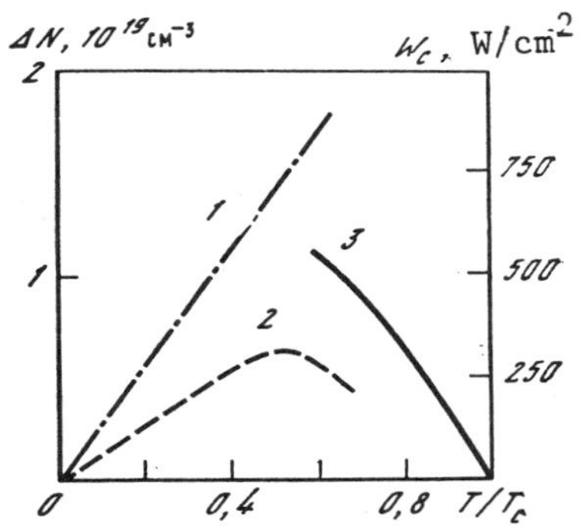

Fig. 20. A comparison of experimental results for Pb to theory
1 - $\Delta N_c(T/T_c)$ relation [37];
2 - $\Delta N_c(T/T_c)$ relation [31];
3 - experimental relation $W_c(T/T_c)$

Recently a number of theoretical studies [30-32] have demonstrated that under laser irradiation a superconductor in certain conditions becomes unstable with respect to the transition to the state with a spatially-inhomogeneous excitation distribution. The transition occurs when the average concentration of excitations in the sample reaches a certain critical value N_c. In experiment, however, N_c is not determined but rather a related quantity of W_c. This allows only a qualitative comparison of experimental results to studies [30-32] where the temperature dependence of the critical excess concentration $\Delta N_c = N_c - N_T$ is obtained (N_T is the concentration of excitations in an equilibrium superconductor at temperature T).

Fig. 20 gives the dependencies of ΔN_c on T/T_c for Pb. Curve 1 was obtained based on the results from study [30] and curve 2 was obtained based on the results from study [31]. Here we employed the values of the state density on the Fermi surface $N(0) = 2.2 \cdot 10^{22}$ $ev^{-1} \cdot cm^{-3}$ and an energy gap value $\Delta_0 = 1.4$ mev.

Studies [30, 31] employed equilibrium excitation distribution functions. If we ignore the temperature dependence of the quasi-particle recombination time τ_R (which clearly is valid for high pumping levels which are achieved in our experiment), then the critical power value obtained from the experiment $W_c \propto \Delta N_c$ with a temperature-independent proportionality factor. Hence the experimental relation $W_c(T)$ with scale factor accuracy reflects the experimental relation $\Delta N_c(T)$. The temperature dependence of W_c for Pb is also given in Fig. 20 (curve 3). It follows from Fig. 20 that the temperature dependence of ΔN obtained in study [30] has a different sign of the first derivative compared to the experimental relation. The dependence of ΔN_c on T/T_c obtained in study [31] is a nonmonotonic relation. In the range of temperatures where we carried out the measurements the signs of the first derivatives coincide, although curve 2 is broken at $T/T_c \approx 0.7$, i.e, according to theory the inhomogeneous state is possible only when $T/T_c < 0.7$, so experiment reveals the existence of the spatially-inhomogeneous state up through $T = T_c$. We should also note that evidently the experiment did not confirm the nonmonotonic nature of the $\Delta N_c(T/T_c)$ relation obtained in study [31]. It follows from curve 2 that at low temperatures the spatially-inhomogeneous state may be obtained at as low pumping levels as desired. In the experiment using Pb films we did not reach temperatures corresponding to a change in the sign of the first derivative on theoretical curve 2. However in experiments with V_3Si where the measurements were carried out in an adjusted temperature range from 0.25 to 1, the laser power (70 w) was sufficient to drive the V_3Si into an inhomogeneous state for $T/T_c >$ 0.915 only. At lower adjusted temperatures no inhomogeneous state arose.

Study [33] which used the nonequilibrium excitation distribution function demonstrated the instability of the irradiated superconductor with respect to the transition to the spatially-inhomogeneous for all $T/T_c < 1$.

According to the "coherent instability" model developed in study [37] this transition occurs at a certain pumping power determined by the dimensionless quantity β_c. Physically the transition is related to the vanishing of the order parameter when the critical quasi particle concentration N_c is reached in the superconductor. The coefficient relating β_c and W_c depends on the kinetic characteristics determining the relaxation and quasi particle recombination processes.

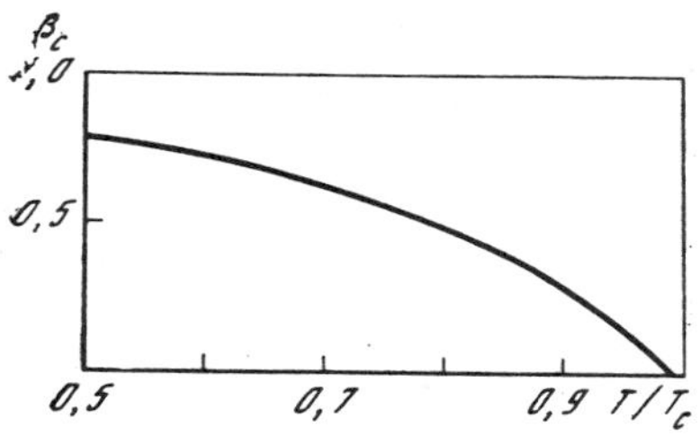

Fig. 21. The theoretical dependence of the dimensionless pumping parameter β_c on adjusted temperature [38]

For the region $T_c - T \ll T_c$ study [33] derived an analytic expression for the temperature dependence of W_c. It takes the form $W_c = A\,(1 - T/T_c)$. However the significant uncertainty in the value of a number of coefficients in A made it impossible to carry out a quantitative comparison of experimental results and theory. From the experiment a coefficient value for Pb of $A = 1.8 \cdot 10^3$ w/cm^2 was obtained, and for V_3Si a value of $A = 7.5 \cdot 10^3$ w/cm^2 was derived.

Study [38] uses the results from studies [33-36] in a numerical calculation of the temperature dependence $\beta_c(T)$ in the interval $0 < T < T_c$ to derive an expression for W_c as a function of temperature and film width.

The expression relating the measured power W_c to the quantity β_c takes the form

$$W_c(T,d) = \beta_c(T)\,\frac{N_c \hbar\omega}{4\tau_R^{\mathrm{eff}} R\,(1-r)}\,\frac{d}{1-e^{-\alpha d}}\,. \qquad (2.1)$$

Here $N_c = p_0 \Delta_0 m/\pi^2 \hbar^3$; m, p_0 are the mass and Fermi momentum of the electron; α is the coefficient of absorption; r is the coefficient of reflection; R is the "multiplication" factor of the quasi particles that accounts for the increase in their concentration due to the absorption of relaxation phonons, τ_R^{eff} is the effective recombination time that differs from the true recombination time τ_R due to the reabsorption of recombination phonons by the factor $(1 + \tau_{es}/\tau_B)$ [23], where τ_{es} is the exit time of the recombination phonons from the film, while τ_B is their lifetime with respect to pair breaking. The universal function β_c depends on temperature and takes the form [38]

$$\beta_c = \frac{1}{\Delta_0^4} \int_0^\infty d\xi \int_0^\infty d\xi' (\xi + \xi')^2 \{n(\xi)\, n(\xi') - N_{\xi'+\xi} [1 - n(\xi) - n(\xi')]\},$$

where $n(\xi)$, N_ξ is the distribution function of the nonequilibrium quasi particles and phonons when $\Delta = 0$. The temperature dependence $\beta_c(T)$ from study [38] obtained by a numerical technique is given in Fig. 21. When $T \to T_c$ $W_c \propto (1 - T/T_c)$, which coincides with the relation obtained in study [33]. With reduction of T the value of β_c increases and approaches a certain limit $\beta_c(0) = 0.84$. The coefficients in equation (2.1) characterize the specific parameters of the superconductor (τ_R, Δ_0, etc.) and the sample (d, r, etc.).

The issue of determining τ_R^{eff} and its dependence on film thickness requires special consideration. The problem is that $\tau_R^{eff} \neq \tau_R$ even in the limit of exceedingly small values of d due to the existence of the total internal reflection angle φ_{max} for the phonons. Thus, for example, in the case of a Pb and Al_2O_3 boundary the value of $\tau_{max} \simeq 11°$ and for incidence angles of $\varphi < \varphi_{max}$ the transmission coefficient is $T \simeq 1$ and $T = 0$ for $\varphi > \varphi_{max}$. In the limit of small values of d with specular reflection of the phonons at the film-substrate boundary $\tau_R^{eff}/\tau_R = (1 - \cos \varphi_{max})^{-1}$ [5]. In the general case ignoring phonon scattering processes compared to the process of new quasi-particle generation from the condensate, the $\tau_R^{eff}(d)$ relation takes the form [5]

$$\tau_R^{eff} = \tau_R(0) \frac{2d}{l_B} \frac{1}{1 - \exp(2d/l_B)}, \tag{2.2}$$

where $\tau_R(0) = \lim_{d \to 0} \tau_R$; $l_B = S\tau_B$ is the free path length of the phonons, S is the speed of sound.

We will analyze the dependence of W_c on film thickness. If we account for the fact that in the test materials the coefficient of absorption exceeds $1/d$, and we assume that the multiplication factor R is independent[1] of d, then we obtain from (2.1) and (2.2)

$$W_c = A l_B [1 - \exp(-2d/l_B)].$$

Here A is the factor independent of d. For a thin film ($d \ll l_B$) W_c is linearly dependent on the width: $W_c \simeq 2Ad$, since as the film volume increases the pumping level necessary to obtain the critical concentration of excitations also increases. For sufficiently large thicknesses ($d \gg l_b$) the $W_c(d)$ relation goes to saturation $W_c \simeq Al_B$, which is related to the reabsorption of recombination phonons. We should note that the asymptotic expression for $W_c(d)$ for the cases $d \to 0$ and $d \to \infty$ are independent of the specific model.

By comparing the experimentally measured values of W_c for the limiting cases of large and small thicknesses we may estimate l_B for the recombination phonons by the formula $l_B = 2W_{c2}d/W_{c1}$ (here the indices 1 and 2 relate to the thick and thin films, respectively).

In order to determine l_B based on results from measurements of W_c in the intermediate range of thicknesses we must select the parameter l_B and the scale factor $R\tau_R(0)$ to assure the best correlation between the experimental dependence $W_c(d)$ and (2.1). The same parameters determine the $W_c(T)$ relation.

The theoretical $W_c(T)$ relations for various values of d and $W_c(d)$ for $T = 6$ K were formulated for the case of Pb according to formula (2.1) and are shown in Figs. 16 and 17 by the solid line curves. For our adjustment parameters we employed the values $R\tau_R(0) = 3.7 \cdot 10^{-8}$ seconds and $l_B = 500$ Å selected to provide best correlation with experiment. A value of $\tau_B = l_B/S = 2.5 \cdot 10^{-11}$ s derived in this manner is in good agreement with the value $\tau_B = 3 \cdot 10^{-11}$ s obtained in study [39].

It is clear from a comparison of the experimental dependencies of W_c on T and d and (2.1) that the "coherent instability" model provides a correct qualitative and quantitative picture of the transition to the spatially-inhomogeneous state. The observed deviation of the $W_c(T)$ relation for $d = 300$ Å may be explained to some degree by the dependence of the coefficient R on d, which was ignored.

We will now consider how, within the scope of the "coherent instability" model we may explain the remaining experimental facts established in the course of this study. As indicated by the model when the critical power is exceeded the superconductor is decomposed into

---

[1] Evidently approximately the same situation exists in reality. Satisfaction of the conditions $\tau_{es} \ll \tau_B$ for $\hbar\omega_{ph} \simeq 2\Delta$ and $\tau_{eS} \gg \tau_B$ for $\hbar\omega_{ph} \gg 2\Delta$ derive from: 1) the reduction in τ_B with an increase in $\hbar\omega_{ph}$ [39] and 2) the increase in the value of S entering into the expression for τ_{es} due to the dispersion for large values of $\hbar\omega_{ph}$.

normal and superconducting regions with the boundary between these regions traveling at a velocity [40]

$$S = \frac{L}{\tau_R} \frac{\delta - \delta_0}{\delta_0} \zeta, \tag{2.3}$$

where $\delta = (\beta/\beta_c)^{1/2} - 1$; $\delta_0 = \delta(\beta_0)$, $\beta_0 = 1{,}37\,\beta_c$; $L = v_F^2 \tau_{imp} \tau_R / 3$ is the diffusion length; τ_{imp} is the impurity scattering time; ζ is the parameter in the order of unity. When $\beta > \beta_0$ this corresponds to an increase in the volume of the normal phase due to the superconducting phase. The amplitude of the recorded signal, i.e., the level of voltage that arises in the sample by the end of the laser pulse will be determined primarily by the velocity of the normal phase boundary and will increase with τ. Here for sufficiently narrow pulses the magnitude of the shift of the phase interface will be much less than the characteristic dimensions of the normal regions that first occur in the transition to the spatially-inhomogeneous state. This means that up to a certain value $\tau = \tau_0$ the value of U will be virtually independent of τ. The boundary value of τ_0 may be determined from the state $S\tau_0 \sim 0.1\ \bar{d}$, where $\bar{d}$ is the characteristic dimensions of the normal region. When $\tau > \tau_0$ an increase in U with growth of τ will be observed.

It was precisely this behavior of the $U(\tau)$ relation that was observed in experiment (Fig. 11). The experimental value is $\tau_0 \sim 10^{-6}$ seconds. By using the derived value of τ_0 and assuming $\bar{d} \sim 10^{-1}$ cm (film width) we may estimate the rate of propagation of the normal phase: $S \sim 0.1\bar{d}/\tau_0 \simeq 10^4$ cm/s. This value of S was obtained for the case where $W/W_c \simeq 1.5$.

The existence of a signal delay with respect to the laser pulse at durations $\tau \sim 10^{-5}$ s (Fig. 10) within the scope of this model may also be related to the motion of the normal region boundary during the laser pulse. Here the initial normal regions increase in size at a rate determined by (2.3). When the dimensions of $\bar{d}$ become equal to the film width, resistance ensues. With growth of temperature at a fixed irradiation power W the W/W_c ratio increases (since W_c drops), which will serve to increase the velocity of the boundary and will reduce the delay $\Delta\tau$. This corresponds to the experimentally observed relation $\Delta\tau(T)$ (Fig. 10). The value of $\Delta\tau$ may also be used to estimate the rate of propagation of the normal phase boundary. If we select the same W/W_c ratio equal to 1.5 (this is the case when $T =$ 5.75 K, as indicated by Fig. 16), we find from Fig. 10 that $\Delta\tau \simeq 4$ μs. If we assume $\bar{d} \simeq 0.1$ cm, we obtain $S = \bar{d}/\Delta\tau = 0.25 \cdot 10^5$ cm/s, which coincides with the estimate of the propagation rate obtained on the basis of the $U/U_N(\tau)$ relation. On the other hand the estimate of this velocity based on (2.3) yields a value of $S \sim L/\tau_R \sim 10^5$ cm/s. Here if we set $\tau_{imp} \simeq 10^{-13}$ s, $V_F = 10^8$ cm/s, $\tau_R \simeq 10^{9}$ s. This is more than an order of magnitude greater than the experimentally observed propagation velocity. Its origin may be related to the reflection of quasi particles off the S-N-boundary [41] which is ignored by the theory. This effect will hinder quasi-particle diffusion, thereby reducing the propagation velocity of the normal phase.

We should note that the propagation of the wavefront is not related to the increase in the volume of the normal phase resulting from Joulean heat, since the magnitude of the observed resistance in the resistive state is independent of the measurement current I up through $I \simeq 0.8\ I_c$. In addition we cannot explain the observed increase in the volume of the normal phase by the absorbed energy of the electromagnetic wave as indicated from an estimate of the propagation velocity of the front for this case. Such an estimate may be obtained on the basis of study [42] which provides an expression for the velocity S of the normal domain boundary:

$$S \approx V_F\ (l/\lambda)(W - W_0)/W.$$

Here l is the free path length; λ is the electron cooling length; W_0 is the irradiation power at which $S = 0$. Assuming $l \sim d \sim 10^{-5}$ cm, $\lambda \sim 10^{-4}$ cm and $(W - W_0)/W \simeq 0.1$-1, we obtain

$$S \sim (10^{-2} - 10^{-1})\ V_F \sim 10^6 - 10^7\ \text{cм/c},$$

which significantly exceeds the experimental estimate.

4. The kinetic aspect of spatially-inhomogeneous state formation

We will now consider the kinetics of the transition of the superconductor to the spatially-inhomogeneous state and the properties of this state based on the experimental results obtained in this study and theoretical concepts provided by the "coherent" instability model.

The kinetics of the nonequilibrium state that arises in thin superconducting films under optical excitation may be represented in the following manner. Quasi particles created from the absorption of light quanta with energy $\hbar\omega \gg \theta_D \gg 2\Delta$ (θ_D is the Debye energy, Δ is the energy gap) have energy in the interval $\Delta < E < \hbar\omega - \Delta$. The high-energy quasi particles with $E \gg \theta_D$ in interacting with the lattice relax over time $\sim 10^{-14}$ s [43] emitting the entire spectrum of phonons [44, 44a]. If the film is sufficiently thick so that the average escape time of the phonons τ_{es} exceeds their lifetime with respect to the breaking of the superconducting pair - τ_B, phonons with energy $\hbar\omega_{ph} \geqslant 2\Delta$ will again create quasi particles thereby increasing their concentration. For these secondary quasi particles $E \gg \hbar\omega$. In relaxing they emit phonons which in turn may again create quasi particles although with a lower energy, etc. Another relaxation channel - electron-electron interaction - may also produce additional quasi particles with lower energy.

In the case of stationary pumping at temperature T an increase in the case of quasi particles in the film over their equilibrium value N_T is limited by the recombination process whose characteristic time is τ_R. In recombination the two excitations produce a Cooper pair, and emit a recombination phonon with $\hbar\omega_{ph} \geqslant 2\Delta$. If τ_R for a phonon with such an energy is less than τ_{es}, the phonon may again break the pair. This results in an effective increase in the excitation lifetime which may be described by introducing the effective recombination time $-\tau_R^{eff} > \tau_R$. As a result of these processes quasi particles with energies $E \simeq \Delta$ are accumulated; these have a certain influence on the order parameter of a nonequilibrium superconductor. The form of the distribution function $n(E)$ of nonequilibrium quasi particles generated by the quasi particle scattering and recombination processes associated with the acoustical phonons is significant. We should note that an equilibrium distribution function with a power-dependent effective temperature will not result in the formation of an inhomogeneous state [45]. Theoretical investigations have revealed [33-37, 46] that $n(E)$ may indeed be significantly out of equilibrium. It is not the equilibrium nature of the distribution function that produces an ambiguous dependence of the order parameter Δ on pumping power. This ambiguity corresponds to the instability of the homogeneous distribution and the possibility for the existence of an intermediate state with alternating normal and superconducting phase regions. The physical reason for instability is related to the dependence of the recombination rate on the order parameter (proportional to Δ^2). Hence if as a result of fluctuation the gap drops, this will result in a reduction in the recombination rate and, as a consequence, will increase the number of quasi particles. This in turn will further reduce the gap, etc. This instability will arise when the critical concentration of excitations N_c is reached in the sample, corresponding to the experimentally-determined critical pumping power W_c. The critical power depends on the temperature and the film thickness.

The structure of the spatially-inhomogeneous state arising when $\beta = \beta_0$ appears, as indicated by the experiment, as normal film sections much less than 100 μm with an approximately constant density along the superconductor film. At a constant power $\beta > \beta_0$ this state is not a steady state. The boundary of the normal regions travels at a velocity determined by the difference $(\beta - \beta_0)$. The experimental estimate yields a value of $S \simeq 10^4$ cm/s. Hence it is possible to observe the spatially-inhomogeneous state only during the time needed to fill the entire sample with the normal phase.

For low durations ($\tau < 1$ μs) we may ignore the increase in the size of the normal regions resulting from boundary motion and at a constant power the volume of the normal phase is constant. However, as the pumping power increases the density of the normal regions that arise from the transition to the spatially-inhomogeneous state grows, while they increase themselves in the volume due to their surrounding transition layer. The observed critical power W_c corresponds to the time when these normal sections make contact and form a structure analogous to a Mendelson clamp [47]. For the case $\tau > 1$ μs the resistive state occurs by the end of the pulse at lower powers due to the increase in the size of the normal regions during the pulse.

Chapter 3

CURRENT INJECTION IN SUPERCONDUCTORS

1. The spatially-inhomogeneous state with tunnel injection in films

This section presents results from an investigation of the properties of superconductors under intense tunnel injection when the concentration of nonequilibrium quasi particles n becomes comparable to unity. Interest in this research is related to the possibility for observing the transition of the superconductor to a spatially-inhomogeneous state in this case which in the case of tunnel injection is stabilized by the current flowing through the tunnel junction, and will be steady-state [48]. This makes it possible to investigate this phenomenon using stationary techniques. One such method is the use of the E-I characteristic of S-I-S tunnel junctions in the injection region, i.e., with voltage across the junction of $V > 2\Delta/e$.

In this study we use the regular four-point configuration of the tunnel junction for measurement applications. The injection current flowing through the junction was generated by a linear current source. The voltage that arises in the junction was injected to the X input of X-Y recording potentiometer PDP-4. A voltage proportional to the injection current at this time was injected to the Y input. The junction was submerged in liquid helium. The measurements were carried out at a temperature below λ: the helium point.

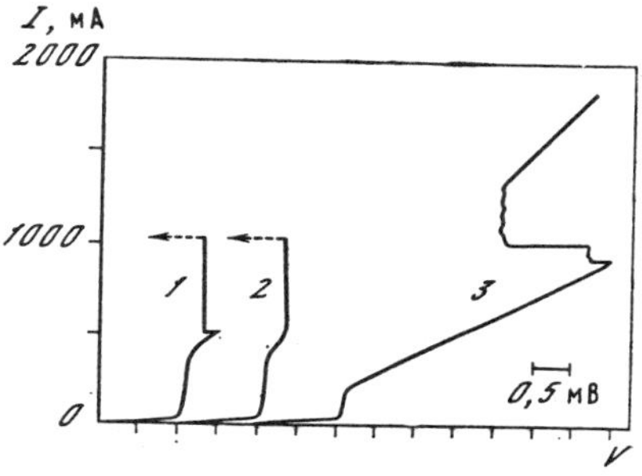

Fig. 22. The E-I characteristics of tunnel junctions
1 - $R_T = 2.8\cdot10^{-3}$ Ohms, d = 1200 Å, S = 0.16 mm^2, T = 1.7 K;
2 - Same as above, T = 2.05 K;
3 - $R_T = 5.5\cdot10^{-3}$ Ohms, d = 2600 Å, S = 0.16 mm^2

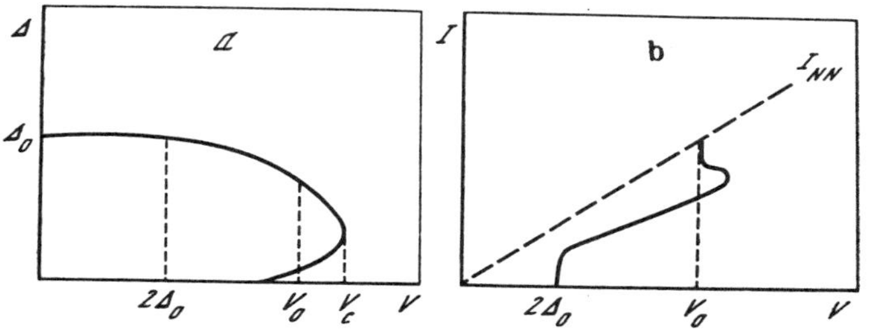

Fig. 23. The energy gap of the superconductor plotted as a function of the voltage across the tunnel junction (a) and the anticipated form of the E-I characteristic of the tunnel junction [48] (b)

These investigations were carried out on $Sn-SnO_x-Sn$ tunnel junctions. The tunnel resistance of the junctions was $R_T \simeq (4-8)\cdot 10^{-4}$ Ohms$\cdot$mm^2. Such resistances were necessary according to estimates in order to pass injection currents through the junction sufficient to generate the necessary concentration of quasi particles. The experiment investigated the E-I characteristic features of the tunnel junctions that occur with a voltage across the junction of $V > 2\Delta/e$.

Fig. 22 shows the E-I characteristics of two junctions that have a different value of R_T and a different relation between thicknesses of films forming the tunnel junction. For curves 1 and 2 $R_T = 3\cdot 10^{-3}$ Ohms with film thicknesses of 1200 Å and 2000 Å; for curve 3 $R_T = 5\cdot 10^{-3}$ Ohms with a film width of 2600 Å. It is clear from the diagram that in the E-I characteristic (curves 1 and 2) at a certain critical injection current level $I_c^{инж}$ corresponding to the voltage V_c there is a negative voltage jump (to $V = V_0$) across the junction after which there is a vertical section corresponding to a rise in current with a fixed voltage across the junction. Two such features were observed in the E-I characteristic of the second junction with similar values of the critical injection currents. We should note that the values of $I_c^{инж}$ for both junctions were lower than the corresponding values of the critical currents I_c of the films. The values of I_c were determined independently and for 1200 Å (the first junction films) equal 0.8 amps and for 2600 Å (the second junction films) equal 1.3 and 1.4 amps. The E-I characteristics (curves 1 and 3) were obtained at a temperature of $T = 1.7$ K. When the temperature was increased other types of E-I characteristics were occasionally observed. One such characteristic recorded at $T = 2.05$ K for the first junction is shown in the figure (curve 2). It is clear that there is no voltage jump in the E-I characteristic at $T = 2.05$ K and the vertical section begins immediately when $V = V_0$. Hysteresis was observed in curves 1 and 3 as the injection current dropped in the return direction.

According to the "coherent" instability model [48] and in the case of optical excitation, the dependence of Δ on the voltage V across the junction in a superconductor with tunnel quasi particle injection near the phase transition point manifests an ambiguous nature, and three possible solutions exist: one describes a drop in Δ with an increase in V, the second describes an increase in Δ and in the third $\Delta = 0$ (Fig. 23). As in the case of optical pumping there is a breakdown into normal and superconducting phases when $V > V_0$. This transition to the spatially-inhomogeneous state results in the formation of an S-shaped E-I characteristic of the tunnel junction. However unlike the case of optical pumping the phase interface in this case is stabilized by the injection current. With an increase in current the region of the superconducting phase drops, while the area of the normal phase grows. Theory provides the relationship between the critical parameters and the characteristics of the sample. Fig. 23 gives the dependence of the gap on the voltage across the junction and the anticipated form of the E-I characteristic for a tunnel injection superconductor obtained in the study. It is clear that beginning at V

$= 2\Delta_0/e$ the gap is reduced, which is related to an increase in the injection intensity. The transition to the inhomogeneous state occurs when $V = V_c$ is reached. Here the voltage across the junction jumps down to $V = V_0$. With an increase in current the voltage across the junction remains approximately constant and is equal to V_0. When there are bare normal phase regions in the sample, the transition to the inhomogeneous state may occur when $V < V_c$.

The derived experimental results are in good qualitative agreement with theory [48]. This is clear from a comparison of curves 1 and 3 (Fig. 22) to the theoretical relation $I(V)$ (Fig. 23) that was obtained in study [48]. All characteristic features are visible in the experimental curves, including the negative voltage spike and the subsequent vertical section of the E-I characteristic. The latter fact is particularly significant since voltage jumps in the E-I characteristic of tunnel junctions have also been observed in a number of other studies [49, 50] although they were related to heating of the film in the micro short circuit region and the transition of some portion of the film to a normal state. The redistribution of current in one of the films forming the tunnel junction that arises in this case when the condition $R_T < R_F$ is satisfied (R_F is the resistance of the film section within the tunnel junction) produced the observed negative voltage spike [49]. This feature vanished when the junction was cooled below λ: the liquid helium point ($T_\lambda = 2.17$ K). The differential resistances of the junction before and after the voltage spike in these cases were approximately equal.

Within the scope of the model utilized [48] the voltage jump is related to the development of the "coherent" instability and the transition of the superconductor from a homogeneous superconducting state to a spatially-inhomogeneous state characterized by the simultaneous coexistence of normal and superconducting phases. This transition occurs when a certain critical quasi-particle concentration in the film is achieved corresponding to the measured value of $I_c^{\text{ИНЗ}}$. The steady-state voltage $V_0 < V_c$ (Fig. 23) corresponds to equality of the energies of the normal and superconducting phases. A further increase in current when $V \simeq V_0$ corresponds to an increase in the normal phase volume at the expense of the superconducting phase until the sample completely converts to a normal state. The voltage interval before the transition $V_0 < V < V_c$ corresponds to the instability region. When there are no normal seed regions the transition occurs when $V = V_c$. The appearance of this seed region results in the development of instability and the transition of the sample to a spatially-inhomogeneous state when $I < I_c^{\text{ИНЗ}}$. If such normal seeding exists from the very beginning (for example, due to local heating), then when $V = V_0$ is achieved the normal phase begins to grow with a fixed voltage across the junction. This is observed in experiment (Fig. 22, curve 2). Here when $T = 2.05$ K close to T_λ there is evidently local heating related to the increase in T_λ due to the energy liberated in the transition. This heat causes a certain small portion of the film to convert to the normal state which then causes a continuous transition to a spatially-inhomogeneous state and associated change in the nature of the features of the E-I characteristic of the junction. For tunnel junctions formed by films of identical thickness, two voltage spikes are observed that are related to the two successive transitions of these films to the spatially-inhomogeneous state with similar values of $I_c^{ИНЗ}$.

In order to test whether or not the transition occurs when a specific quasi-particle concentration level is reached we investigated the dependence of the critical injection current on film volume v within the area of the tunnel junction. In the case of various film thicknesses forming the junction we selected a film volume of smaller thickness. The derived $I_c^{ИНЗ}$ (υ) relation is shown in Fig. 24. It is clear from the figure that $I_c^{ИНЗ} \sim \upsilon$ regardless of R_T, which is consistent with the assumption made here. Using the relation $I_c^{ИНЗ}/e \sim n_c\upsilon/\tau_R$, we may determine the value of n_c. Assuming $\tau_R \sim 10^{-8}$ s for Sn [39] we obtain $n_c \sim 0.4 \cdot 10^{19}$ cm^{-3} which is within an order of magnitude of the theoretical estimate of $1.3 \cdot 10^{19}$ cm^{-3} [48].

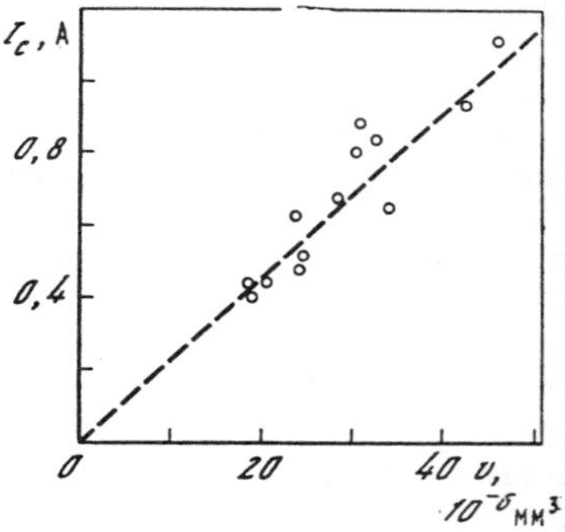

Fig. 24. The critical injection current $I_c^{ИНЗ}$ plotted as a function of volume υ of the film within the junction region

Thus, the derived results may serve as an experimental confirmation of the "coherent" instability model and allows us to conclude that in both tunnel injection and optical pumping there is a range of quasi-particle concentrations in which the superconductor assumes a spatially-inhomogeneous state.

This effect may find practical application. Such a tunnel junction biased in the instability region may serve as a capable detector of high energy particles. In this range between V_0 and V_c the quasi particles in the superconductor are analogous to supercooled vapor in a cloud chamber. When the particle makes impact a portion of the film makes the transition to a normal state due to the liberated energy. This in turn causes development of instability and splits the entire superconductor into phases. Thus, the impact of a microscopic particle may appear as a finite voltage jump in the E-I characteristic.

2. Enhancement of superconductivity with quasi particle tunneling

The idea of possibly expanding the energy gap of a superconductor by passing current through a S-I-S'-I-S tunnel system ($\Delta_S > \Delta_{S'}$) was first offered by Parmenter [51]. He demonstrated that in this situation the reduction in the gap Δ_S in the far wafers would be accompanied by an increase in the gap $\Delta_{S'}$ in the central film. An analogous situation for *n-I-S-I-p* tunnel junctions (*n*, *p* are the corresponding superconductors) was examined in the study by Aronov and Gurevich [52]. It was demonstrated that in this case both the total number of quasi particles in the superconductor and their energy distribution change (the former drops), thereby causing an increase in the superconductor gap. An increase in the gap in tunneling may be observed in *S-I-S* systems. In this case the enhancement of superconductivity will be observed in both superconducting films. This effect was first predicted in study [53].

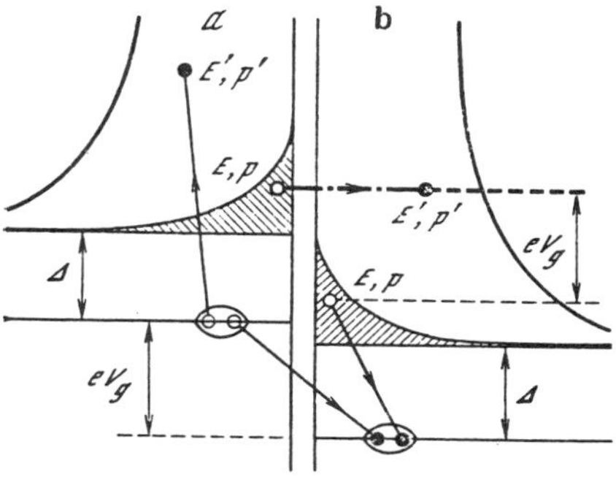

Fig. 25. Tunneling processes described by the operators and for an initial energy E and bias eV_g. The total result of these two processes is an increase in the excitation energy by eV_g on both sides of the barrier

We note that the quasi particles localized near the gap have a determinant influence on the order parameter of a superconductor. If some external source changes the quasi particle distribution (without changing their total number) so that the states near the gap are freed, the components of the superconducting pairs may scatter to free states. This will be equivalent to increasing the binding energy and will serve to increase superconductivity. This model was proposed by Eliashberg et al. [54, 54a] in order to explain experiments on enhancement of superconductivity by microwave emission. Under microwave irradiation the field heats the quasi particles and increases their average energy. Here no additional quasi particles are produced since the emission frequency $\hbar\omega < 2\Delta$. This same effect, obviously, may be achieved by using a photon flux with $\hbar\omega_{ph} < 2\Delta$. This was done in study [55] where the authors observed an increase in the critical current of the superconducting film irradiated by low-frequency phonons.

It turns out [53] that an analogous phenomenon may be observed in quasi particle tunneling when the voltage across the tunnel junction is $V < 2\Delta/e$. In order to explain how this occurs we will consider the processes that make a contribution to the tunnel current between two identical superconductors *a* and *b* when $V < 2\Delta/e$ (Fig. 25). The tunneling processes shown in Fig. 25 are described by quasi particle Bogolubov-Valatin operators $u_{p'}u_p\gamma^+_{p'\uparrow}\gamma_{p\uparrow}$ and $v_{p'}v_p\gamma_{-p'\downarrow}\gamma^+_{-p\downarrow}$, where v and u are the coherent factors and $\gamma^+(\gamma)$ is the quasi particle creation (destruction) operator.

The first operator describes the process shown by the dotted line in the figure. In this process a single quasi particle in a is destroyed and a single quasi particle in b is created. The second operator describes the quasi particle destruction process in b and the quasi particle creation process in a. Here in each process a single electron is transported through the barrier from a to b. In the superconductor model the first of these processes corresponds to electron transport from a to b while the second corresponds to hole transport from b to a. The clear result of these two processes will be increasing quasi particle energy by eV on both sides of the barrier without changing the total number of quasi particles. Thus, it is analogous to the absorption of phonons or microwave quanta.

The branch imbalance that arises in tunneling [56] does not change the gap since the pair states are blocked identically by both electrons and holes.

In order to provide a noticeable increase in the gap due to tunnel heating it is, obviously, necessary to maintain the high excitation rate Γ_T comparable to the thermalization rate Γ_S (for tin $\Gamma_S \sim 10^9\ s^{-1}$). The tunnel heating rate is $\Gamma_T = I/2e\upsilon N(T)$ where I is the tunnel current, υ is the film volume within the tunnel junction, $N(T)$ is the equilibrium quasi particle concentration. It is clear that the most significant effect would be expected in superconductors with long thermalization times (less than Γ_S).

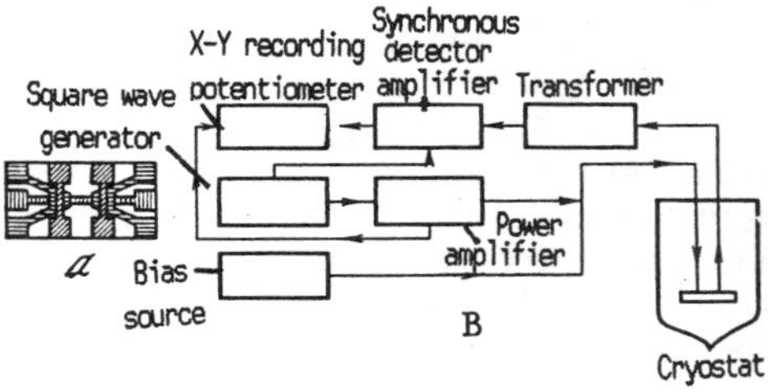

Fig. 26. The geometry of the double tunnel junction. The shaded area represents the deposited metallic layers (a) and the block diagram of the device for investigating the influence of current tunnel injection on the superconducting gap is shown in (b)

In this study we measured the change in the gap $\delta\Delta$ in one of the superconducting wafers of a tunnel junction in both the quasi-particle tunneling region and in the injection region. For this purpose we utilized the double tunnel junction technique [57-59]. Such a junction consists of three overlapping superconductor films separated by oxide barriers. One of the films - the generator - injects quasi particles, while the other functions as a passive detector recording the change in the gap $\delta\Delta$ in the central film which is simultaneously the upper wafer of the generator and the lower wafer of the detector. The geometry of these samples is shown in Fig. 26a. We should note that there is a significant difference in the tunnel resistance of the generator and the detector. The generator must have a low tunnel resistance in order to obtain sufficiently high rates of quasi particle excitation while the detector, on the other hand, must have a significantly higher resistance in order to avoid significant changes in the quasi particle concentration due to intrinsic bias current.

In order to improve measurement sensitivity periodic injection of unipolar square-wave current pulses (with a frequency of ~1 kHz) was used; the amplitude of these pulses increased over time from 0 to I_{max}. The value of I_{max} corresponded to a voltage of V_{max} across the junction at the end of the test section of the E-I characteristic of the generator.

A block diagram of the experimental set up is shown in Fig. 26b. The square wave unipolar current pulses with a frequency of ~1 kHz and an amplitude slowly increasing over time originate at square wave generator G6-15 and pass through the power amplifier and are injected to the generator junction of the sample in the cryostat. The operating point on the E-I characteristic of the tunnel junction-detector was shifted by means of bias current from the bias source to a higher voltage range at a voltage of $V \simeq 2\Delta/e$. The a.c. across the detector corresponding to the change in the gap under the influence of current injection was sent through the input transformer to the input of synchronous detector amplifier USD-1. A signal proportional to the level of $\delta\Delta$ was taken from the output of the amplifier and injected to the Y input of X-Y recorder PDP-4. A signal from the power amplifier proportional to the amplitude of the injection current pulses was input to the X input. The reference signal for the USD-1 originated from an auxiliary output of master oscillator G6-15.

The tunnel junction was submerged in liquid helium. In order to improve heat entrainment all measurements were conducted at a temperature below λ: the helium point.

The error in determining the a.c. component of the voltage across the junction was ~30% for low signal levels (~1 nv) and ~5% for high signals. However the accuracy in determining the change in the gap is significantly lower since significant uncertainty exists in the estimate of the detection area.

For the investigation we employed a Sn-SnO_x-Sn-SnO_x-Sn double tunnel junction. In the experiment we used a junction-detector to measure the change in the gap $\delta\Delta$ in the central Sn film that arises from the passage of current I_g through the junction-generator as a function of I_g. A typical $\delta\Delta(I_g)$ relation is shown in Fig. 27. It is clear from the diagram than $\delta\Delta > 0$ in a certain range of currents I_g which corresponds to an enhancement of superconductivity in this range of I_g. For small values of I_g, $\delta\Delta$ increases with an increase in current. At a certain value of I_g the level of $\delta\Delta$ reaches a maximum after which it decays and enters a region where $\delta\Delta < 0$, which corresponds to the transition from enhancement of the gap to suppression of the gap. The interval of I_g currents corresponding to a reduction in $\delta\Delta$ and the passage of $\delta\Delta(I_g)$ through zero appears in the current rise area on the E-I characteristic of the generator with a voltage of $V_g \simeq 2\Delta/e$ across the generator. This is clear from Fig. 28 where the value of $\delta\Delta$ is plotted as a function of V_g. When formulating the $\delta\Delta(V_g)$ relation the corresponding voltage values were determined on the basis of the E-I characteristic of the generator recorded at the

same temperature. The corresponding E-I characteristic of the generator is shown in Fig. 28.

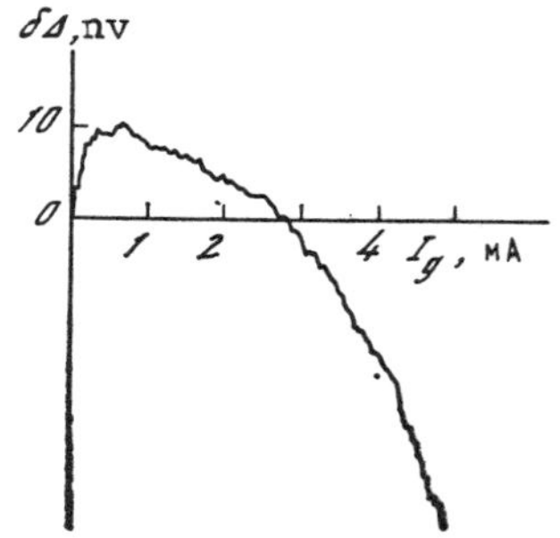

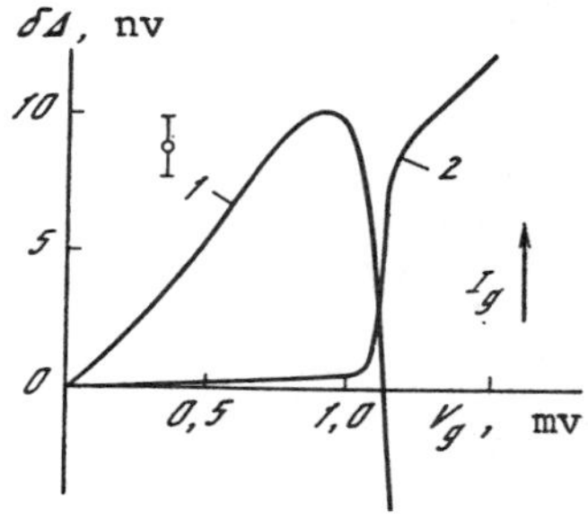

Fig. 27. The experimental plot of the change in the gap in the cente
Sn film as a function of generator current.

Fig. 28. The change in the gap plotted as a function of the voltag
across the generator (1) and the E-I characteristic of the generato
(2)
The line shows absolute measurement error

The influence of temperature on the change in the gap was als investigated. Fig. 29 gives the $\delta\Delta(V_g)$ relations obtained at variou temperatures in the junction with generator and detector resistance equal to 0.02 and 0.04 ohms respectively. The thickness of th central film was 300 Å. It is clear from the diagram that with a re duction in temperature $\delta\Delta$ in the $V_g < 2\Delta/e$ voltage range drops. Th dependence of the maximum gap change $\delta\Delta_m$ on temperature T/T_c is show in the insert. The highest value of $\delta\Delta_m$ obtained here was ≃40 n which corresponds to a relative gap increase of $\delta\Delta/\Delta \simeq 0.8\cdot10^{-4}$. Ex trapolation of the $\delta\Delta_m(T/T_c)$ relation to zero $\delta\Delta_m$ shows that the en hancement effect vanishes at temperatures of $T \sim 0.4\cdot T_c$.

The derived results are in good agreement with the Eliashber model. The increase in $\delta\Delta$ with low voltages (currents) is related t an increase in the injection rate with growth of $V_g(I_g)$, since her the populations of the high-energy states by quasi particles in creases. The sharp decay of $\delta\Delta$ when $V_g \simeq 2\Delta/e$ is related to an in crease in the quasi particle concentration due to the predominance o tunneling processes with pair breaking when $V \geqslant 2\Delta/e$. Here we shoul note that the observed superconductivity enhancement effect is no related to the possible action of Josephson radiation, which could b generated in the tunnel junction (i.e., resulting in regular microwav enhancement). This follows from the fact that $\delta\Delta > 0$ in the voltag range $\Delta/e < V_g < 2\Delta/e$ as well for which the value of the Josephso quantum is $\hbar\omega_J = 2eV_g > 2\Delta$. Such a quantum is capable of pair break ing while radiation at this frequency would cause an increase in quas particle concentration and a decrease in Δ.

The reduction in $\delta\Delta_m$ observed in the experiment with a reduction in temperature is consistent with the fact that the role of the increase in the "center of gravity" of the distribution function in quasi particle tunneling is reduced to weakening of the temperature factor, i.e., Δ^* is always less than Δ_0. (Here Δ^* is the expanded gap when $T > 0$, while Δ_0 is the gap when $T = 0$.) Hence $\delta\Delta \to 0$ when $\Delta(T) \to \Delta_0$. This result is also in agreement with the data from study [60] in which vanishing of the superconductivity enhancement effect in films under microwave irradiation was observed by cooling them below a certain critical temperature.

We will now carry out a quantitative comparison of experiment and theory. The anticipated magnitude of the enhancement effect in Sn may be estimated from the simple relation obtained in study [53] for $T \sim T_c$ and $eV = 2\Delta/5$:

$$\delta\Delta/\Delta_0 \simeq 0{,}5\tau_0 I_0. \qquad (3.1)$$

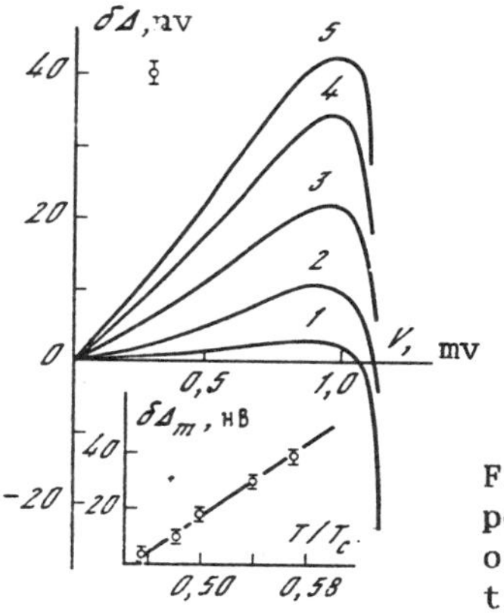

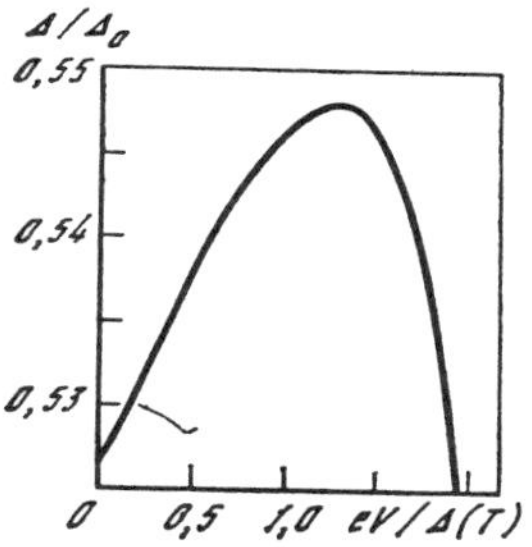

Fig. 30. The energy gap of the superconductor plotted as a function of voltage across the tunnel junction [61]

Fig. 29. The magnitude of the gap change plotted as a function of the voltage across the generator for various temperatures
1 - 1.74 K; 2 - 1.81 K; 3 - 1.91 K; 4 - 2.06 K; 5 - 2.13 K. The insert gives the dependence of the maximum gap change on the adjusted temperature

Taking R_T = 0.02 Ohms, $\tau_0 = 2.3\cdot10^{-9}$ s [39], $N(0) = 1.4\cdot10^{22}$ ev$^{-1}\cdot$cm^{-3}, d = 300 Å, $S_T = 3\cdot10^{-3}$ cm^2, Δ_0 = 0.5 mev, we obtain $\delta\Delta \approx$ 1.7 mev. This result exceeds the experimentally measured value of $\delta\Delta \approx$ 10 nv by approximately two orders of magnitude (for $eV = 2\Delta/5$ and $T/T_c = 0.57$). The reason for this divergence is evidently related to the assumption in study [53] of an equilibrium nature of the phonon distribution. A more realistic case is considered in study [61] where the nonequilibrium distribution functions of quasi particles and phonons are determined on the basis of the solution of a system of coupled kinetic equations, and numerical methods are used to obtain the $\delta\Delta/\Delta(V_g)$ relation. Fig. 30 gives the $\Delta/\Delta_0(V_g)$ relation taken from study [61] for $T = 0.9T_c$ and the coupling parameter

$$A = [\tau_0/2RN(0)e^2d]\left(\frac{T}{\Delta_0}\right)^3 = 0,1. \qquad (3.2)$$

Here R is the specific resistance of the junction and d is the film thickness.

It is clear from a comparison of Figs. 28 and 30 that the overall form of the experimental relation $\delta\Delta(V_g)$ is in qualitative agreement with the theoretical curve. According to study [61] when reducing the coupling parameter A the maximum of the $\delta\Delta(V_g)$ curve will shift towards higher voltages, while the enhancement range may stretch across the entire interval from 0 to $V = 2\Delta/e$. We will estimate the value of A for our case. According to (3.2) using the values of R, τ_0, and $N(0)$ obtained above, we obtain $A = 1\cdot10^{-3}$. The maximum value of $\delta\Delta/\Delta$ in the conditions of Fig. 30 is $4\cdot10^{-2}$. In order to recalculate this value for $A = 10^{-3}$ (here $T = 0.9\ T_c$), we may use relation (3.1), accounting for the fact that A is proportional to the product $\tau_0\ I_0$ and, consequently, $\delta\Delta/\Delta_0 \sim A$. Thus, we obtain for $A = 10^{-3}$ the value $\delta\Delta/\Delta \sim 4\cdot10^{-4}$ when $T/T_c = 0.9$. For a temperature $T = 2.13$ K ($T/T_c = 0.57$) at which the measurements were conducted, we may estimate the value of $\delta\Delta/\Delta$ by assuming a linear dependence of $\delta\Delta/\Delta$ on T/T_c across the entire range $0.4 < T/T_c < 0.9$. This yields for the case $T/T_c = 0.57$ a value of $\delta\Delta/\Delta \simeq 1.4\cdot10^{-4}$ which is in satisfactory agreement with the experimental value of $\delta\Delta/\Delta \simeq 0.8\cdot10^{-4}$.

It is also interesting to consider the results obtained using a double junction whose generator E-I characteristic had a subharmonic structure. There exist several models today [62] that predict such a subharmonic structure: 1) multiparticle tunneling; 2) the influence of Josephson radiation generated in the junction on the tunnel current; 3) multiple Andreev reflection of tunneling quasi particles. Multiple experiments have not yet clarified the true cause of subharmonic features.

The idea of the experiment was to attempt to differentiate the mechanisms that produce the subharmonic structure by its influence on the superconducting gap. This is possible in principle since the first two mechanisms are accompanied by an increase in the concentration of quasi particles (i.e., suppression of the gap), while in the third mechanism the total concentration remains unchanged on both sides of the barrier. In this case in the quasi particle tunneling state gap enhancement will be observed.

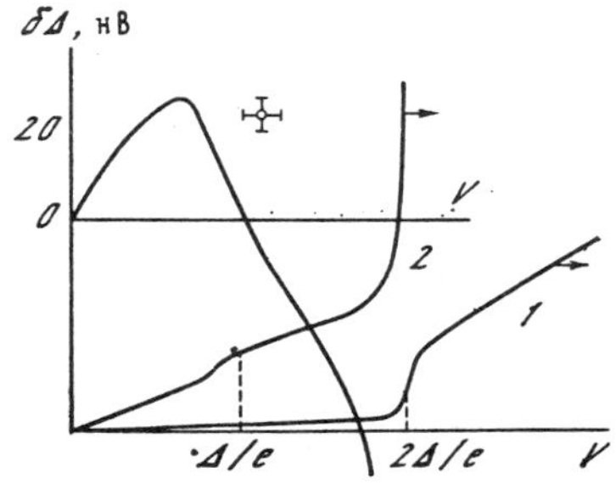

Fig. 31. The gap change plotted as a function of voltage across the generator for a junction with a subharmonic structure (1) and the E-I characteristic of the generator (2)

Fig. 31 gives measurement results for a double Sn-Sn-Sn junction (curve 1 is the E-I characteristic of the test junction-generator, curve 2 is the same E-I characteristic on a magnified (X10) current scale). The subharmonic structure when $eV_g \simeq 2\Delta/2$ is visible and the appearance of this structure may be related to one of the three mechanisms discussed above. It is clear from Fig. 31 that the value of $\delta\Delta$ is positive and increases with an increase in current in the initial section. In the subharmonic region a decay of $\delta\Delta$ is observed together with a further transition to the negative region corresponding to the transition from gap enhancement to gap suppression. On the other hand it was demonstrated previously that gap enhancement is observed across the entire interval $0 < V_g < 2\Delta$ in tunnel junctions whose E-I characteristics do not have a subharmonic structure. Consequently the mechanism that produces the subharmonic structure in this case causes a simultaneous increase in the concentration of quasi particles since the gap is narrowed. Thus, we may conclude that multiple Andreev reflection in this case does not make a determinant contribution to the formation of the subharmonic structure in the E-I characteristics of tunnel junctions.

BIBLIOGRAPHY

1. Kopaev, Yu.V. "Electron and structural transformations and superconductivity" TR. FIAN, 1975, V. 86, p. 3-100.

2. Aronov, A.G., Spivak, B.Z. "Nonequilibrium distributions in superconductors" FNT, 1978, V. 4, No. 1, p. 1365-1403.

3. Chang, J.J., Scalapino, D.J. "Nonequilibrium superconductivity" J. LOW TEMP. PHYS., 1978, V. 31, No. 1/2, p. 1-32.

4. Elesin, V.F., Kopaev, Yu.V. "Superconductors with excess quasi particles" UFN, 1981, V. 133, No. 2, p. 259-307.

5. Eisenmenger, W., Daym, A. "Quantum generation and detection of incoherent phonons in superconductors" PHYS. REV. LETT., 1967, V. 18, No. 4, p. 125-127.

6. Kinder, H., Lassman, K., Eisenmenger, W. "Phonon emission by quasi particle decay in superconducting tunnel junction" PHYS. LETT. A, 1970, V. 31, No. 8, p. 475-476.

7. Daym, A., Miller, B., Weigand, J. "Phonon generation and detection in superconducting lead diodes. PHYS. REV. B - SOLID STATE, 1971, V. 3, No. 9, p. 2949-2961.

8. Wyatt, A.F.G., Dmitriev, V.M., Moore, W.S., Sheard, F.W. "Microwave-enhanced critical supercurrents in constricted tin films" PHYS. REV. LETT., 1966, V. 16, No. 25, p. 1166-1169.

9. Dayem, A.H., Wiegand, J.J. "Behavior of thin-films superconducting bridges in a microwave field" PHYS. REV., 1967, V. 155, No. 2, p. 419-428.

10. Latyshev, Yu.I., Nad', F.Ya. "Microwave-enhanced superconductivity mechanism" ZhETF, 1976, V. 71, No. 6, p. 2158-2167.

11. Dmitriev, V.M., Khristenko, Ye.V. "External electromagnetic emission-induced and stimulated superconductivity" FNT, 1978, V. 4, No. 7, p. 821-856.

12. Testardi, L.R. "Destruction of superconductivity by laser light" PHYS. REV. B, 1971, V. 4, No. 7, p. 2189-2196.

13. Yanson, I.K., Balkashin, O.P., Krasnogorov, A.Yu. "Nonequilibrium filling of quasi-particle states in a superconductor under laser irradiation" PIS'MA V ZhETF, 1976, V. 23, No. 8, p. 458-462.

14. Balkashin, O.P., Yanson, I.K., Kotkevich, A.V. "The tunnel effect in superconductors with nonequilibrium filling of quasi-particle states under laser irradiation" ZhETF, 1977, V. 72, No. 3, p. 1182-1191.

15. Ivlev, V.I. "The influence of light pumping on the tunneling characteristics of superconductors" PIS'MA V ZhETF, 1976, V. 24, No. 2, p. 86-90.

16. Jaworski, F., Parker, W.H. "Experimental determination of the quasi particle energy distribution in optically perturbed superconducting films" PHYS. REV. B, 1979, V. 20, No. 3, p. 945-951.

17. Owen, C.S., Scalapino, D.J. "Superconductivity state under influence of external dynamic pair breaking" PHYS. REV. LETT., 1972, V. 28, No. 24, p. 1559-1562.

18. Vardanyan, R.A., Ivlev, B.I. "The influence of laser emission on superconductivity" ZhETF, 1973, V. 65, No. 6, p. 2315-2326.

19. Parker, W.H., Williams, W.D. "Photoexitation of quasi particles in nonequilibrium superconductors" PHYS. REV. LETT., 1972, V. 29, No. 14, p. 924-927.

20. Fröhlich, H., Müller, F., Weiss, F., Zach, H.G. "Messung der Lebensdauer von optischangeregten Quasiteilchen in supraleitenden Bleischichten" EXP. TECHN. PHYS., 1976, V. 24, No. 4, p. 323-329.

21. Jaworski, F., Parker, W.H., Kaplan, S.B. "Quasi particle and phonon lifetimes in superconductors" PHYS. REV. LETT., 1967, V. 19, No. 1, p. 27-30.

22. Parker, W.H. "Effective quasi particle lifetime in superconducting

Symmetrization" SOLID STATE COMMUN., 1974, V. 15, No. 6, p. 1003-1006.

23. Rothwarf, A., Taylor, B.N. "Measurements of recombination lifetimes in superconductors"PHYS. REV. LETT., 1967, V. 19, No. 1 p. 27-30.

24. Sai-Halasz, G.A., Chi, C.C., Denenstein, A., Langenberg, D.N. "Effect of dynamic external pair breaking of superconducting films" PHYS. REV. LETT., 1974, V. 33, No. 4, p. 215-218.

25. Hu., P., Dynes, P.C., Narayanamurti, W. "Dynamics of quasi particles in superconductors" PHYS. REV. B, 1974, V. 10, No. 7, p. 2786-2788.

26. Golovashkin, A.I., Mitsen, K.V., Motulevich, G.P. "Experimental research on the nonequilibrium state of superconductors under laser excitation" In: 18-3 Vsesoyuz. sobeshch. po fizike nizkikh temperatur [The 18th All Union Conference on Low Temperature Physics] 16-24 September 1974, Topic Paper, Kiev, 1974, p. 477-478.

27. Janik, R., Morelli, L., Cirillo, N.C., et al. "Effects of laser irradiation on weak link devices" IEEE TRANS. MAGN., 1975, V. 11, No. 2, p. 687-699.

28. Buisson, R., Chicault, R., Madeore, F., Romestain, R. "Anomalous resistive states created by light pulses in superconducting films" J. PHYS. LETT., 1978, V. 39, No. 9, p. L130-L133.

29. Golovashkin, A.I., Mitsen, K.V., Motulevich, G.P. "Experimental investigation of the nonequilibrium state of superconductors under laser excitation" ZhETF, 1975, V. 68, No. 4, p. 1408-1412.

30. Baru, V.G., Sukhanov, A.A. "New types of instabilities from nonequilibrium superconductor excitation" PIS'MA V ZhETF, 1975, V. 21, No. 4, p. 209-212.

31. Chang, J.J., Scalapino, D.J. "New instability in superconductors under external dynamic pair breaking" PHYS. REV. B - SOLID STATE, 1974, V. 10, No. 9, p. 4047-4049.

32. Scalapino, D.J., Huberman, B.A. "Onset of an inhomogeneous state in a nonequilibrium superconducting film", PHYS. REV. LETT., 1977, V. 39, No. 2, p. 1365-1368.

33. Elesin, V.F. "The features of the phase transition of optically-pumped nonequilibrium superconductors" ZhETF, 1976, V. 71, No. 4, p. 1490-1501.

34. Elesin, V.F., Kondrashov, V.Ye., Shamraev, B.N. "A theory of opti-

cal pumping breakdown of superconductivity" FTT, 1977, V. 19, No. 8, p. 1525-1527.

35. Elesin, V.F., "Phase transition theory in nonequilibrium superconductors accounting for reabsorption of phonons" FTT, 1977, V. 19, No. 10, p. 2977-2985.

36. Elesin, V.F., Kondrashov, V.Ye., Shatraev, V.N., Sukhikh, A.P. "The dependence of the order parameter of a nonequilibrium superconductor on optical pumping intensity" FTT, 1977, V. 19, No. 12, p. 3702-3704.

37. Elesin, V.F., Kondrashov, V.Ye., Sukhikh, A.S. "Kinetic theory of nonequilibrium superconductors with a broad quasi-particle source" FTT, 1979, V. 21, No. 11, p. 3225-3233.

38. Golovashkin, A.I., Elesin, V.F., Ivanenko, O.M., et al. "The conditions for the onset of a spatially-inhomogeneous state in laser irradiated superconductors" FTT, 1980, V. 22, No. 1, p. 105-109.

39. Kaplan, S.B., Chi, C.C., Langenberg, D.N., et al. "Quasi particles and phonons lifetimes in superconductors" PHYS. REV. B, 1976, V. 14, No. 11, p. 4854-4873.

40. Elesin, V.F. "The nonstationary "intermediate" state of nonequilibrium superconductors" ZhETF, 1977, V. 73, No. 1, p. 355-366.

41. Andreev, V.F. "The thermal conductivity of the intermediate state of superconductors" ZhETF, 1964, V. 46, No. 5, p. 1823-1828.

42. Kashey, V.A. "The resistive domain in a long superconductor of short cross-section in a super high frequency field" FTT, 1978, V. 20, No. 5, p. 1454-1460.

43. Golovashkin, A.I., Motulevich, G.P. "The optical properties of lead in the visible and infrared in a temperature range from room temperature through helium temperatures" ZhETF, 1967, V. 53, No. 5, p. 1526-1532.

44. Gurshi, R.I. "A quantum kinetic equation for electrons in metals" ZhETF, 1957, V. 33, No. 2, p. 451-455.

44a. Gurshi, R.I. "A theory of absorption of electromagnetic waves in metals in the infrared" ZhETF, 1967, V. 33, No. 3, p. 660-664.

45. Parker, W.H. "Modified heating theory of nonequilibrium superconductors" PHYS. REV. B, 1975, V. 12, No. 9, p. 3667-3672.

46. Elesin, V.F. "Nonequilibrium states of superconductors with optical excitation of quasi particles" ZhETF, 1974, V. 66, No. 5, p. 1755-1761.

47. Bean, C.P. "Magnetization of hard superconductors" PHYS. REV. LETT., 1962, V. 8, No. 6, p. 250-253.

48. Elesin, V.F., Lebchenko, Ye.B. "A theory of the phase transition of a nonequilibrium superconductor with tunnel quasi particle injection" FTT, 1978, V. 20, No. 8, p. 2459-2465.

49. Iwanyshyn, O., Leslie, J.D., Smith, H.J.T. "A study of evaporated superconducting weaklink junctions" CANAD. J. PHYS., 1970, V. 48, No. 4, p. 470-476.

50. Jahn, M.T., Kao, Y.H. "Voltage switching effect and microheating in Josephson weak superconductors" J. PHYS. SOC. JAP., 1976, V. 40, No. 2, p. 377-382.

51. Parmenter, R.H. "Enhancement of superconductivity by extraction of normal carriers" PHYS. REV. LETT., 1961, V. 7, No. 7, p. 274-277.

52. Aronov, A.G., Gurevich, V.L. "Tunneling of excitations from a superconductor and growth of T_C" ZhETF, 1972, V. 63, No. 5, p. 1809-1821.

53. Peskovatskiy, S.A., Seminozhenko, V.P. "Enhancement of superconductivity by a tunnel d.c. current" FNT, 1976, V. 2, No. 7, p. 943-945.

54. Eliashberg, G.M. "High frequency field-enhanced superconductivity of films" PIS'MA V ZhETF, 1970, V. 11, No. 3, p. 186-189.

54a. Ivlev, B.I., Lisitsyn, S.G., Eliashberg, G.M. "Nonequilibrium excitation in superconductors in high-frequency fields" J. LOW TEMP. PHYS., 1973, V. 10, No. 3/4, p. 449-468.

55. Tredwell, T.J., Jacobsen, E.H. "Phonon-induced increase in the energy dap of superconducting films" PHYS. REV. B, 1976, V. 13, No. 7, p. 2931-2932.

56. Tinkham, M. "Tunneling generation, relaxation and tunneling detection of hole-electron imbalance in superconductors" PHYS. REV. B, 1972, V. 6, No. 5, p. 1747-1756.

57. Miller, B.I., Dayem, A.H. "Relaxation and recombination times of quasi particles in superconducting Al thin films" PHYS. REV. LETT., 1976, V. 18, No. 23, p. 1000-1004.

58. Levine, J.L., Hsieh, S.Y. "Recombination time of quasi particles in superconducting aluminum" PHYS. REV. LETT., 1968, V. 20, No. 18, p. 994-997.

59. Gray, K.E., Long, A.R., Adkins, C.J. "Measurements of the lifetime of excitations in superconducting aluminum" PHILOS. MAGNETIC., 1969, V. 29, No. 164, p. 273-278.

60. Pals, J.A., Dobben, J. "Measurements of microwave-enhanced super-conductivity in aluminum strips" PHYS. REV. B, 1979, V. 20, No. 3 p. 935-940.

61. Chang, J.J. "Cap enhancement in superconducting thin films due to quasi particle tunnel injection" PHYS. REV. B, 1978, V. 17, No. 5 p. 2137-2140.

62. Octavio, M., Tinkham, M., Blonder, G.E., et al. "Subharmonic energy-gap structure in superconducting constrictions" PHYS. REV. 1983, V. 27, No. 11, p. 6739-6740.

Kinetic Theory of Nonequilibrium Processes in Superconductors

V.G. Valeev, G.F. Zharkov, Yu.A. Kukharenko

Abstract: The Green-Keldysh function technique in the Eliashberg representation is used to formulate a gradient-invariant scheme for describing the kinetics of pure superconductors in external fields that vary slowly in time and are weakly inhomogeneous in space; this scheme is based on kinetic equations for the density matrix of a superconductor with collision integrals that conserve the total particle number. A system of two-fluid hydrodynamic equations is derived for pure superconductors accounting for retardation effects and the frequency-dependent kinetic coefficients are calculated. Collisional relaxation of the order parameter modulus is examined. Relaxation, diffusion and thermal diffusion are investigated and the spectrum of fluctuations in the interbranch imbalance is found. The nonequilibrium shift of the chemical potential in the current state of a superconductor under the influence of thermal inhomogeneities is found.

INTRODUCTION

This study is devoted to an investigation of the nonequilibrium phenomena in superconductors. The quarter century history of the development of the kinetics of superconducting systems - one of the most interesting fields in the modern physics of the low-temperature state - may be traced back to the studies by Geylikman [1] and Bardeen, Rickayzen, and Tewordt [2] who first employed the kinetic equation for the distribution function of normal excitations. Since that time the use of the kinetic equation technique has become an integral attribute of the majority of research on nonequilibrium superconductivity (see monograph [3] containing bibliographic references up through 1970 and survey [4] covering achievements of recent years). Hence the significant attention devoted to the problem of the substantiation of this kinetic scheme proceeding from initial principles [5, 6, 8, 10] in the literature is a natural process.

This approach was formulated in the study by Galayko [6] for application to pure superconductors in external fields that vary sufficiently slowly in time and are weakly inhomogeneous in space. This approach is based on the statement that at times greater than the characteristic formation time of the Cooper pair condensate there is a reduction in the description of the system using the isospin-non-diagonal density matrix in the electron-hole space. As a result the quasi particle description becomes valid and the behavior of the superconductor is completely characterized by the excitation distribution function and the order parameter related by a matching equation whose slow variation is described by a Boltzmann kinetic equation with an interaction constant-squared collision integral. Eliashberg accounted for electron-electron collisions in study [7]. Electron scattering in this case was determined by the matrix element of Coulomb interaction. An analogous formalism was developed by Aronov and Gurevich [8] by means of the Green-Keldysh function technique in the Nambu representation. In study [10] the kinetic equations for the electron excitations and phonons in superconductors were based on a generalization to the nonequilibrium case of the compensation principle of dangerous diagrams in Bogolubov equilibrium perturbation theory [11].

However the collision integrals derived in these and many other studies do not satisfy the conservation law of the particle number. Hence in order to restore the gradient invariance of the theory a continuity equation should be added for the volumetric density of the electrons in this scheme. Only Kulik's study [12] has found the correct electron-phonon collision integral in superconductors, although its primary feature that conserves particle number (its isospin nondiagonal nature in the quasi particle representation) was not taken into account.

An alternative and very effective method of investigating nonequilibrium processes in superconductors is the energy-integrated Green's function technique proposed by Eilenberger [13] and Larkin and Ovchinnikov [14]. Study [14] is based on the Keldysh diagrammatic technique [9] that makes it possible to describe significantly nonequilibrium states of a system that arise from equilibrium with adiabatic switching on of a powerful external field. This approach was later generalized by Eliashberg [7, 15] for the retarded interaction for the case of heavily nonequilibrium states of superconductors. The quasi-classical Green's function technique was further developed in the studies by Gorkov and Kopnin [16, 17, 17a], Larking and Ovchinnikov [18, 20], and Schmid and Schön [19] which also investigated the method of determining quasi particle distributions in this theory which allows description of a situation when the concept of the spectrum of element excitations becomes meaningless. The issue of the penetration of a longitudinal electrical field into the superconductors with an arbitrary impurity concentration was investigated by Ovchinnikov [21]. A detailed description of the primary concepts of this approach may be found in the study by A. Schmid [22] and his later survey [23]. The broad range of application of the quasi-classical Green's function technique satisfying the requirement of gradient invariance makes it a universal apparatus for investigating the kinetics of strongly nonequilibrium superconductors which, however, is not as simple as the kinetic equation method for the quasi particle distribution function and requires very complex calculations.

Shelankov [24]. In the "pure" limit this equation becomes a kinetic equation for the distribution function of normal excitations (see, for example, [8]).

Study [25] has carried out a comparative analysis of results produced in the coincident region of parameters of the system by both these methods. In our opinion the conclusion that these approaches are equivalent in pure superconductors near T_c as cited in study [25] is in error. Specifically this is indicated by a comparison of results from a numerical solution of equations describing the population imbalance of the excitation spectrum in tunnel N-S-structures to experimental material (see survey [26]). The results obtained in this study also point to this conclusion. The reason for this discrepancy is that we cannot ignore the contribution of the components of the density matrix that are nondiagonal in the quasi particle representation nor the collision integrals.

The Nambu representation is used primarily in the investigation of superconductors. The Hamiltonian of the system in this representation contains diagonal field operator-squared terms compared to the exact Hamiltonian. These additions are not significant in describing the dynamics of superconductors, although calculation of the macroscopic quantities in a kinetic equation written in the Nambu representation causes divergent expressions; the spin-index-nondiagonal interactions such as electron-paramagnetic impurity collisions cannot be accounted for correctly in this representation which describes only the isotopic degree of freedom. This difficulty was overcome by Eliashberg [27] who correctly formulated the matrix representation in the spin-isospin space.

The purpose of this study was to explain the reasons why the kinetic equation for the distribution function of normal excitations of a superconductor is not in agreement with the continuity equation for the electron density and to formulate a kinetic gradient-invariant theory of nonequilibrium processes in superconductors.

The material is presented in the following sequence. In the first chapter we use the Green-Keldysh function technique in the Eliashberg representation to derive kinetic equations for a single-particle density matrix of a superconductor with collision integrals that conserve the number of particles. The different symmetry of the true Hamiltonian and the Hamiltonian with a self-consistent Hartree-Fock-Bogolubov field with a nonzero value of the order parameter causes the collision integral for collisions between electrons and any excitations and impurities in the superconductor to not be a diagonal matrix in the quasi particle ν-representation, and it is fundamentally necessary to account for the components of the density matrix that are nondiagonal in the ν-representation together with the collision integral to observe the law of conservation of particle number and gradient invariance of the theory. The structure of the collision integrals obtained here is such that even the existence of deviations from equilibrium in the "ν-diagonal channel" alone over significant times whose scale is given by the scattering mechanism produces ν-nondiagonal components of the density matrix determining the phase of the order parameter. In the first chapter we calculate the electron-electron collision integral in Lenard-Balescu form that accounts for the multiparticle nature of scattering in the low-temperature state of matter as well as the collision integrals for superconductor electron-phonon collisions and collisions with nonmagnetic and paramagnetic impurities. For all these interactions we examine an approximation of the relaxation time and calculate the corresponding collision frequencies (for the electron-electron interaction we use a model with point interaction) near absolute zero and in the vicinity of the phase transition point, yet outside the critical region. For each of the collision integrals calculated in the study we find a complete set of additive invariants and the equilibrium solution. It is demonstrated that by accounting for the components of the kinetic equation that are nondiagonal in the quasi particle representation the conservation law of the number of particles is an automatic consequence of the equation for the density matrix and thus is not an independent equation. This fact provides explicit gradient invariance of the kinetic scheme developed here which consists of a kinetic equation for a single-electron density matrix, and matching equations for the modulus and phase of the order parameter and Maxwell's equations. The matching equation for the phase of the order parameter becomes useful only when accounting for the components of the density matrix that are nondiagonal in the quasi particle representation.

In this chapter the kinetic equation method including collision integrals that conserve the number of particles developed in the first chapter is used to obtain two-fluid hydrodynamic equations of an uncharged superconducting Fermi-fluid in a viscous approximation and for calculating the kinetic coefficients entering into these equations. This problem was considered previously in studies by Shumejko [28] and Soda and Fudzhiki [29] (in the latter study the problem of bulk viscosity was not considered) based on a kinetic equation for the quasi particle distribution function. However the technique for deriving the hydrodynamic equations based on the existence of conservation laws evidently cannot be carried out correctly within the scope of this approach: the Boltzmann equation for the quasi particle distribution function may be obtained from an equation for a single-electron density matrix ignoring the components of the density matrix nondiagonal in the quasi particle representation and the collision integrals and, as stated above, for this reason it contradicts the continuity equation for the electron density. The electron-electron collision integral calculated in Chapter 1 contains a complete set of additive invariants; hence, in the hydrodynamic stage the kinetic equation for the density matrix of a Fermi-fluid may be solved by the Chapman-Enskog classical method [30]. A significant element of the application of this method is the search for a complete set of right eigenvectors of the collision integral corresponding to the zeroth eigenvalue; due to the nonhermitian nature of the collision integrals in the superconductors, their right and left eigenvectors do not coincide. As it turns out the right eigenvectors that cause the collisional terms to vanish and correspond to conservation of the number of particles and energy are nondiagonal in the quasi particle representation. A solution of the kinetic equation has revealed that at significant times the ν-nondiagonal components of the density matrix near T_c make the primary contribution to the coefficient of bulk viscosity and are significant for describing slow equilibrium processes in the excitation gas. Chapter 2 also examines the damping of the first and second sounds in a nonlinear superconducting Fermi-fluid. An analysis of experimental research aimed at measuring dissipative corrections makes it possible to conclude that the calculations carried out in this study have been confirmed experimentally.

In Chapter 3 we examine a number of specific problems of the kinetic theory of relaxation processes in superconductors that are important both in methodological and applied aspects. Electron-electron collisions, electron-phonon collisions and electron-impurity collisions play the predominant role in the relaxation processes of a superconducting system. An important requirement that allows us to correctly account for scattering processes in the formulation of a solution of the kinetic equation is the need to observe solvability conditions of the integrodifferential equation for the nonequilibrium addition to the density matrix. Observing these solvability conditions imparts some meaning to the very procedure of solving the kinetic equation and makes it possible to obtain a physically adequate picture of the processes under consideration. For this purpose in this study we propose a generalization of the Chapman-Enskog method using the solution of Boltzmann's equation valid for all values of the perturbation frequency ω within the scope of applicability of the kinetic equation for the density matrix ($\omega \ll \Delta$) in the range of wave numbers of the perturbation $k \ll v_F^{-1} | i\omega + \nu_p|$ where ν_p is the transport momentum relaxation time of the electrons. The application of this method to the kinetic equation derived above for the density matrix of the superconductor allows us to consistently account for the operator properties of the collision integral and results in a system of two-fluid hydrodynamic equations accounting for retardation. This system of equations, after calculating the kinetic coefficients, becomes closed and allows complete description of the behavior of the superconductor in this region (ω, k). In Section 3 of Chapter 3 (after a brief introduction and a presentation of the method of solving the kinetic equation (Section 1) and calculation of the tunnel electron source (Section 2), which, like the collision integrals, is a nondiagonal matrix in the quasi particle representation) expressions are found for the kinetic coefficients dependent on the perturbation frequency and that account for the retardation effects. In Section 4 we examine the behavior of the modulus of the order parameter and we solve the problem of collisional relaxation of small spatially-homogeneous perturbations of the modulus of the order parameter $\delta\Delta$ near T_c (yet outside the critical temperature range). The asymptotic expression found for the $\delta\Delta(T)$ relation shows that the damping of these perturbations over significant times is determined by inelastic collisions.

Section 5 of this chapter is devoted to an investigation of the evolution of the branch population imbalance of the excitation spectrum of superconductors $\delta\tilde{\mu}$. Here we use two-fluid hydrodynamic retardation equations to derive an equation that describes the spatial and temporal behavior of $\delta\tilde{\mu}$ with tunnel injection and thermal inhomogeneities, and the possibility for an experimental test of the derived expressions is discussed for the corresponding diffusion coefficients and relaxation frequency that differ from those found previously using other techniques.

In Section 6 we examine the thermoelectric effect; the conditions for observing this effect in superconductors were identified by Ginzburg [32]. Using the equation derived in Section 5 for the quantity $\delta\tilde{\mu}$ characterizing the interbranch imbalance in superconductors we examine the nonstationary thermoelectric effect in a bimetallic ring consisting of various superconductors. After analysis of the experimental data obtained to date in experiments to measure the thermoelectric addition to the magnetic flux penetrating the bimetallic ring, the nonequilibrium shift of the chemical potential in the current state of the superconductor with a temperature gradient is found and a method of testing this result is proposed (Section 6).

In Section 7 we derive the spectral function of fluctuations in the population imbalance of the excitation spectrum of a superconductor using hydrodynamics with retardation and the linewidth of Josephson radiation is found.

All results from the third chapter described above were derived in the situation where the external actions do not excite a spin degree of freedom. In the final, ninth section of this chapter by solving a kinetic equation for the density matrix of the superconductor we derive a hydrodynamic retardation equation describing relaxation, diffusion and convective transport of the nonequilibrium spin density in superconductors with a tunnel source of polarized electrons, and the frequency-dependent coefficient of spin diffusion is calculated.

Chapter 1

KINETIC EQUATIONS FOR SUPERCONDUCTORS WITH COLLISION INTEGRALS PRESERVING PARTICLE NUMBER

In this chapter we use the Green-Keldysh function technique in the Eliashberg representation [27] to derive kinetic equations for a single-electron density matrix with collision integrals that conserve the number of particles. We then generalize the Balescu [33] and Lenard [34] result to the case of superconductors and find the expression for the electron-electron collision integral that accounts for the dynamical screening of Coulomb interaction that does not contain divergences with small momenta. The electron-electron collision integral is calculated in a point interaction model together with the collision integrals for electron-phonon, electron-nonmagnetic impurity and electron-paramagnetic collisions. We show that the characteristic feature of the collision integrals for electron collisions with any excitations and impurities in superconductors is their nondiagonal nature in the quasi particle ν-representation. A comprehensive accounting of the ν-nondiagonal components of the density matrix and the collision integrals in investigating nonequilibrium phenomena in superconductors is closely related to conservation of particle number and is fundamentally important for meeting the requirement of gradient invariance.

For all indicated collision integrals we derive an approximation of the relaxation time. The relaxation frequencies that arise here are calculated in two limiting cases: near absolute zero and near the critical point.

1. Derivation of the kinetic equation

We will formulate a scheme with a self-consistent field in the Gorkov model and will account for electron-electron, electron-phonon and electron-impurity collisions as a source of relaxation only [12]. The Hamiltonian with point interaction written for the "vectors" ψ_α, $\alpha = 1$-4, with the components $(\psi_\uparrow, \psi_\downarrow^+, \psi_\uparrow^+, \psi_\downarrow)$, takes the form [27]

$$H = \int d\mathbf{r}\psi^\alpha(\mathbf{r}) N_{\alpha\beta} H_0(\mathbf{r}) \psi_\beta(\mathbf{r}) - \\ - \frac{1}{2}|g| \int d\mathbf{r}\psi^\alpha(\mathbf{r}) N_{\alpha\beta}\psi_\beta(\mathbf{r})\psi^\gamma(\mathbf{r}) N_{\gamma\delta}\psi_\delta(\mathbf{r}), \tag{1.1}$$

where

$$\psi^\alpha = I^{\alpha\beta}\psi_\beta, \quad \tilde{I} = \begin{pmatrix} \hat{0} & \hat{1} \\ \hat{1} & \hat{0} \end{pmatrix}, \quad \tilde{N} = \begin{pmatrix} \hat{\sigma}_\uparrow & \hat{0} \\ \hat{0} & \hat{\sigma}_\downarrow \end{pmatrix},$$
$$\hat{\sigma}_\uparrow = \frac{1}{2}(\hat{1} + \hat{\sigma}_z), \quad \hat{\sigma}_\downarrow = \frac{1}{2}(\hat{1} - \hat{\sigma}_z), \quad H_0 = p^2/2m - \lambda,$$

where the wavy line prime designates matrices in the spin-isospin space, differentiated by the isospin index from two-dimensional matrices which we will designate in the regular manner; σ_i are standard Pauli matrices in the real particle representation (the s-representation). We will consider a nonferromagnetic system when external actions are diagonal with respect to the spin index. We will determine the Green-Keldysh function in the normal manner:

$$(\tilde{G})_\beta^{\alpha, ik}(1,1') = -i \langle T_c \psi_\beta(1_i)\ \psi^\alpha\ (1'_k)\rangle, \tag{1.2}$$

where $(1) = (\mathbf{r}_1 t_1)$, the indices i and k characterize which of the two branches of the Keldysh contour [9] the time coordinate belongs to; $\langle\ldots\rangle$ is averaging with the density matrix taken at time $t = t_0$ in which the Heisenberg operators coincide with the Schrodinger operators. Functions (1.2) have a block structure of the type

$$\begin{pmatrix} \hat{A} & \hat{0} \\ \hat{0} & \hat{B} \end{pmatrix},$$

where the diagonals contain the Green-Keldysh functions in the Nambu representation. Summation of the diagrams results in the Dyson equation:

$$\tilde{G}_0^{-1}(1)\tilde{G}(1,1') = \tilde{E}\hat{\sigma}_z^{(k)}\delta(1-1') + \int d2\tilde{E}\hat{\sigma}_{(z)}^{(k)}\tilde{\Sigma}(1,2)\tilde{G}(2,1'), \tag{1.3}$$

$$\tilde{G}(1,1')[\tilde{G}_0^{-1}(1')]^{+} = \tilde{E}\hat{\sigma}_z^{(k)}\delta(1-1') + \int d2\tilde{G}(1,2)\tilde{\Sigma}(2,1')\tilde{E}\hat{\sigma}_z^{(k)}. \tag{1.4}$$

In (1.3)-(1.4) $\tilde{\Sigma}$ is the self-energy matrix where the regular graphical technique is used to calculate the matrix. Each solid or wavy line corresponds to the matrix electron Green's function or the Green's function of the field/interaction carrier (determined analogous to (1.2) accounting for the statistics), multiplied by (i); the simple tip corresponds to the matrix $\tilde{\Gamma}_0$:

$$\tilde{\Gamma}_0 = \tilde{N} - \tilde{I}\tilde{N}\tilde{I} = \begin{pmatrix} \hat{\sigma}_z & \hat{0} \\ \hat{0} & \hat{\sigma}_z \end{pmatrix}, \tag{1.5}$$

while each closed loop with a single tip is assigned an additional (with respect to the multiplier in the Nambu technique) factor one-half;

$$\tilde{G}_0^{-1}(1) = i\tilde{E}\frac{\partial}{\partial t_1} + \tilde{\Gamma}_0 H_0(1), \tag{1.6}$$

where $\tilde{E}$ is the identity matrix in the spin-isospin space; $\sigma_z^{(k)}$ is the matrix with the Keldysh index.

According to study [35] in order to satisfy a differential law of preservation of the number of true particles it is necessary for the approximation of the Green's function $\tilde{G}$ to satisfy equations (1.3) and (1.4) simultaneously. Drafting the difference of these equations we obtain for G_{-+}

$$i\tilde{E}\left(\frac{\partial}{\partial t_1} + \frac{\partial}{\partial t_2}\right)\tilde{G}^{-+}(1,2) = \tilde{\Gamma}_0 H_0(1)\tilde{G}^{-+}(1,2) - \\ - H_0(2)\hat{G}^{-+}(1,2)\tilde{\Gamma}_0 + \tilde{J}^{-+}(1,2), \tag{1.7}$$

where

$$\tilde{J}^{-+}(1,2) = \int d3\,\{\tilde{\Sigma}^{-i}(1,3)\tilde{G}^{i+}(3,2) + \tilde{G}^{-i}(1,3)\tilde{\Sigma}^{i+}(3,2)\}; \tag{1.8}$$

summation is implied with respect to all repeating indices here and henceforth; in (1.8) I = (-; +) are the traditional values of the Keldysh indices.

In order to calculate the field part of the kinetic equation we must find the integral term in (1.7) in the first interaction order. In the Gorkov model $D_0^{++}(1, 2) = -D_0^{--}(1, 2) = |g|\delta(1-2)$, $D_0^{-+} = D_0^{+-} = 0$. By including the subtraction result in the energy operator of the self-consistent field $\mathcal{E}(\mathbf{r}t)$ we may write (1.7) as

$$iE\left(\frac{\partial}{\partial t_1}+\frac{\partial}{\partial t_2}\right)\tilde{G}^{-+}(1,2)=\tilde{\varepsilon}(1)\tilde{G}^{-+}(1,2)-\tilde{G}^{-+}(1,2)\tilde{\varepsilon}(2)+\mathcal{J}'^{+}(1,2), \qquad (1.9)$$

where

$$\tilde{\varepsilon}(1)=\begin{pmatrix}\hat{\varepsilon}(1) & \hat{0}\\ \hat{0} & -\hat{\varepsilon}^T(1)\end{pmatrix}, \qquad (1.10)$$

$$\hat{\varepsilon}(1)=\xi(1)\hat{\sigma}_z-\Delta(1)\hat{\sigma}_+-\Delta^*(1)\hat{\sigma}_--{}^1/_2|g|n(1)\hat{\sigma}_z+{}^1/_2|g|ns_z(1); \qquad (1.11)$$

in the s-representation

$$\hat{\sigma}_x=\begin{pmatrix}0 & 1\\ 1 & 0\end{pmatrix}, \quad \hat{\sigma}_y=\begin{pmatrix}0 & -i\\ i & 0\end{pmatrix}, \quad \hat{\sigma}_\pm={}^1/_2(\hat{\sigma}_x\pm i\hat{\sigma}_y),$$
$$\xi(1)=p_1^2/2m-\lambda, \qquad (1.12)$$
$$n(1)=\mathrm{Sp}\,(\hat{\sigma}_z\hat{\rho}(\mathbf{r}_1\mathbf{r}_2t_1)+{}^1/_2\delta(\mathbf{r}_1-\mathbf{r}_2)\}|_{\mathbf{r}_2\to\mathbf{r}_1}$$

is the volumetric electron density;

$$s_z(1)=\mathrm{Sp}\,\{\hat{\rho}(\mathbf{r}_1\mathbf{r}_2t_1)-{}^1/_2\delta(\mathbf{r}_1-\mathbf{r}_2)\}|_{\mathbf{r}_2\to\mathbf{r}_1} \qquad (1.13)$$

is the density of the projection of the electron spin onto the quantization axis (O_z);

$$\Delta(1)=|g|\ \mathrm{Sp}\,\{\hat{\sigma}_+\hat{\rho}(\mathbf{r}_1\mathbf{r}_1t_1)\} \qquad (1.14)$$

is the complex order parameter; $J'^{-+}(1,2)$ is the contribution of diagrams of order higher than one in g. We employed the relation between the nonequilibrium density matrices in the Eliashberg and Nambu representations:

$$\tilde{\rho}(\mathbf{r}_1\mathbf{r}_2t)=\begin{pmatrix}\hat{\rho}(\mathbf{r}_1\mathbf{r}_2t), & \hat{0}\\ \hat{0}, & \hat{1}-\hat{\rho}^T(\mathbf{r}_2\mathbf{r}_1t)\end{pmatrix}. \qquad (1.15)$$

The gradient invariance of equation (1.9) makes it possible to identify the phase of the order parameter in the normal manner: carry out the transformation of all matrices in (1.9) as

$$\tilde{\Pi}(1,2)=\exp\left(\frac{i}{2}\tilde{\Gamma}_0\chi(1)\right)\tilde{\Pi}'(1,2)\exp\left(-\frac{i}{2}\tilde{\Gamma}_0\chi(2)\right) \qquad (1.16)$$

and require that after this the order parameter Δ becomes real: $\mathrm{Im}(\Delta')=0$. We will drop the prime in all cases below and will remember that transformation (1.16) has already been done. The function $\chi(\mathbf{r}t)$ infers the phase of the order parameter; the equation for this quantity will take the form

$$\mathrm{Sp}\,\{\hat{\sigma}_y\rho(\mathbf{r},\mathbf{r},t)\}=0. \qquad (1.17)$$

Converting to the variables $\mathbf{r}=1/2(\mathbf{r}_1+\mathbf{r}_2)$, $\mathbf{R}=\mathbf{r}_1-\mathbf{r}_2$, $t=1/2(t_1+t_2)$, $\tau=T_1-T_2$ and using the identity

$$\tilde{G}^{-+}(\mathbf{r}_1t_1,\mathbf{r}_2t_2)|_{t_2=t_1}=i\tilde{\rho}(\mathbf{r}_1\mathbf{r}_2t_1), \qquad (1.18)$$

we obtain in the mixed Wigner representation for the function $\rho(\mathbf{r}\mathbf{p}t)$:

$$\frac{\partial}{\partial t}\tilde{\rho}+\frac{1}{2}\left\{\left[\frac{\partial\tilde{\varepsilon}_p}{\partial\mathbf{p}},\frac{\partial}{\partial\mathbf{r}}\tilde{\rho}\right]_+-\left[\frac{\partial\tilde{\varepsilon}_p}{\partial\mathbf{r}},\frac{\partial}{\partial\mathbf{p}}\tilde{\rho}\right]_+\right\}+i\,[\tilde{\varepsilon}_{pr}\tilde{\rho}]_-=-\tilde{J}'^{-+}(\mathbf{r}\mathbf{p}t). \quad (1.19)$$

Here $[\tilde{A},\ \tilde{B}]_{\pm} = \tilde{A}\tilde{B} \pm \tilde{B}\tilde{A}$; $\mathbf{p}$ is the momentum that is Fourier-conjugate to the difference coordinate $\mathbf{r}$. We have accounted for the fact that the energy operator $\varepsilon_p(\mathbf{r}t)$:

$$\tilde{\varepsilon}_p=\begin{pmatrix}\tilde{\varepsilon}_p(\mathbf{r}t) & \hat{0}\\ \hat{0} & -\hat{\varepsilon}_{-p}(\mathbf{r}t)\end{pmatrix}, \quad (1.20)$$

where

$$\hat{\varepsilon}_p=\tilde{\xi}_p\hat{\sigma}_z-\Delta\hat{\sigma}_x+\mathbf{p}\mathbf{v}_s+\frac{1}{2}|g|\,s_z,$$

$$\tilde{\xi}_p=p^2/2m-\tilde{\mu},\quad \tilde{\mu}=\lambda-\frac{1}{2}n\,|g|-\Phi-\frac{1}{2}mv_s^2, \quad (1.21)$$

$$\Phi(\mathbf{r}t)=\frac{1}{2}\frac{\partial}{\partial t}\chi(\mathbf{r}t)+e\varphi(\mathbf{r}t) \quad (1.22)$$

is the gradient-invariant potential,

$$\mathbf{v}_s=\frac{1}{2m}\frac{\partial}{\partial\mathbf{r}}\chi(\mathbf{r}t)-\frac{e}{m}\mathbf{A} \quad (1.23)$$

is the superfluid velocity; $e = -|e|$ is the electron charge; $c = \hbar = 1$; $\tilde{\varepsilon}_p$ is not transposed with the density matrix; $(\varphi,\ \mathbf{A})$ is the external field that we included in (1.20). In deriving equation (1.19) we accounted for the fact that $\tilde{\varepsilon}p$ and ρ vary slowly at distances in the order of the coherence length $\xi(T)$ although we have not assumed slowness of the change in these quantities in time.

In order to find the collision integral we will consider the function $\tilde{J}'^{-+}(\mathbf{r},t)$. We may show that this quantity contains two types of contributions. The first results from electron interaction with the virtual quanta of the field/interaction carrier and causes corrections to appear in the excitation spectrum; qualitatively this effect may be accounted for by renormalizing the constant g; we will assume that such renormalization is complete and will henceforth limit our examination to terms of the second type resulting from the interaction of electrons and the real field quanta and corresponding to scattering processes. If all quantities vary slowly over lengths of $\sim\xi(T)$ and over times $\sim 1/\Delta$, the desired collision integral $\tilde{R}^{(2)}$ may be written as

$$\tilde{R}^{(2)}(\mathbf{r}\mathbf{p}t)=\frac{1}{2}\int_{-\infty}^{\infty}\frac{d\omega}{2\pi}\{\tilde{G}^{-+}(\mathbf{r}\mathbf{p}t\omega)\,\tilde{\Sigma}^{+-}(\mathbf{r}\mathbf{p}t\omega)-\tilde{G}^{+-}(\mathbf{r}\mathbf{p}t\omega)\,\tilde{\Sigma}^{-+}(\mathbf{r}\mathbf{p}t\omega)\}^{(2)}+(\text{h. c.}) \quad (1.24)$$

(here we have accounted for diagrams $\sim g^2$). Obviously (1.24) becomes the regular expression for the collision term of the normal system, if $\tilde{G}$ and Σ commute. Here ω is the frequency that is Fourier-conjugate to the difference time τ.

We will show that in superconductors as in normal metals the total number of electrons is conserved in the collisions. For this we will consider the quantity $\mathcal{F}(\tilde{A}) = \langle \tilde{A}\tilde{R}^{(2)}(\mathbf{rp}t)\rangle$, where $\tilde{A}$ is a random matrix, $\langle\ldots\rangle = \sum_{\mathbf{p}} \mathrm{Sp}(\ldots)$. The corrections to the vertices in the superconductors are small, as in normal metals [36, 37]. In the simple vertex approximation we will have:

$$\mathcal{F}(\tilde{A}) = \frac{1}{2}\,\mathrm{Sp}\int \frac{1}{(2\pi)^8}\, d^4P\, d^4Q\, [\tilde{A}, \tilde{\Gamma}_0]_- \times \\ \times \tilde{G}^{-+}(P+Q)\tilde{\Gamma}_0\tilde{G}^{+-}(P)\,[iD^{+-}(Q)] + (\text{h. c.}). \tag{1.25}$$

We utilize the identity

$$D^{+-}(Q) = D^{-+}(-Q), \qquad Q = (\mathbf{q},\ \omega_q)$$

and for simplicity we have dropped the sum coordinates in the arguments of the Green's functions. Formula (1.25) is suitable not only for electron-electron and electron-phonon interactions, but also for electron-impurity interactions in the Abrikossov-Gorkov model [38, 39] (the squared modulus of the Fourier-component of the electron-impurity interaction potential plays the role of the D functions here). In the case of paramagnetic impurities $\tilde{\Gamma}_0$ should be replaced by $\tilde{\Gamma}_S$ (see below). Obviously $[\tilde{N}, \tilde{\Gamma}_0]_- = 0$. Below we will see that $[\tilde{N}, \tilde{\Gamma}_S]_- = 0$. Hence $\mathcal{F}(\tilde{N}) = 0$ or

$$\langle \tilde{N}\tilde{R}^{(2)}(\mathbf{rp}\ t)\rangle = 0. \tag{1.26}$$

We may easily show that the continuity equation for the volumetric electron density follows automatically from identity (1.26) and kinetic equation (1.19) with collision integral (1.24).

Further,

$$[\tilde{\sigma}_z, \tilde{\Gamma}_0]_- = 0$$

corresponds to the projection of the electron spin onto the O_z axis in the Eliashberg representation (see below). This means that in electron-electron, electron-phonon, and electron-nonmagnetic impurity collisions the spin projection is conserved, and scattering by paramagnetic impurities causes relaxation of the spin density. Analysis reveals that in a nonferromagnetic system when external actions are diagonal in the spin space, the collision integral $\hat{r}^{(2)}$ has a diagonal block structure. We then by definition set

$$\tilde{R}^{(2)}(\mathbf{rp}t) = \begin{pmatrix} \hat{R}^{(2)}(\mathbf{rp}t) & \hat{0} \\ \hat{0} & -\hat{R}^{(2)T}(\mathbf{r}, -\mathbf{p}, t) \end{pmatrix}. \tag{1.27}$$

It is clear that in this situation we may limit our examination to the kinetic equation for the density matrix in the Nambu representation $\hat{\rho}(\mathbf{rp}t)$. Thus, the desired kinetic equation takes the form

$$\frac{\partial}{\partial t}\hat{\rho} + \frac{1}{2}\left\{\left[\frac{\partial \hat{\varepsilon}_p}{\partial \mathbf{p}}, \frac{\partial}{\partial \mathbf{r}}\hat{\rho}\right]_+ - \left[\frac{\partial \hat{\varepsilon}_p}{\partial \mathbf{r}}, \frac{\partial}{\partial \mathbf{p}}\hat{\rho}\right]_+\right\} + i\,[\hat{\varepsilon}_p, \hat{\rho}]_- = \hat{R}^{(2)}(\mathbf{rp}t). \tag{1.28}$$

Kinetic equation (1.28) together with the matching equations for the modulus of the order parameter

$$\Delta = {}^1/_2 \mid g \mid \langle \hat{\sigma}_x \hat{\rho} \rangle \tag{1.29}$$

and its phase

$$0 = \langle \hat{\sigma}_y \hat{\rho} \rangle \tag{1.30}$$

and Maxwell's equations form a complete set of equations for pure superconductors if we express its collision integral through the density matrix.

2. Collision integrals and relaxation frequencies

In order to calculate the collision integrals we must find the Green-Keldysh functions in the approximation of a self-consistent field limiting our examination to the locally-homogeneous solution of the Dyson equations for g. For example, for $\tilde{G}^{-+}$ in the approximation of the local homogeneity of the system we may easily obtain Cauchy's problem in τ:

$$\begin{gathered} 2i \frac{\partial}{\partial \tau} \tilde{G}^{-+}(\mathbf{rp}t\tau) = [\tilde{\varepsilon}_p(\mathbf{r}t), \tilde{G}^{-+}(\mathbf{rp}t\tau)]_+, \\ \tilde{G}^{-+}(\mathbf{rp}t0) = i\tilde{\rho}(\mathbf{rp}t), \end{gathered} \tag{1.31}$$

where $\tilde{\rho}$ is the nonequilibrium single-particle density matrix and from here

$$\tilde{G}^{-+}(\mathbf{rp}t\tau) = i \exp\left(-\frac{i}{2}\tilde{\varepsilon}_p\tau\right)\tilde{\rho}(\mathbf{rp}t)\exp\left(-\frac{i}{2}\tilde{\varepsilon}_p\tau\right). \tag{1.32}$$

Analogously

$$\tilde{G}^{+-}(\mathbf{rp}t\tau) = i \exp\left(-\frac{i}{2}\tilde{\varepsilon}_p\tau\right)[\hat{E} - \tilde{\rho}(\mathbf{rp}t)]\exp\left(-\frac{i}{2}\tilde{\varepsilon}_p\tau\right). \tag{1.33}$$

The approximation of a self-consistent field which is used to calculate Green's functions (1.32)-(1.33) is justified if the collision frequencies are small compared to the order parameter; otherwise, such as the case of "dirty" superconductors, it is necessary to account for the renormalization of the vertex and corrections to the spectrum related to the electron-impurity interaction. The assumption of local homogeneity imposes a constraint on the values of the wave numbers k and frequencies ω of the external perturbations acting on the system:

$$k\xi(T) \ll 1, \quad \omega \ll \Delta. \tag{1.34}$$

Relations (1.34) have the range of applicability of kinetic equation (1.28) with the collision integral found in this section.

It is convenient to examine equation (1.28) in the quasi particle (ν^-) representation. This is the representation in which the matrix $\hat{\varepsilon}(\mathbf{r}t)$ is diagonal.

We will make the transformation from the s-representation to the ν-representation by means of the Bogolubov transform with the matrix $\hat{U}_p$:

$$\hat{U}_p = U_p\hat{1} - v_p i\hat{\sigma}_y, \tag{1.35}$$

where

$$\left.\begin{matrix} u_p \\ v_p \end{matrix}\right\} = \frac{1}{\sqrt{2}}\left(1 \pm \frac{\tilde{\xi}_p}{\varepsilon_p}\right)^{1/2}, \qquad \varepsilon_p = \sqrt{\xi_p^2 + \Delta^2}. \tag{1.36}$$

We will introduce the matrices $\{\hat{\tau}\}$ having the form of standard Pauli matrices in the ν-representation. Then

$$\hat{\varepsilon}_p = \varepsilon_p\hat{\tau}_3 + \mathbf{p}\mathbf{v}_s + {}^1/_2\,|\,g\,|\,s_z. \tag{1.37}$$

For convenience we incorporate one additional matrix $\hat{E}_p$ by the formula

$$\hat{E}_p = \hat{\varepsilon}_p + (\mathbf{p}\mathbf{v}_s + {}^1/_2\;|\,g\,|\,s_z)\;\hat{\tau}_1. \tag{1.38}$$

Identity (1.26) in condition (1.27) is equal to

$$\langle\hat{\sigma}_z\hat{R}^{(2)}\{\tilde{\rho}\}\rangle = 0. \tag{1.39}$$

The matrix $\hat{\sigma}_x$ is nondiagonal in the ν-representation and takes the form

$$\hat{\sigma}_z = \frac{\tilde{\xi}_p}{\varepsilon_p}\hat{\tau}_3 + \frac{\Delta}{\varepsilon_p}\hat{\tau}_1. \tag{1.40}$$

We may easily see that identity (1.39) reflecting conservation of the number of particles from the collisions is satisfied for any $\hat{\rho}$ only in accounting for the components of the collision integral that are non-diagonal in the ν-representation. Without any additional stipulations we will write out the expressions for the function $\hat{R}^{(2)}\{\hat{\rho}\}$.

1. **The Lenard-Balescu collision integral in superconductivity theory.** Here we will generalize the result from Balescu [33] and Lenard [34] to the case of superconductors and will obtain the electron-electron collision integral that accounts for the dynamical screening of Coulomb interaction and is free from divergences with low momenta.

We will write the Dyson equation for the Green's functions $D(Q)$ in the loop approximation limiting ourselves to simple vertices. For example,

$$D^{++}(Q) = D_0^{++}(Q) + D_0^{++}(Q)\,\pi^{++}(Q)\,D^{++}(Q) + D_0^{++}(Q)\,\pi^{+-}(Q)\,D^{-+}(Q); \tag{1.41}$$

$$D^{+-}(Q) = D_0^{++}(Q)\,\pi^{++}(Q)\,D^{+-}(Q) + D_0^{++}(Q)\,\pi^{+-}(Q)\,D^{--}(Q), \tag{1.42}$$

where

$$\pi^{++}(Q) = i\int \frac{d^4L}{(2\pi)^4}\,\mathrm{Sp}\,\{\hat{\sigma}_z \hat{G}^{++}(L+Q)\,\hat{\sigma}_z \hat{G}^{++}(L)\},$$
$$\pi^{+-}(Q) = i\int \frac{d^4L}{(2\pi)^4}\,\mathrm{Sp}\,\{\hat{\sigma}_z \hat{G}^{+-}(L+Q)\,\hat{\sigma}_z \hat{G}^{-+}(L)\}; \tag{1.43}$$
$$D_0^{++}(Q) = V(q) = 4\pi e^2/q^2 \tag{1.44}$$

is the Green's function of the free Coulomb field.

It is easy to see from (1.42) and the analogous equation for $D^{-+}(Q)$ that diagrams with D^{-+} and D^{+-} have additional interaction smallness with respect to the contribution to the self-energy parts from diagrams with D^{++} and D^{--} and correspond to the accounting of ternary collisions. Limiting our examination to pair collisions only we obtain, for example, for Σ^{-+}:

$$-i\hat{\Sigma}^{-+}(P) = -\int \frac{d^4P'}{(2\pi)^4}\,\hat{\sigma}_z i\hat{G}^{-+}(P')\,\hat{\sigma}_z\,\mathrm{Sp}\int \frac{d^4P_1}{(2\pi)^4}\,\{\hat{\sigma}_z i\hat{G}^{-+}(P_1)\,\hat{\sigma}_z i\hat{G}^{+-}(P_1')\}\,\times$$
$$\times\,\delta(P + P_1 - P_1' - P')\,iD_0^{--}(P'-P)\,iD^{++}(P'-P) +$$
$$+\int \frac{d^4P'd^4P_1}{(2\pi)^8}\,\hat{\sigma}_z i\hat{G}^{-+}(P_1')\,\hat{\sigma}_z i\hat{G}^{+-}(P_1)\,\hat{\sigma}_z i\hat{G}^{-+}(P')\,\hat{\sigma}_z\delta(P + P_1 - P_1' - P')\,\times$$
$$\times\,iD_0^{--}(P' - P_1)\,iD^{++}(P' - P). \tag{1.45}$$

The second term in (1.45) represents the contribution of the exchange diagram. The following relation exists between the Green's function $D^{++}(Q)$ and the dielectric constant of the system $\epsilon(Q)$

$$\mathrm{Re}\,D^{++}(Q) = D_0(Q)\,\mathrm{Re}\,\{\epsilon^{-1}(Q)\}. \tag{1.46}$$

Finally we obtain the desired expression for the electron-electron collision integral in the ν-representation:

$$\hat{R}^{(2)}_{e-e} = \hat{R}^{(2)}_a + \hat{R}^{(2)}_b, \tag{1.47}$$

where the first component corresponds to direct scattering and the second corresponds to exchange scattering:

$$(\hat{R}^{(2)}_a)_{\nu\nu'} = \pi\sum A_{\substack{p\,p'\\ \nu\,\sigma'}} A_{\substack{p'\,p\\ \sigma'\,\alpha}} A_{\substack{p_1\,p_1'\\ \sigma_1\,\sigma_1'}} A_{\substack{p_1'\,p_1\\ \alpha_1'\,\alpha_1}} \{(\hat{1} - \hat{\rho}_p)_{\alpha\nu'}\,(\hat{1} - \hat{\rho}_{p_1})_{\alpha_1\sigma_1}\,\hat{\rho}_{\substack{p_1'\\ \sigma_1'\alpha_1'}}\,\hat{\rho}_{\substack{p'\\ \sigma'\alpha'}} -$$
$$-\,\hat{\rho}_{\substack{p\\ \alpha\nu'}}\,\hat{\rho}_{\substack{p_1\\ \alpha_1\sigma_1}}\,(\hat{1} - \hat{\rho}_{p_1'})_{\sigma_1'\alpha_1'}\,(\hat{1} - \hat{\rho}_{p'})_{\sigma'\alpha'}\}\,[V(\mathbf{p}' - \mathbf{p})]^2\,\times$$
$$\times\,|\epsilon(\mathbf{p}' - \mathbf{p}, \hat{E}_{\substack{p'\\ \sigma'\alpha'}} - \hat{E}_{\substack{p\\ \alpha\nu'}})|^{-2}\,\delta(\mathbf{p} - \mathbf{p}_1 - p_1' - \mathbf{p}')\,\times$$
$$\times\,\delta(\hat{E}_{\substack{p\\ \alpha\nu'}} + \hat{E}_{\substack{p_1\\ \alpha_1\sigma_1}} - \hat{E}_{\substack{p_1'\\ \sigma_1'\alpha_1'}} - \hat{E}_{\substack{p'\\ \sigma'\alpha'}}) + (\text{h. c.}); \tag{1.48}$$

$$
\begin{aligned}
(\hat{R}_b^{(2)})_{\nu\nu'} = & -\pi \sum A_{\substack{p\,p_1'\\ \nu\,\sigma_1'}} A_{\substack{p_1'\,p_1\\ \alpha_1'\,\alpha_1}} A_{\substack{p_1\,p'\\ \sigma_1\,\sigma'}} A_{\substack{p'\,p\\ \alpha'\,\alpha}} V(\mathbf{p}_1 - \mathbf{p}')\, V(\mathbf{p}' - \mathbf{p}) \times \\
& \times \left| \epsilon\left(\mathbf{p}' - \mathbf{p}, \hat{E}_{\substack{p'\\ \sigma'\alpha'}} - \hat{E}_{\substack{p\\ \alpha\nu'}}\right)\right|^{-2} \{(\hat{1} - \hat{\rho}_p)_{\alpha\nu'} (\hat{1} - \hat{\rho}_{p_1})_{\alpha_1\sigma_1} \hat{\rho}_{\substack{p_1'\\ \sigma_1'\alpha_1'}} \hat{\rho}_{\substack{p'\\ \sigma'\alpha'}} - \\
& - \hat{\rho}_{\substack{p\\ \alpha\nu'}} \hat{\rho}_{\substack{p_1\\ \alpha_1\sigma_1}} (\hat{1} - \hat{\rho}_{p_1'})_{\sigma_1'\alpha_1'} (\hat{1} - \hat{\rho}_{p'})_{\sigma'\alpha'}\}\, \delta(\mathbf{p} + \mathbf{p}_1 - \mathbf{p}_1' - \mathbf{p}') \times \\
& \times \delta\left(\hat{E}_{\substack{p\\ \alpha\nu'}} + \hat{E}_{\substack{p_1\\ \alpha_1\sigma_1}} - \hat{E}_{\substack{p_1'\\ \sigma_1'\alpha_1'}} - \hat{E}_{\substack{p'\\ \sigma'\alpha'}}\right) + (\text{h. c.}),
\end{aligned}
\tag{1.49}
$$

where we use the convention

$$
\begin{aligned}
A_{\substack{p\,p'\\ \sigma\,\sigma'}} = (\hat{U}_p \hat{\sigma}_z \hat{U}_{p'}^{+})_{\sigma\sigma'} = & \frac{1}{2}\left\{\left(1 + \sigma\frac{\xi_p}{\varepsilon_p}\right)^{1/2} \left(1 + \sigma'\frac{\xi_{p'}}{\varepsilon_{p'}}\right)^{1/2} - \right. \\
& \left. - \sigma\sigma' \left(1 - \sigma\frac{\xi_p}{\varepsilon_p}\right)^{1/2} \left(1 - \sigma'\frac{\xi_{p'}}{\varepsilon_{p'}}\right)^{1/2}\right\}.
\end{aligned}
\tag{1.50}
$$

As we have seen in Section 2, the particle number and spin projection of the electrons are conserved in electron-electron collisions; we may show by direct calculation that the cross-term in (1.47) preserves momentum and energy. The corresponding identities take the form of (1.39)

$$\langle \hat{1}\hat{R}^{(2)}\{\hat{\rho}\}\rangle = 0, \tag{1.51}$$

$$\langle \mathbf{p}\hat{R}^{(2)}\{\hat{\rho}\}\rangle = 0, \tag{1.52}$$

$$\langle \hat{\varepsilon}_p = \hat{R}^{(2)}\{\hat{\rho}\}\rangle = 0 \tag{1.53}$$

and are satisfied for any density matrix.

We may easily verify that collision integral (1.47) vanishes by the distribution

$$\hat{\rho}_0 = n_0\,[(\varepsilon_p \hat{\tau}_3 + \mathbf{p}\,(\mathbf{v}_s - \mathbf{v}_n) - \nu_s)/T] \tag{1.54}$$

for any ν_s. In (1.54) $\hat{\varepsilon}_p$ is the real nonequilibrium energy; $\mathbf{v}_n$ is the velocity of the normal component; $n_0(x) = (e^x + 1)^{-1}$. Substituting (1.54) into the definition of s_z when ν_s, $p_F\nu_s \ll \Delta$, we may obtain

$$\mathbf{v}_s = \left\langle \frac{1}{T}\left[-n_0'\left(\frac{\varepsilon_p}{T}\right)\right]\right\rangle^{-1} s_z. \tag{1.55}$$

The relation between the collision integral and the self-consistently determined dielectric constant of the system reflects the fact that collisions in a low-temperature medium are, essentially, many-particle collisions.

If we set $\epsilon(Q) = 1$ in (1.48), (1.49) we will see that the diagonal part of our result coincides with the Eliashberg electron-electron collision integral [7].

2. Electron-electron collision frequencies in the Gorkov model. The collisional term of the kinetic equation in a model with point interaction may be obtained from the result of the preceding section by replacing all Green's functions of the Coulomb field with the constant g. Writing $\hat{\rho}$ as

$$\hat{\rho} = \hat{\rho}_0(\hat{\varepsilon}_p) + \delta\tilde{\rho}, \tag{1.56}$$

where $\hat{\varepsilon}_p$ is the real nonequilibrium value of the energy operator of the self-consistent field, $\delta\tilde{\rho}$ is a small nonequilibrium addition, we may easily obtain the linearized collision integral $\hat{\mathscr{L}}^{(2)}\{\delta\tilde{\rho}\}$ which we do not provide in its entirety in view of its cumbersome size. Factoring out the cofactors representing frequency collisions in the normal manner and dropping integral terms we obtain in the ν-representation:

$$\hat{\mathscr{L}}^{(2)}_{(\nu)}\{\delta\tilde{\rho}\} = -\begin{pmatrix} \left[\nu_\alpha\delta\rho_{11} - \frac{\nu_\beta}{2}\operatorname{Sp}(\hat{\tau}_1\delta\tilde{\rho})\right] & \left[\nu_\gamma\delta\rho_{12} - \frac{1}{2}\nu_\delta\operatorname{Sp}(\hat{\tau}_3\delta\tilde{\rho})\right] \\ \left[\nu_\gamma\delta\rho_{21} - \frac{1}{2}\nu_\delta\operatorname{Sp}(\hat{\tau}_3\delta\tilde{\rho}\right] & \left[\nu_\alpha\delta\rho_{22} + \frac{\nu_\beta}{2}\operatorname{Sp}(\hat{\tau}_1\delta\tilde{\rho})\right] \end{pmatrix}, \tag{1.57}$$

where we designate

$$\begin{aligned} \nu_\alpha^{ee} = {} & 2\pi g^2 n_0^{-1}\left(\frac{\varepsilon_p}{T}\right)\sum\{(A_{\substack{p\,p'\\1\,\sigma'}})^2 - (A_{\substack{p_1\,p_1'\\ \sigma_1\,\sigma_1'}})^2 - A_{\substack{p\,p_1'\\1\,\sigma_1'}}A_{\substack{p_1'\,p_1\\ \sigma_1'\,\sigma_1}}A_{\substack{p_1\,p'\\ \sigma_1\,\sigma'}}A_{\substack{p'\,p\\ \sigma'\,1}}\}\times \\ & \times n_0(-\sigma_1\varepsilon_p/T)\,n_0(\sigma_1'\varepsilon_{p_1'}/T)\,n_0(\sigma'\varepsilon_{p'}/T)\times \\ & \times\delta(\mathbf{p}+\mathbf{p}_1'-\mathbf{p}_1'-\mathbf{p}')\,\delta(\varepsilon_p+\sigma_1\varepsilon_{p_1}-\sigma_1'\varepsilon_{p_1'}-\sigma'\varepsilon_{p'}); \end{aligned} \tag{1.58}$$

$$\begin{aligned} \nu_\beta^{ee} = {} & 2\pi g^2\sum\{A_{\substack{p\,p'\\1\,\sigma'}}A_{\substack{p'\,p\\ \sigma'\,2}}(A_{\substack{p_1\,p_1'\\ \sigma_1\,\sigma_1'}})^2 - A_{\substack{p\,p_1'\\1\,\sigma_1'}}A_{\substack{p_1'\,p_1\\ \sigma_1'\,\sigma}}A_{\substack{p_1\,p'\\ \sigma_1\,\sigma'}}A_{\substack{p'\,p\\ \sigma'\,2}}\}\times \\ & \times(n_0(-\sigma_1\varepsilon_{p_1}/T)\,n_0(\sigma_1\varepsilon_{p_1}/T)\,n_0(\varepsilon_{p'}\sigma'/T)\,\delta(\mathbf{p}+\mathbf{p}_1-\mathbf{p}_1'-\mathbf{p}')\times \\ & \times\delta(\sigma_1\varepsilon_{p_1}-\sigma_1'\varepsilon_{p_1'}-\sigma'\varepsilon_{p'}); \end{aligned} \tag{1.59}$$

$$\begin{aligned} \nu_\gamma^{ee} = {} & 2\pi g^2\sum\{(A_{\substack{p_2\,p'\\1\,\sigma'}})^2(A_{\substack{p_1\,p_1'\\ \sigma_1\,\sigma_1'}})^2 - A_{\substack{p\,p_1'\\1\,\sigma_1'}}A_{\substack{p_1'\,p_1\\ \sigma_1'\,\sigma_1}}A_{\substack{p_1\,p'\\ \sigma_1\,\sigma'}}A_{\substack{p'\,p\\ \sigma'\,1}}\}\times \\ & \times n_0(-\sigma_1\varepsilon_{p_1}/T)\,n_0(\sigma_1'\varepsilon_{p_1'}/T)\,n_0(\sigma'\varepsilon_{p'}/T)\,\delta(\mathbf{p}+\mathbf{p}_1-\mathbf{p}_1'-\mathbf{p}')\times \\ & \times\delta(\sigma_1\varepsilon_{p_1}-\sigma_1'\varepsilon_{p_1'}-\sigma'\varepsilon_{p'}); \end{aligned} \tag{1.60}$$

$$\begin{aligned} \nu_\delta^{ee} = {} & 2\pi g^2\sum\{A_{\substack{p\,p'\\1\,\sigma'}}A_{\substack{p'\,p\\ \sigma'\,2}}(A_{\substack{p_1\,p_1'\\ \sigma_1\,\sigma_1'}})^2 - A_{\substack{p\,p_1'\\1\,\sigma_1}}A_{\substack{p_1'\,p_1\\ \sigma_1'\,\sigma_1}}A_{\substack{p_1'\,p'\\ \sigma_1\,\sigma'}}A_{\substack{p'\,1\\ \sigma'\,2}}\}\times \\ & \times n_0(-\sigma_1\varepsilon_{p_1}/T)\,n_0(\sigma_1'\varepsilon_{p_1'}/T)\,n_0(\sigma'\varepsilon_{p'}/T)\,\delta(\mathbf{p}+\mathbf{p}_1-\mathbf{p}_1'-\mathbf{p}')\times \\ & \times\delta(\varepsilon_p+\sigma_1\varepsilon_{p_1}-\sigma_1'\varepsilon_{p_1'}-\sigma'\varepsilon_{p'}). \end{aligned} \tag{1.61}$$

In formulae (1.58)-(1.61)

$$(A_{\substack{p\,p'\\ \sigma\,\sigma'}})^2 = \frac{1}{2}\left(1 + \sigma\sigma'\,\frac{\tilde{\xi}_p\tilde{\xi}_{p'} - \Delta^2}{\varepsilon_p\varepsilon_{p'}}\right), \tag{1.62}$$

$$A_{\substack{p\,p'\\ 1\,\sigma'}} A_{\substack{p'\,p\\ \sigma'\,2}} = \frac{\sigma'}{2}\,\frac{\Delta}{\varepsilon_p\varepsilon_p'}\,(\tilde{\xi}_p + \tilde{\xi}_{p'}),$$

$$|\mathbf{V}_s - \mathbf{V}_n| \ll \Delta/p_F, \tag{1.63}$$

where ν_α^{ee} is the regular "quasi particle collision frequency" resulting from point interaction. We may show that $\nu_\beta^{ee} = 0$, ν_α^{ee} and ν_γ^{ee} are even while ν_δ^{ee} is an odd function of the variable ξ_p with a high degree of accuracy. For the functions ν_α^{ee}, ν_γ^{ee}, ν_δ^{ee} we may obtain general estimates when $T \to 0$ and $T \to T_c - 0$ near the Fermi surface: the behavior of collision frequencies when $\xi_p \sim 0$ are determined by the kinetic properties of the metal in a weakly-nonequilibrium state.

Thus,

a) $T \to 0$:

$$\nu_\alpha^{ee} \simeq c_\alpha (mp_F g)^2 \frac{\Delta_0^2}{\mu_F}\left(\frac{T}{\Delta_0}\right)^{1/2} \exp\left(-\frac{\Delta_0}{T}\right),$$

$$\nu_\gamma^{ee} \simeq \nu_\alpha^{ee} \exp\left(-\frac{\Delta_0}{T}\right), \qquad \nu_\delta^{ee} \simeq \frac{1}{2}\,\frac{\tilde{\xi}_p}{\varepsilon_p}\,\nu_\alpha^{ee}, \tag{1.64}$$

where $c_\alpha > 0$ is the dimensionless constant;

b) $T \to T_c - 0$:

$$\nu_\alpha^{ee} \simeq \nu_\gamma^{ee} \simeq \nu_0 (1 + b_\alpha \Delta/T_c), \qquad \nu_\delta^{ee} \simeq b_\delta \frac{\xi_p}{\varepsilon_p}\left(\frac{\Delta}{T_c}\right)^{1/2} \nu_0, \tag{1.65}$$

where ν_0 is the electron-electron collision frequency in a normal system when $T = T_c$:

$$\nu_0 \sim T_c^2/\mu_F,$$

where $b_\alpha > 0$, $b_\delta > 0$ are the dimensionless constants.

3. Electron-phonon collision integral. The Frohlich Hamiltonian [40] written in the Eliashberg representation

$$H_{e\text{-}ph} = \int d\mathbf{r}\,\psi^\alpha(\mathbf{r})\,N_{\alpha\beta}\psi_\beta(\mathbf{r})\,\varphi_{ph}(\mathbf{r}), \tag{1.66}$$

where φ_{ph} is the phonon field operator. Solving Dyson's equation for the phonon Green's functions, we obtain for the locally-homogeneous system

$$iD^{-+}(\mathbf{k}\omega) = |m_k|^2 \{N_\mathbf{k}\delta(\omega - \omega_k) + (1 + N_\mathbf{k})\,\delta(\omega + \omega_k)\}, \tag{1.67}$$

where $|m_k|^2 = \alpha^2 k^2/2\omega_k \bar\rho_0 \Omega$, α is the electron-phonon interaction constant, $\bar\rho_0$ is the density of matter, Ω is the volume of the system; for acoustic phonons $\omega_k = s_0 k$ where s_0 is the speed of sound and for simplicity we assume an isotropic value; $N_k(\mathbf{r}t)$ is the distribution function of phonons that is, generally speaking, a nonequilibrium function [41, Chapter 10, Section 92]. Using (1.67) and (1.24) we obtain in the ν-representation

$$(\hat R^{(2)}_{e-ph})_{\nu\nu'} = \pi \sum A_{\underset{\nu}{p}\,\underset{\sigma'}{p'}} A_{\underset{\alpha'}{p'}\,\underset{\alpha}{p}} |m_{\mathbf{q}}|^2 \{(\hat 1 - \hat\rho_p)_{\alpha\nu'} \hat\rho_{\underset{\sigma'\alpha'}{p'}} \times$$
$$\times [N_{\mathbf{q}} \delta(\hat E_{\underset{\alpha\nu'}{p}} - \hat E_{\underset{\sigma\alpha'}{p'}} - \omega_q) + (1 + N_{-\mathbf{q}}) \delta(\hat E_{\underset{\alpha\nu'}{p}} - \hat E_{\underset{\sigma'\alpha'}{p'}} + \omega_{-q})] -$$
$$- \hat\rho_{\underset{\alpha\nu'}{p}} (\hat 1 - \hat\rho_{p'})_{\sigma'\alpha'} [(1 + N_{\mathbf{q}}) \delta(\hat E_{\underset{\alpha\nu'}{p}} - \hat E_{\underset{\sigma'\alpha'}{p'}} - \omega_q) +$$
$$+ N_{-\mathbf{q}} \delta(\hat E_{\underset{\alpha\nu'}{p}} - \hat E_{\underset{\sigma\alpha'}{p'}} + \omega_{-q})]\} \delta(\mathbf{p}' - \mathbf{p} - \mathbf{q}) + (\text{h. c.}). \tag{1.68}$$

In the collisions described by formula (1.68) the electron number is conserved together with the projection of the electron spin and the total momentum and energy of the electron-phonon system. (It is also necessary to account for the U-processes in the crystals). $\hat R^{(2)}_{e-ph}$ describes the relaxation of the electron subsystem to a state of the type in (1.54) when $V_n = 0$, while the phonons relax to the state $N_q = N_0(\omega_q/T)$, where $N_0(x) = (e^x - 1)^{-1}$. In the relaxation time approximation when $p_F v_s \ll \Delta$, $\hat R^{(2)}_{e-ph}$ takes the form of (1.57), where $\nu^{ph}_\beta = 0$,

$$\nu^{ph}_\alpha = 2\pi \sum |m_q|^2 (A_{\underset{1}{p}\underset{\sigma'}{p'}})^2 \left\{\left[n_0\left(-\frac{\sigma'\varepsilon_{p'}}{T}\right) + N_q\right] \delta(\varepsilon_p - \sigma'\varepsilon_{p'} - \omega_q) +\right.$$
$$\left. + \left[n_0\left(\frac{\sigma'\varepsilon_{p'}}{T}\right) + N_{-q}\right] \delta(\varepsilon_p - \sigma'\varepsilon_{p'} + \omega_q)\right\} \delta(\mathbf{p}' - \mathbf{p} - \mathbf{q}); \tag{1.69}$$

$$\nu^{ph}_\gamma = 2\pi \sum |m_q|^2 (A_{\underset{1}{p}\underset{\sigma'}{p'}})^2 \left\{\left[n_0\left(-\frac{\sigma'\varepsilon_{p'}}{T}\right) + N_q\right] \delta(\sigma'\varepsilon_{p'} + \omega_q) +\right.$$
$$\left. + \left[n_0\left(\frac{\sigma'\varepsilon_{p'}}{T}\right) + N_q\right] \delta(\sigma'\varepsilon_{p'} - \omega_q)\right\} \delta(\mathbf{p}' - \mathbf{p} - \mathbf{q}); \tag{1.70}$$

$$\nu^{ph}_\delta = 2\pi \sum |m_q|^2 A_{\underset{1}{p}\underset{\sigma'}{p'}} A_{\underset{\sigma'}{p'}\underset{2}{p}} \left\{\left[n_0\left(-\frac{\sigma'\varepsilon_{p'}}{T}\right) + N_q\right] \delta(\varepsilon_p - \sigma'\varepsilon_{p'} - \omega_q) +\right.$$
$$\left. + [n_0(\sigma'\varepsilon_{p'}/T) + N_q] \delta(\varepsilon_p - \sigma'\varepsilon_{p'} + \omega_q)\right\} \delta(\mathbf{p}' - \mathbf{p} - \mathbf{q}). \tag{1.71}$$

It is convenient by following study [42] to introduce the characteristic frequency $\nu_c = 2\pi\beta T_c^3$, where β is the dimensional constant dependent on the properties of the matter. For frequencies $\nu^{ph}_\alpha - \nu^{ph}_\delta$ we may obtain general formulae on the Fermi surface where $\varepsilon_p = \Delta$. For the frequency ν^{ph}_α such a formula was obtained in study [42] and takes the following form:

$$\nu^{ph}_\alpha(\Delta) = \nu_c \left(\frac{\Delta}{T}\right)^3 \sum_{n=1}^{\infty} \left\{(-1)^{n+1} 4\left[2 + \left(\frac{T}{\Delta}\right)^2\right] K_1\left(\frac{n\Delta}{T}\right) + 2\frac{T}{\Delta} K_0\left(\frac{n\Delta}{T}\right) +\right.$$
$$+ \exp\left(-\frac{n\Delta}{T}\right)\left[4 + \frac{3T}{n\Delta} + 2\left(\frac{T}{\Delta}\right)^2\right) K_1\left(\frac{n\Delta}{T}\right) +$$

$$+\left[\left(4-\frac{3T}{n\Delta}+2\left(\frac{T}{\Delta}\right)^2\right)K_1\left(\frac{n\Delta}{T}\right)+\exp\left(\frac{n\Delta}{T}\right)\times\right. \tag{1.72}$$
$$\left.\times\left[\left(4-\frac{3T}{n\Delta}+2\left(\frac{T}{\Delta}\right)^2\right)K_1\left(\frac{n\Delta}{T}\right)+\left(4-\frac{T}{n\Delta}\right)K_0\left(\frac{n\Delta}{T}\right)\right]\right\},$$

where $K_n(x)$ are the Bessel functions of the second order imaginary argument.

The frequency ν_γ^{ph} is energy-independent and is equal to [43]

$$\nu_\gamma^{ph}=2\nu_c\left(\frac{T}{T_c}\right)^3\left(\frac{\Delta}{T}\right)^2\sum_{n=0}^{\infty}\left\{\frac{\Delta}{T}K_1\left((2n+1)\frac{\Delta}{T}\right)+\right.$$
$$\left.+(2n+1)^{-1}K_2\left((2n+1)\frac{\Delta}{T}\right)\right\}. \tag{1.73}$$

We may show that the quantity ν_δ^{ph} takes the form

$$\nu_\delta^{ph}=\nu_c\frac{\tilde{\xi}_p}{\varepsilon_p}\frac{\Delta}{T}\left(\frac{T}{T_c}\right)^3 Z\left(\frac{\Delta}{T},\frac{\varepsilon_p}{T}\right).$$

On the Fermi surface for the function $Z(\Delta/T,\ \varepsilon_p/T)$ the following formula is valid

$$Z\left(\frac{\Delta}{T},\frac{\Delta}{T}\right)=\sum_{n=1}^{\infty}\left[(-1)^{n+1}+\exp\left(n\frac{\Delta}{T}\right)\right]\times$$
$$\times\left\{\frac{1}{n}\frac{\Delta}{T}K_1\left(\frac{n\Delta}{T}\right)+2\left(\frac{\Delta}{T}\right)^2\left[K_0\left(\frac{n\Delta}{T}\right)-K_1\left(\frac{n\Delta}{T}\right)\right]\right\}+$$
$$+\sum_{n=1}^{\infty}\left[(-1)^{n+1}+\exp\left(-\frac{n\Delta}{T}\right)\right]\left\{\frac{1}{n}\frac{\Delta}{T}K_1\left(\frac{n\Delta}{T}\right)+\right.$$
$$\left.+2\left(\frac{\Delta}{T}\right)^2\left[K_0\left(\frac{n\Delta}{T}\right)+K_1\left(\frac{n\Delta}{T}\right)\right]\right\}. \tag{1.74}$$

At low temperatures the series in (1.72)-(1.74) rapidly converge: the leading terms of the expansions in terms of the parameter T/Δ_0 take the form

$$\nu_\alpha^{ph}\simeq\frac{15}{8}\sqrt{\frac{\pi}{2}}\zeta\left(\frac{7}{2}\right)\left(\frac{T_c}{\Delta_0}\right)^{1/2}\left(\frac{T}{T_c}\right)^{7/2}\nu_c,$$
$$\nu_\gamma^{ph}\simeq\sqrt{2\pi}\left(\frac{T}{T_c}\right)^3\left(\frac{\Delta_0}{T}\right)^{5/2}\exp\left(-\frac{\Delta_0}{T}\right)\nu_c,$$
$$\nu_\delta^{ph}\simeq\frac{3}{4}\sqrt{\frac{\pi}{2}}\zeta\left(\frac{5}{2}\right)\left(\frac{T}{T_c}\right)^3\left(\frac{\Delta_0}{T}\right)^{1/2}\frac{\tilde{\xi}_p}{\varepsilon_p}\nu_c, \tag{1.75}$$

where $\Delta_0 = \Delta(0)$; $\zeta(x)$ is the Riemann zeta-function. When $T \to T_c - 0$ we may obtain the following maximum values of the relaxation frequencies:

$$\nu_\alpha^{ph}=7\zeta(3)\nu_c,\qquad \nu_\gamma^{ph}=\frac{7}{2}\zeta(3)\nu_c,$$
$$\nu_\beta^{ph}=0,\qquad \nu_\delta^{ph}=\frac{\pi^2}{2}\frac{\Delta}{T_c}(\operatorname{sign}\tilde{\xi}_p)\nu_c. \tag{1.76}$$

the quantity ν_γ^{ph} coincides with the recombination frequency ν_r calculated on the Fermi surface for the Cooper pairs given in study [44]. However unlike ν_r the quantity ν_γ^{ph} is energy-independent and its meaning is different. In addition to the order parameter $\Delta(T)$ the quantity ν_γ^{ph} determines the temporal evolution of the components of the density matrix that are nondiagonal in the ν-representation. It is clear that when $T_c - T \ll T_c$ relaxation of the excitation distribution function and of the ν-nondiagonal components of the density matrix is characterized by the same time scale $\nu_\alpha^{ph}(T_c) \sim \nu_\gamma^{ph}(T_c)$.

4. **Electron-nonmagnetic impurity collision integral.** We will assume that the atoms of a nonmagnetic impurity with a low concentration N_i are distributed chaotically in a pure superconductor, and we will assume that the potential-electron-impurity interaction $U(\mathbf{r})$ satisfies the applicability condition of Born's approximation

$$\int U(\mathbf{r})\, d\mathbf{r} \ll \mu_F/p_F^3. \tag{1.77}$$

In Eliashberg's representation

$$H_{ei}^n = \sum_\alpha \int d\mathbf{r} U(\mathbf{r} - \mathbf{r}_\alpha)\, \psi^\beta(\mathbf{r}) N_{\beta\gamma} \psi_\gamma(\mathbf{r}). \tag{1.78}$$

After averaging over the impurity positions we obtain in the ν-representation

$$\begin{aligned}(\hat{R}^{(2)}_{(n)i})_{\nu\nu'} = \pi N_i \sum_{\mathbf{q}} |U(\mathbf{q})|^2 \{A_{\substack{\mathbf{p}\,\mathbf{p}'\\ \nu\,\alpha'}} A_{\substack{\mathbf{p}'\,\mathbf{p}\\ \sigma'\,\sigma}} [(\hat{1} - \hat{\rho}_{\mathbf{p}})_{\sigma\nu'} \hat{\rho}_{\substack{\mathbf{p}'\\ \alpha'\,\sigma'}} - \hat{\rho}_{\substack{\mathbf{p}\\ \sigma\,\nu'}} (\hat{1} - \hat{\rho}_{\mathbf{p}'})_{\alpha'\sigma'}]\} \times \\ \times\, \delta(\hat{E}_{\substack{\mathbf{p}'\\ \alpha'\,\sigma'}} - \hat{E}_{\substack{\mathbf{p}\\ \sigma\,\nu'}})\, \delta(\mathbf{p}' - \mathbf{p} - \mathbf{q}) + (\text{h. c.});\end{aligned} \tag{1.79}$$

$$U(\mathbf{q}) = \int d\mathbf{r} U(\mathbf{r}) \exp(i\mathbf{q}\mathbf{r}).$$

Elastic collision integral (1.79) conserves the total number of particles, the spin projection and the energy; it describes the relaxation of the system to the state $f(E_p + \nu_s)$, where f is any real function. The τ-approximation for (1.79) takes the form of (1.57) where the frequencies are determined by the formulae

$$\begin{aligned}\nu_\alpha^i = \pi N_i \sum |U(\mathbf{q})|^2 \left(1 + \sigma' \frac{\tilde{\xi}_p \tilde{\xi}_{p'} - \Delta^2}{\varepsilon_p \varepsilon_{p'}}\right) \times \\ \times\, \delta(\varepsilon_p - \sigma' \varepsilon_{p'} + (\mathbf{p} - \mathbf{p}')\mathbf{v}_s)\, \delta(\mathbf{p}' - \mathbf{p} - \mathbf{q});\end{aligned} \tag{1.80}$$

$$\nu_\beta^i = \pi N_i \sum |U(\mathbf{q})|^2 \frac{\Delta(\tilde{\xi}_p + \tilde{\xi}_{p'})}{\varepsilon_p \varepsilon_{p'}} \sigma' \delta(\sigma' \varepsilon_{p'} + (\mathbf{p}' - \mathbf{p})\mathbf{v}_s)\, \delta(\mathbf{p}' - \mathbf{p} - \mathbf{q}); \tag{1.81}$$

$$\nu_\gamma^i = \pi N_i \sum |U(\mathbf{q})|^2 \left(1 + \sigma' \frac{\tilde{\xi}_p \tilde{\xi}_{p'} - \Delta^2}{\varepsilon_p \varepsilon_{p'}}\right) \delta(\sigma' \varepsilon_{p'} + (\mathbf{p}' - \mathbf{p})\mathbf{v}_s)\, \delta(\mathbf{p}' - \mathbf{p} - \mathbf{q}); \tag{1.82}$$

$$\nu_\delta^i = \pi N_i \sum |U(\mathbf{q})|^2 \frac{\Delta(\tilde{\xi}_p + \tilde{\xi}_{p'})}{\varepsilon_p \varepsilon_{p'}} \sigma \delta(\varepsilon_p - \sigma' \varepsilon_{p'} + (\mathbf{p}' - \mathbf{p})\mathbf{v}_s)\, \delta(\mathbf{p}' - \mathbf{p} - \mathbf{q}). \tag{1.83}$$

We may show that on the Fermi surface when $2p_F v_s < \Delta^i_\beta = \nu^i_\gamma = 0$; the principal terms of the expansion ν^i_α and n^i_δ in terms of the parameter $p_F v_s/\Delta$ take the form $\nu^i_\alpha = \nu_i \mid \tilde{\xi}_p \mid / \varepsilon_p$, $\nu^i_\delta = -\frac{\Delta}{\varepsilon_p} \operatorname{sign}(\tilde{\xi}_p)\, \nu_i$, where ν_i is the electron-impurity collision frequency in the normal metal when $T = T_c$.

5. Electron-paramagnetic impurity collision integral. The Hamiltonian of the exchange interaction of conduction electrons and the electrons of incomplete shells of impurity paramagnetic ions [45] cannot be written in the Nambu representation:

$$H^m_{e-i} = \sum_\alpha \int d\mathbf{r} J(\mathbf{r} - \mathbf{r}_\alpha)\, \psi^\beta(\mathbf{r})\, (\tilde{\sigma}\mathbf{s}_\alpha)_{\beta\gamma}\, \psi_\gamma(\mathbf{r}), \tag{1.84}$$

where $J(\mathbf{r})$ is the s-d-exchange integral, $1/2\tilde{\sigma}$ is the electron spin operator; in the Eliashberg representation:

$$\tilde{\sigma} = \tilde{\sigma}_l \mathbf{e}_l; \quad |\mathbf{e}_l| = 1, \qquad l = x, y, z,$$
$$\tilde{\sigma}_z = \begin{pmatrix} \hat{\sigma}_+ & 0 \\ 0 & -\hat{\sigma}_- \end{pmatrix}; \quad \tilde{\sigma}_y = \begin{pmatrix} \hat{0} & -i\hat{\sigma}_+ \\ i\hat{\sigma}_- & \hat{0} \end{pmatrix}; \quad \tilde{\sigma}_x = \begin{pmatrix} \hat{0} & \hat{\sigma}_+ \\ \hat{\sigma}_- & \hat{0} \end{pmatrix};$$

$\mathbf{s}_\alpha$ is the ion spin operator at the node $\mathbf{r}_\alpha$. The interaction vertex is proportional to the matrix

$$\tilde{\Gamma}_s = (\tilde{\sigma}\mathbf{s}) - \tilde{I}\, (\tilde{\sigma}\mathbf{s})^T \tilde{I} = \begin{pmatrix} s_z\hat{1} & is_-\hat{\sigma}_y \\ -is_+\hat{\sigma}_y & -s_z\hat{1} \end{pmatrix}, \tag{1.85}$$

where $s_\pm = s_x \pm is_y$.

We will consider the Abrikossov-Gorkov model [38]. Then we average over the impurity spins

$$\overline{s_i s_k} = \frac{1}{3}\delta_{i3}\delta_{ik}\overline{s^2}, \qquad \overline{s_+ s_-} = \frac{2}{3}\overline{s^2}, \qquad \overline{s_+^2} = \overline{s_-^2} = 0. \tag{1.86}$$

Calculations using formula (1.24) after averaging over the spins and positions of the impurities produce a result of the type (1.27); consequently,

$$\begin{aligned}(\hat{R}^{(2)}_i(m))_{\nu\nu'} = {} & \pi N_m \sum |J(\mathbf{q})|^2 s(s+1) \Big\{ \frac{1}{3} B_{\substack{p\,p'\\ \nu\,\sigma'}} B_{\substack{p'\,p\\ \alpha\,\sigma}} \times \\ & \times [(\hat{1} - \hat{\rho}_p)_{\sigma\nu'} \hat{\rho}_{\substack{p'\\ \sigma'\alpha'}} - \hat{\rho}_{\substack{p\\ \sigma\nu'}} (\hat{1} - \hat{\rho}_{p'})_{\sigma'\alpha'}]\, \delta(\hat{E}_{\substack{p\\ \sigma\alpha'}} - \hat{E}_{\substack{p'\\ \sigma'\nu'}}) + \\ & + \frac{2}{3} C_{\substack{p\,p'\\ \nu\,\sigma'}} C_{\substack{p'\,p\\ \alpha'\,\sigma}} [(\hat{1} - \hat{\rho}_p)_{\sigma\nu'} (\hat{1} - \hat{\rho}^T_{-\mathbf{p}'})_{\sigma'\alpha'} - \hat{\rho}_{\substack{p\\ \sigma\nu'}} \hat{\rho}^T_{-\mathbf{p}'}] \times \\ & \times \delta(\hat{E}_{\substack{p\\ \sigma\nu'}} + \hat{E}_{\substack{-\mathbf{p}'\\ \sigma'\alpha'}}) \Big\}\, \delta(\mathbf{p}' - \mathbf{p} - \mathbf{q}) + (\text{h. c.}),\end{aligned} \tag{1.87}$$

where

$$\hat{B}_{pp'} = \hat{U}_p \hat{1} \hat{U}_{p'} = (u_p u_{p'} + v_p v_{p'})\, \hat{1} - i\hat{\sigma}_y (v_p u_{p'} - u_p v_{p'}); \tag{1.88}$$
$$\hat{C}_{pp'} = \hat{U}_p i\hat{\sigma}_y \hat{U}_{p'} = -i\hat{\sigma}_y \hat{B}_{pp'}; \tag{1.89}$$

N_m is the concentration of paramagnetic impurities.

Collision integral (1.87) describes the relaxation of this system to a state described by an arbitrary Hermitian matrix of the type $f(\hat{E}_p)$ and it conserves the number of particles and the energy; obviously this is the only cross-term considered here that produces relaxation of the spin degree of freedom (which is described by the part of collision integral (1.78) proportional to matrices $C_{pp'}$). In the τ-approximation (1.57)

$$\nu_\alpha^m = \pi N_m \sum |J(\mathbf{q})|^2 \frac{2}{3} s(s+1) \{ |B_{\underset{1\sigma'}{pp'}}|^2 \delta(\hat{E}_{\underset{\sigma'\sigma'}{p'}} - \hat{E}_{\underset{11}{p}}) + 2|C_{\underset{1\sigma'}{pp'}}|^2 \delta(\hat{E}_{\underset{11}{p}} + \hat{E}_{\underset{\sigma'\sigma'}{p'}})\} \delta(\mathbf{p}' - \mathbf{p} - \mathbf{q}); \tag{1.90}$$

$$\nu_\beta^m = \pi N_m \sum |J(\mathbf{q})|^2 \frac{2}{3} s(s+1) \{ B_{\underset{1\sigma'}{pp'}} B_{\underset{\sigma'2}{p'p}} \delta(\hat{E}_{\underset{\sigma'\sigma'}{p'}} - \mathbf{p}\mathbf{v}_s) + 2C_{\underset{1\sigma'}{pp'}} C_{\underset{\sigma'2}{p'p}} \delta(\hat{E}_{\underset{\sigma'\sigma'}{-\mathbf{p}'}} + \mathbf{p}\mathbf{v}_s)\} \delta(\mathbf{p}' - \mathbf{p} - \mathbf{q}); \tag{1.91}$$

$$\nu_\gamma^m = \pi N_m \sum |J(\mathbf{q})|^2 \frac{2}{3} s(s+1) \{ |B_{\underset{1\sigma'}{pp'}}|^2 \delta(\hat{E}_{\underset{\sigma'\sigma'}{p'}} - \mathbf{p}\mathbf{v}_s) + 2|C_{\underset{1\sigma'}{pp'}}|^2 \delta(\hat{E}_{-\mathbf{p}'} + \mathbf{p}\mathbf{v}_s)\} \delta(\mathbf{p}' - \mathbf{p} - \mathbf{q}); \tag{1.92}$$

$$\nu_\delta^m = \pi N_m \sum |J(\mathbf{q})|^2 \frac{2}{3} s(s+1) \{ B_{\underset{1\sigma'}{pp'}} B_{\underset{\sigma'2}{p'p}} \delta(\hat{E}_{\underset{\sigma'\sigma'}{p'}} - \hat{E}_{\underset{11}{p}}) + 2C_{\underset{1\sigma'}{pp'}} C_{\underset{\sigma'2}{p'p}} \delta(\hat{E}_{-\mathbf{p}'} + \hat{E}_{\underset{11}{p}})\} \delta(\mathbf{p}' - \mathbf{p} - \mathbf{q}). \tag{1.93}$$

Thus, the collision integrals for superconductor electron collisions with any excitations and impurities are nondiagonal in the quasi particle representation; accounting for the ν-nondiagonal elements of the density matrix and the collision integrals automatically satisfies the continuity equation for the particle density and observance of the gradient invariance. The contribution of ν-nondiagonal components of the density matrix to the macroscopic quantities is significant not only in examining high-frequency phenomena in superconductors [46, 12] but also in the hydrodynamic frequency range.

The ν-diagonal components of the collision integrals calculated here coincide with the cross-terms of kinetic equations for the distribution function of quasi particles, naturally ignoring the contribution of the ν-nondiagonal elements of the density matrix to them.

Chapter 2

THE HYDRODYNAMICS OF A NONIDEAL SUPERCONDUCTOR FLUID. FIRST AND SECOND SOUND DAMPING

In this chapter we will provide a derivation of two-fluid hydrodynamic equations of a Fermi superfluid in a viscous approximation. We will proceed from the kinetic equation for the density matrix of the superconductor electrons with a collision integral that conserves particle number.

Hydrodynamic equations for an ideal superfluid were first derived phenomenologically by Landau [47]. Bogolubov [48] based on equations of motion for the field operators has demonstrated that the Landau equations describe the behavior of an ideal Bose superfluid. Khalatnikov [49] investigated a wide range of issues in superfluidity theory; specifically, based on a kinetic equation for the distribution function of normal excitations he calculated the dissipative terms in the Landau equations and found their kinetic coefficients. Galysevich [50] applied the Bogolubov method to the problem of a Fermi superfluid and considered the issue of the form of the dissipative corrections. In the study by Hohenberg and Martin [51] the linear response method [52] was used to derive relations that relate the kinetic coefficients and the hydrodynamic asymptotics of the Green's functions. The Landau equations for an ideal Fermi superfluid were derived in studies [53, 54]. Morozov [55] derived hydrodynamic equations of a viscous Fermi superfluid using the Zubarev method [56] of formulating a nonequilibrium statistical operator and derived expressions for the kinetic coefficients through the temporal correlation functions. In the studies by Shumeyko [28] and Soda and Fudjiki [29] the kinetic coefficients of the hydrodynamic equations of a neutral Fermi superfluid with point interaction are found by means of a kinetic equation for the excitation distribution function [4-6, 8, 10]. This equation may be obtained from kinetic equation (1.28) ignoring the density matrix components and collision integral nondiagonal in the ν-representation. Here the equation for the phase of order parameter (1.30) becomes meaningless and the collision integral becomes the source of particles; in order to restore gradient invariance the continuity equation for the electron density must be imposed in addition which results in additional limitations on the space-time behavior of the macroscopic quantities. Moreover, as indicated by an analysis of equation (1.28) deviation from equilibrium in a ν-diagonal channel over a significant time period whose scale is determined by the collision integral produces ν-nondiagonal components of the density matrix and it is necessary to account for these in order to satisfy the law of preservation of particle number. We may easily see that the quantity $\left\langle \frac{\Delta}{\varepsilon_p} \hat{\tau}_1 \delta\hat{\rho} \right\rangle$, providing the correction of the nondiagonal components $\hat{\rho}$ to the total particle number does not oscillate in time, but rather manifests a relaxation behavior with frequency ν_γ.

It is interesting to investigate how these features are manifest in the last, hydrodynamic stage of system evolution.

As in study [28] we will consider a neutral superconducting Fermi fluid with direct interaction between particles assuming that there are no external fields. The primary step in deriving the hydrodynamic equations is obtaining the conservation laws for local quantities.

1. Conservation laws

The conservation laws for the densities of particle number, spin projection, current and energy follow automatically from kinetic equation (1.28) and identities (1.39), (1.51)-(1.53). The expressions for the macroscopic quantities here are correct if we account for the relation between the Nambu and Eliashberg representations as examined in Chapter 1. Equation (1.28) allows us to find the flux terms with second order gradient accuracy sufficient for calculating the transport coefficients.

1. Continuity equation for the particle number density. This equation is not the result of equation (1.30) for the phase of the order parameter as we may show (such a statement, for example, is made in study [57]); it is the direct consequence of the kinetic equation. This may be seen if we follow directly from the equation for the density matrix prior to eliminating the phase of the order parameter (equation (1.9) when $t_1 = t_2$). From (1.15) we have

$$n(\mathbf{r}, t) = \langle \tilde{N}\tilde{\rho} \rangle = \langle \hat{\sigma}_+ \hat{\rho}(\mathbf{r}\mathbf{p}t) + \hat{\sigma}_-(\hat{1} - \hat{\rho}^T(\mathbf{r}, -\mathbf{p}, t)\rangle.$$

Writing an analogous equation for the function $\hat{\rho}^T(\mathbf{r}, -\mathbf{p}, t)$ in addition to (1.28) we draft the combination $\hat{\sigma}_+\hat{\rho} + \hat{\sigma}_-(1 - \hat{\rho}^T)\rangle$; by virtue of identity (2.39) we obtain from the right $\langle \hat{\sigma}_z \hat{R}^{(2)}\{\hat{\rho}\}\rangle = 0$. After transformation of the convective terms we obtain

$$\frac{\partial}{\partial t} n + \Delta \langle \hat{\sigma}_J \hat{\rho} \rangle + \operatorname{div} \mathbf{j} = 0.$$

In place of $\Delta\langle\hat{\sigma}_y\hat{\rho}\rangle$ we obtain from (1.9) the following contribution to the left half of the commutator $[\hat{\varepsilon}_p, \hat{\rho}]_-$:

$$\langle \Delta (\hat{\rho})_{12} - \Delta^* (\hat{\rho})_{21} \rangle \equiv 0.$$

Thus, the continuity equation

$$\frac{\partial}{\partial t} n + \operatorname{div} \mathbf{j} = 0, \tag{2.1}$$

where

$$\mathbf{j} = \mathbf{j}_0 + n\mathbf{v}_s, \quad \mathbf{j}_0 = \left\langle \frac{\mathbf{p}}{m}\left(\hat{\rho} - \frac{1}{2}\right)\right\rangle, \tag{2.2}$$

follows directly from the kinetic equation. We note that the kinetic equation for the excitation distribution function [4-6, 8, 10] produces an incorrect expression for the current which was initially used in Geylikman's first study [1] (see also [2]); hence, for example, in study [58] formula (2.2) was obtained by averaging the operator expression for the electron current.

2. Continuity equations for the density of the electron spin projection. As noted above in electron-electron collisions the projection of electron spin is conserved. The formal expression of this fact takes the form

$$\frac{\partial}{\partial t} s_z + \operatorname{div} \mathbf{j}_s = 0, \tag{2.3}$$

where the spin projection flux is determined by the formula

$$\mathbf{j}_s = \mathbf{j}_s^{(0)} + s_z \mathbf{v}_s, \quad \mathbf{j}_s^{(0)} = \left\langle \frac{\mathbf{p}}{m} \left(\hat{\sigma}_z \hat{\rho} + \frac{1}{2} \right) \right\rangle \tag{2.4}$$

and follows from equation (1.28) and identity (1.51).

3. Momentum conservation law. For the current density

$$\mathbf{j} = \left\langle \frac{\mathbf{p}}{m} \{ \hat{\sigma}_+ \hat{\rho}(\mathbf{r}\mathbf{p}t) - \hat{\sigma}_- (\hat{1} - \hat{\rho}^T(\mathbf{r}, -\mathbf{p}, t)) \} \right\rangle + n\mathbf{v}_s$$

we obtain from equations (1.28) and (1.28.1) and identity (1.52) the equation of motion

$$\frac{\partial}{\partial t} j_\alpha + \frac{\partial}{\partial r_{\alpha'}} \pi_{\alpha'\alpha} = 0. \tag{2.5}$$

The momentum flux tensor $\pi_{\alpha\alpha'}$ takes the form

$$\pi_{\alpha\alpha'} = \pi_{\alpha\alpha'}^{(0)} + j_{0\alpha} v_{s\alpha'} + j_{0\alpha'} v_{s\alpha} + n v_{s\alpha} v_{s\alpha'}, \tag{2.6}$$

where

$$\pi_{\alpha\alpha'}^{(0)} = \left\langle \frac{p_\alpha p_{\alpha'}}{m^2} \left(\hat{\sigma}_z \hat{\rho} + \frac{1}{2} \right) \right\rangle - \frac{1}{2m} |g| \left\{ \frac{n^2}{2} + \frac{2\Delta^2}{g^2} + \frac{s_z^2}{2} \right\} \delta_{\alpha\alpha'}. \tag{2.7}$$

4. Energy balance equation. Identity (1.53) after some cumbersome transformations of equations (1.28) for $\hat{\rho}$ and $\hat{\rho}^T$ produces a continuity equation for the energy density $\mathcal{E}(\mathbf{r}t)$:

$$\frac{\partial}{\partial t} \mathcal{E} + \operatorname{div} \mathbf{j}_\mathcal{E} = 0, \tag{2.8}$$

where

$$\mathcal{E} = \mathcal{E}_0 + m \mathbf{v}_s \mathbf{j}_0 + \frac{1}{2} n m v_s^2, \tag{2.9}$$

while the vector of the energy flux density is

$$\mathbf{j}_\mathcal{E} = j_\mathcal{E}^{(0)} + m \overleftrightarrow{\pi}^{(0)} \mathbf{v}_s + \frac{1}{2} m v_s^2 \mathbf{j}_0 + \mathcal{E} \mathbf{v}_s. \tag{2.10}$$

The index (0) everywhere designates quantities that after the substitution $\mathbf{v}_s \to 0$, $\mathbf{v}_n \to \mathbf{v}_n - \mathbf{v}_s$ yield values of the corresponding macroscopic fields in the coordinate system K', where $\mathbf{v}_s = 0$. In (2.9), (2.10)

$$\mathcal{E}_0 = \left\langle \frac{p^2}{2m} \left(\hat{\sigma}_z \hat{\rho} + \frac{1}{2} \right) \right\rangle - \frac{1}{2} |g| \left(\frac{n^2}{2} + 2 \frac{\Delta^2}{|g|^2} - \frac{s_z^2}{2} \right), \tag{2.11}$$

$$\mathbf{j}_\mathcal{E}^{(0)} = \left\langle \frac{\mathbf{p}}{m} \frac{p^2}{2m} \left(\hat{\rho} - \frac{1}{2} \right) \right\rangle. \tag{2.12}$$

It is clear that ν are the nondiagonal components of the density matrix and they make a contribution to the quantities ν, $\pi_{\alpha\alpha'}^{(0)}$, $\mathcal{E}$, and Δ.

5. **Equations for superfluid velocity.** Bogolubov [48] demonstrated that in a Bose superfluid the equation for $\mathbf{v}_s$ is due to the reality condition of the order parameter after elimination of the phase. Such a situation also exists for a Fermi superfluid.

It follows from our definitions of the parameter $\tilde{\mu}$ and the superfluid velocity (1.21), (1.23) that

$$m\frac{\partial}{\partial t}\mathbf{v}_s + \frac{\partial}{\partial \mathbf{r}}\left(\tilde{\mu} + \frac{1}{2}|g|n + \frac{1}{2}mv_s^2\right) = 0. \qquad (2.13)$$

Following the arguments of study [48] we will identify the meaning of the quantity $\tilde{\mu}$. In the total thermodynamic equilibrium state in the frame where $\nu_n = 0$, the parameter λ of the large canonical ensemble is equal to the chemical potential of the system $\mu_0(P, T, u)$, where P is pressure, T is temperature, $\mathbf{u} = \mathbf{v}_s - \mathbf{v}_n$. Hence in equilibrium

$$\tilde{\mu} = \mu_0(P, T, u) - \frac{1}{2}|g|n - \frac{1}{2}mu^2. \qquad (2.14)$$

We will assume that equality (2.14) remains valid in local equilibrium as well when $P = P(\mathbf{r}, t)$, $T = T(\mathbf{r}t)$, $\mathbf{u} = \mathbf{u}(\mathbf{r}, t)$. Then for states similar to local-equilibrium, we obtain from (2.13) and (2.14)

$$m\frac{\partial}{\partial t}\mathbf{v}_s + \frac{\partial}{\partial \mathbf{r}}\left(\mu_0(P, T, u) + \frac{1}{2}mv_s^2 - \frac{1}{2}mu^2 + \delta\tilde{\mu}\right) = 0, \qquad (2.15)$$

where $\delta\tilde{\mu}$ is the nonequilibrium addition to the self-consistent parameter $\tilde{\mu}$; in order to determine this addition we will use the equation for the phase of the order parameter (1.30). The applicability condition of equation (2.15) takes the form $\delta\tilde{\mu}/\mu_F \ll 1$.

In study [28] the quantity $\tilde{\mu}$ is identified with the chemical potential of the system in the frame where $\mathbf{v}_n = 0$, while in an equation analogous to (2.15) there is no function μ_0 (P, T, u).

Equations (2.1), (2.3), (2.5) and (2.8) represent the conservation laws and account for the existence of the spin degree of freedom. In the nonferromagnetic system in the case of external actions that are spin-index nondiagonal it is natural to consider states of fermions with a compensated spin. By selecting the initial conditions for equation (1.28) in this class of states, we may eliminate relaxation of spin projection. Hence we may then assume $s_z = 0$ (see study [59]).

2. Hydrodynamic equations of an ideal superconducting fluid

With large times the leading term of kinetic equation (1.28) is the collision integral and hence the equilibrium solution is determined by the condition

$$\hat{R}^{(2)}\{\hat{\rho}_0\} = 0. \qquad (2.16)$$

It then follows from (1.28) that $\hat{\rho}_0$ commutes with the equilibrium energy operator of the self-consistent field. We may easily show that the general expression itself for $\hat{\rho}_0$ satisfying these requirements when $s_z = 0$ takes the form

$$\hat{\rho}_0 = n_0 \left[(\varepsilon_p \hat{\tau}_3 + \mathbf{p}(\mathbf{v}_s - \mathbf{v}_n))/T\right], \tag{2.17}$$

where $\mathbf{v}_n$ is the velocity of the normal component. It is obvious that collision integral $R^{(2)}_{e-e}$ vanishes by the distribution $\hat{\rho}_0$ as a function of the real nonequilibrium energy.

We will call the state of the system described by the density matrix $\hat{\rho}_l$ in (2.17) the local-equilibrium state when all parameters of this matrix depend on the coordinates and time. $\hat{\rho}_l$ depends on five independent functions; we normally use $n(\mathbf{r}t)$, $\mathbf{j}(\mathbf{r}t)$ and $\mathscr{E}(\mathbf{r}t)$ as such functions. We will assume that these quantities coincide with the true nonequilibrium values of the corresponding macroscopic quantities.

It is natural to assume that with large times in the zeroth approximation the state of a weakly-nonequilibrium system is described by the distribution $\hat{\rho}_l$. This assumption underlies the Chapman-Enskog scheme for solving the kinetic equation that allows us, in conditions of smallness of the gradients of the hydrodynamic fields, to express the fluxes of macroscopic quantities through their gradients, and to calculate the transport coefficients that arise here and at the same time to close the system of hydrodynamic equations.

We may show that the solution of the zeroth approximation of $\hat{\rho}_l$ does not contain thermal fluxes or voltages and results in the hydrodynamic equations of an ideal superfluid [48, 53, 54]. In the absence of external fields when $s_z = 0$ this system takes the form

$$\frac{\partial}{\partial t} n + \operatorname{div} \mathbf{j} = 0,$$

where

$$\mathbf{j} = n_n \mathbf{v}_n + n_s \mathbf{v}_s, \qquad n_s = n - n_n, \tag{2.18}$$

$n_n = \frac{1}{u^2}\left\langle \frac{\mathbf{p}\mathbf{u}}{m} \hat{\rho}_l \right\rangle$ is the density of the normal component;

$$\frac{\partial}{\partial t} j_\alpha + \frac{\partial}{\partial r_\beta} \pi_{\beta\alpha} = 0, \tag{2.19}$$

where

$$\pi_{\alpha\beta} = n_n v_{n\alpha} v_{n\beta} + n_s v_{s\alpha} v_{s\beta} + \frac{1}{m} P \delta_{\alpha\beta}, \tag{2.20}$$

P is pressure:

$$P = TS - \mathscr{E}_0 + \mu_0 n + n_n u^2, \tag{2.21}$$

S is entropy:

$$S = -\langle \rho_l \ln \rho_l + (1 - \rho_l) \ln (1 - \rho_l) \rangle, \tag{2.22}$$

$$\frac{\partial}{\partial t} \mathscr{E} + \operatorname{div} \mathbf{j}_{\mathscr{E}} = 0,$$

where the energy density is determined by the expressions (2.9), (2.11) while the energy flux is determined by (2.10), (2.12) calculated with the solution of $\hat{\rho}_l$; a BCS equation is valid for the order parameter Δ:

$$m \frac{\partial}{\partial t} \mathbf{v}_s + \frac{\partial}{\partial \mathbf{r}} (\mu_0 + m v_s^2/2 - m u^2/2) = 0. \tag{2.23}$$

We will provide one additional useful relation. From the self-consistent equation and thermodynamic relations in local equilibrium we may easily find that in the zeroth approximation

$$\left(\frac{\partial}{\partial t} \Delta\right)_l = 0. \tag{2.24}$$

Below we will use (2.24) to calculate the left half of the first approximation equation.

2. Application of the Chapman-Enskog method to solving the kinetic equation for the density matrix of a superconductor

The Chapman-Enskog method [30] contains two important assumptions. It is assumed that all hydrodynamical variables are quantities in the order of unity; an exact solution of $\hat{\rho}$ has a time dependence only through the hydrodynamic quantities. Corrections to $\hat{\rho}_l$ in these assumptions take the form of series in terms of the gradients of the macroscopic quantities.

In order to find the equilibrium correction to $\hat{\rho}_l$ in the first approximation in the gradients we will represent the density matrix as

$$\hat{\rho} = \hat{\rho}_l + \delta\hat{\rho}. \tag{2.25}$$

It is clear that all derivatives in the left half of (1.28) may be calculated with $\hat{\rho}_l$. In order to transform the commutator $[\hat{\varepsilon}_p, \hat{\rho}]_-$ and the collision integral it is convenient to write $\hat{\rho}$ somewhat differently

$$\hat{\rho} = n_0 \left(\frac{\hat{\varepsilon}_p - \mathbf{p}\mathbf{v}_n}{T}\right) + \delta\hat{\rho}, \tag{2.26}$$

where ε_p is the nonequilibrium energy operator. In accordance with the discussion above

$$[\hat{\varepsilon}_p, \hat{\rho}]_- = [\hat{\varepsilon}_p, \delta\hat{\rho}]_-, \tag{2.27}$$

$$\hat{R}^{(2)}_{e-e}\{\hat{\rho}\} = \hat{R}^{(2)}_{e-e}\{\delta\hat{\rho}\}. \tag{2.28}$$

It is clear that after factoring out terms of the second order in $\delta\hat{\rho}$ in (2.27), (2.28) we may consider the spectrum to be an equilibrium spectrum. The operator obtained from (1.48), (1.49) by conservation of quantities that are linear with respect to $\delta\hat{\Delta}$ only (called the linearized collision integral) will be designated as $\hat{\mathscr{L}}^{(2)}$.

We obtain for the first order correction to the density matrix a linear inhomogeneous integral equation

$$\frac{\partial}{\partial t}\hat{\rho}_l + \frac{1}{2}\left\{\left[\frac{\partial\hat{\varepsilon}_p}{\partial \mathbf{p}}, \frac{\partial}{\partial \mathbf{r}}\hat{\rho}_l\right]_+ - \left[\frac{\partial\hat{\varepsilon}_p}{\partial \mathbf{r}}, \frac{\partial}{\partial \mathbf{p}}\hat{\rho}_l\right]_+\right\} = \hat{L}\{\delta\tilde{\rho}\}. \tag{2.29}$$

Naturally we select as the region for determining the nonHermitian operator

$$\hat{L} = -i\,[\hat{\varepsilon}_p, \ldots]_- + \hat{\mathscr{L}}^{(2)}\{\ldots\} \tag{2.30}$$

the set of two-dimensional Hermitian matrices having the property of (1.30), i.e., $\langle\sigma_y\delta\tilde{\rho}\rangle = 0$.

In order to solve equation (2.29) we must require the left half of its part to be orthogonal to all left eigenvectors of the operator $\hat{L}$ responsible for the zeroth eigenvalue. It is obvious that identities (1.39), (1.51)-(1.53) remain valid for $\hat{\mathscr{L}}^{(2)}\{\delta\tilde{\rho}\}$ as well with any $\delta\rho$. Accounting for the limitation on the range of determination of the operator $\hat{L}$ we may easily see that the matrices $\hat{\sigma}_z$, 1, $\mathbf{p}$, and $\xi_p\hat{\sigma}_z - \Delta\hat{\sigma}_x + \mathbf{p}\mathbf{v}_s$ form the "zero subspace" of the left eigenvectors of the operator L (property (1.30) satisfies for L the identity (1.39)) completely analogous to the selection of the additive collision invariants of the normal Fermi-system.

Thus, the solution conditions of equation (2.29) are nothing other than zero approximation equations for the macroscopic quantities. Hence in calculating the left half of (2.29) we must replace all time derivatives with spatial gradients in accordance with equations for an ideal Fermi superfluid as in the normal case.

For ease of calculations we will represent $\hat{\rho}_l$ as

$$\rho_l = {}^1/_2\,\{n_0\,[(\varepsilon_p + \mathbf{p}\mathbf{u})/T] + n_0\,(-\varepsilon_p + \mathbf{p}\mathbf{u})/T]\} + {}^1/_2\hat{\tau}_3\,\{n_0\,[(\varepsilon_p + \mathbf{p}\mathbf{u})/T - \\ -n_0\,[(\varepsilon_p + \mathbf{p}\mathbf{u})/T)\}, \tag{2.31}$$

where $\hat{\tau}_3$ in the s-representation is a variable matrix:

$$\tau_3 = (\hat{\xi}_p/\varepsilon_p)\,\hat{\sigma}_z - (\Delta/\varepsilon_p)\hat{\sigma}_x.$$

As in study [28] we will limit our case to the situation when $u \ll \Delta/p_F$ and we may consider the problem to be spherically symmetrical. For any parameter

$$\frac{\partial}{\partial x}\hat{\rho}_l = n_0'\left(\frac{\varepsilon_p}{T}\right)\left\{\hat{\tau}_3\frac{\partial}{\partial x}\left(\frac{\varepsilon_p}{T}\right) + \frac{\partial}{\partial x}\left(\frac{\mathbf{p}\mathbf{u}}{T}\right)\right\} +$$

$$+\frac{\operatorname{th}(\varepsilon_p/2T)}{2\varepsilon_p}\left\{-\frac{\Delta}{\varepsilon_p}\frac{\partial}{\partial x}(\tilde{\xi}_p)+\frac{\tilde{\xi}_p}{\varepsilon_p}\frac{\partial}{\partial x}\Delta\right\}\hat{\tau}_1. \tag{2.32}$$

The matrix $\hat{\tau}_1$ in the s-representation takes the form

$$\hat{\tau}_1=\frac{\Delta}{\varepsilon_p}\hat{\sigma}_z+\frac{\tilde{\xi}_p}{\varepsilon_p}\hat{\sigma}_x.$$

It is convenient to select n, Δ and $\mathbf{u}$ as the independent thermodynamic variables; in view of (2.24) terms with $\delta\Delta/\Delta t$ drop out.

Further:

$$\frac{\partial}{\partial t}\hat{\rho}_l=n_0'\left(\frac{\varepsilon_p}{T}\right)\left\{\hat{\tau}_3\left[\frac{\tilde{\xi}_p}{\varepsilon_p}\frac{1}{T}\left(\frac{\partial\mu}{\partial n}\right)_\Delta+\frac{\varepsilon_p}{T^2}\left(\frac{\partial T}{\partial n}\right)_\Delta\right]\operatorname{div}\mathbf{j}+\right.$$
$$\left.+\frac{1}{T}\frac{n}{n_n}S\frac{\mathbf{p}}{m}\nabla T\right\}-\hat{\tau}_1\frac{\operatorname{th}\varepsilon_p/2T}{2\varepsilon_p}\frac{\Delta}{\varepsilon_p}\left(\frac{\partial\mu}{\partial n}\right)_\Delta\operatorname{div}\mathbf{j}. \tag{2.33}$$

After some rather cumbersome transformations of the convective terms and neglecting components of the second order of smallness of the $\mathbf{u}\nabla\tilde{\mu}$ type, we finally obtain the following equation for $\delta\tilde{\rho}$ which may easily be written in the coordinate system K', where $\mathbf{v}_s'=0$, $\mathbf{v}_n'=\mathbf{v}_n-\mathbf{v}_s$, $\mathbf{j}'=\mathbf{j}_0$:

$$n_0'\left(\frac{\varepsilon_p}{T}\right)\left\{\left(\frac{1}{T}\frac{n}{n_s}S-\frac{\tilde{\xi}_p}{T^2}\right)\frac{\mathbf{p}}{m}\nabla T+\frac{\varepsilon_p}{T^2}\left(\frac{\partial T}{\partial n}\right)_\Delta\operatorname{div}\mathbf{j}'+\right.$$
$$+\frac{1}{T}\frac{\tilde{\xi}_p}{\varepsilon_p}\left[\left(\frac{\partial\mu}{\partial n}\right)_\Delta\operatorname{div}\mathbf{j}'-\frac{p^2}{3m}\operatorname{div}\mathbf{v}_n'\right]-\frac{1}{T}\frac{\tilde{\xi}_p}{\varepsilon_p}\frac{p_\alpha p_\beta}{2m}\left(\frac{\partial v_{n\beta}'}{\partial r_\alpha}+\frac{\partial v_{n\alpha}'}{\partial r_\beta}-\right.$$
$$\left.\left.-\frac{2}{3}\delta_{\alpha\beta}\operatorname{div}\mathbf{v}_n'\right)\right]\hat{\tau}_3\Big\}-n_0'\left(\frac{\varepsilon_p}{T}\right)\frac{\hat{\tau}_1}{T}\frac{\Delta}{\varepsilon_p}\left\{\frac{p^2}{3m}\operatorname{div}\mathbf{v}_n'+\right.$$
$$\left.+\frac{p_\alpha p_\beta}{2m}\left(\frac{\partial v_{n\beta}'}{\partial r_\alpha}+\frac{\partial v_{n\alpha}'}{\partial r_\beta}-\frac{2}{3}\delta_{\alpha\beta}\operatorname{div}\mathbf{v}_n'\right)\right\}-\frac{\operatorname{th}\varepsilon_p/2T}{2\varepsilon_p}\frac{\Delta}{\varepsilon_p}\left(\frac{\partial\mu}{\partial n}\right)_\Delta\operatorname{div}\mathbf{j}'=$$
$$=-i\,[\hat{\varepsilon}_p,\delta\tilde{\rho}]_-+\hat{\mathscr{L}}^{(2)}\{\delta\tilde{\rho}\}. \tag{2.34}$$

The operator $\hat{\mathscr{L}}^{(2)}$ which we will not provide here due to its cumbersome size is not Hermitian; its left and right eigenvectors, generally speaking, do not coincide. Nonetheless the subspace of the right eigenvectors of this operator corresponding to the zeroth eigenvalue may be found in the same manner as in the normal case (see, for example, [41, Chapter 1, Section 6]). Thermodynamically the equilibrium solution of $\hat{\rho}_l$ depends on three independent parameters, two scalar parameters, and a single vector parameter. For convenience we will select the following set: $(\tilde{\mu}, \Delta, \mathbf{u})$. We may easily show that the collision integral $\hat{R}^{(2)}\{\hat{\rho}\}$ may be written as a certain multidimensional integral of the linear combination of matrix products. Hence the following statement is valid: since $\hat{R}^{(2)}\{\hat{\rho}\}$ vanishes by the equilibrium density matrix with any coefficients $(\tilde{\mu}, \Delta, \mathbf{u})$, its zeros will also be the small additions

$$\delta\tilde{\rho}_{\tilde{\mu}}=\frac{\partial}{\partial\tilde{\mu}}(\hat{\rho}_l)_{\Delta,\mathbf{u}}\delta\tilde{\mu},\qquad \delta\tilde{\rho}_\Delta=\frac{\partial}{\partial\Delta}(\hat{\rho}_l)_{\tilde{\mu},\mathbf{u}}\delta\Delta,\qquad \delta\tilde{\rho}_{\mathbf{u}}=\frac{\partial}{\partial\mathbf{u}}(\hat{\rho}_l)_{\tilde{\mu},\Delta}\delta\mathbf{u}. \tag{2.35}$$

This statement may be checked directly by varying the collision integral $\hat{R}^{(2)}\{\hat{\rho}\}$ and transforming the integrals that arise here with the products of the δ-functions using the action rules on the generalized functions.

Thus, if we consider states with a compensated electron spin, the operator $\hat{\mathscr{L}}^{(2)}$ has three independent "parasitic" solutions:

$$\delta\tilde{\rho}_{\mu} = \left\{ -n_0'\left(\frac{\varepsilon_p}{T}\right)\left[\frac{1}{T}\frac{\xi_p}{\varepsilon_p} + \frac{\varepsilon_p}{T^2}\left(\frac{\partial T}{\partial \mu}\right)_{\Delta \mathbf{u}}\right]\hat{\tau}_3 + \frac{\operatorname{th}\varepsilon_p/2T}{2\varepsilon_p}\frac{\Delta}{\varepsilon_p}\tau_1\right\}\alpha_{\mu}, \tag{2.36}$$

$$\delta\tilde{\rho}_{\Delta} = \left\{ -n_0'\left(\frac{\varepsilon_p}{T}\right)\left[-\frac{1}{T}\frac{\Delta}{\varepsilon_p} + \frac{\varepsilon_p}{T^2}\left(\frac{\partial T}{\partial \Delta}\right)_{\mu \mathbf{u}}\right]\hat{\tau}_3 + \frac{\operatorname{th}\varepsilon_p/2T}{2\varepsilon_p}\frac{\Delta}{\varepsilon_p}\frac{\xi_p}{\Delta}\hat{\tau}_1\right\}\frac{d\Delta}{d\mu}\alpha_{\Delta}, \tag{2.37}$$

$$\delta\tilde{\rho}_{\mathbf{u}} = -n_0'\left(\frac{\varepsilon_p}{T}\right)\frac{1}{T}\frac{\mathbf{p}}{m}\hat{1}\alpha_{\mathbf{u}}. \tag{2.38}$$

As in the normal case in order to determine the constants α_{μ}, α_{δ}, and $\alpha_{\mathbf{u}}$ we must use the fact that the solution of the zeroth approximation of $\hat{\rho}_l$ is normalized to the true macroscopic values of the volumetric density of fermion number, current, and energy density. The corresponding conditions in the coordinate system (K') take the form

$$\delta n = 0, \tag{2.39}$$

$$\delta \mathbf{j}_0 = 0, \tag{2.40}$$

$$\delta \mathscr{E}_0 = 0. \tag{2.41}$$

For simplicity we will designate the inhomogeneous term of the equation by the symbol Left$\{\hat{\rho}_l\}$. Accounting for (2.36)-(2.38) we may rewrite the equation for $\delta\tilde{\rho}$:

$$\text{Left}\,\{\hat{\rho}_l\} = -i\,[\hat{\varepsilon}_p, \delta\tilde{\rho}]_- + \hat{\mathscr{L}}^{(2)}\{\delta\tilde{\rho} - \delta\tilde{\rho}_{\mu} - \delta\tilde{\rho}_{\Delta} - \delta\tilde{\rho}_{\mathbf{u}}\}. \tag{2.42}$$

We will solve equation (2.42) using for the linearized collision integral the approximation of the relaxation time so that we may account for the existence of "parasitic" solutions (2.36)-(2.38). For classical gases this approximation was developed in study [60]. Estimates of the characteristic frequencies of electron-electron collisions (1.64), (1.65) allows us to calculate the kinetic coefficients in two limiting cases: near zero temperature and the vicinity of the phase transition point. For this we will write a complete system of equations in order to determine the nonequilibrium additions. Here, as in calculating Left$\{\hat{\rho}_l\}$ the features associated with the existence of an isotopic degree of freedom are clearly manifest.

The appearance of the nonequilibrium addition to the density matrix $\delta\tilde{\rho}$ containing three unknown constants (α_{μ}, α_{Δ}, and $\alpha_{\mathbf{u}}$) is accompanied by the appearance of self-consistent nonequilibrium additions ($\delta\tilde{\mu}$) and ($\delta\Delta$) in the excitation spectrum. In order to determine the five independent quantities we have five equations: two matching equations (for the phase and modulus of the order parameter) and three conditions (2.39)-(2.41) deriving from the selection of the normalization of $\hat{\rho}_l$. We will write the system of these equations in variations. For this we will represent

$$\hat{\rho} = n_0\left(\frac{\hat{\varepsilon}_p^{(0)} - \mathbf{p}\mathbf{v}_n}{T}\right) + \delta\left[n_0\left(\frac{\hat{\varepsilon}_p - \mathbf{p}\mathbf{v}_n}{T}\right)\right] + \delta\tilde{\rho}. \tag{2.43}$$

Using (2.32) we will have

$$\hat{\rho} = n_0\left(\frac{\hat{\varepsilon}_p^{(0)} + \mathbf{p}\mathbf{v}_n}{T}\right) + \delta\tilde{\rho} +$$
$$+\left\{-n_0'\left(\frac{\varepsilon_p^{(0)}}{T}\right)\frac{1}{T}\frac{\xi_p^{(0)}}{\varepsilon_p^{(0)}}\hat{\tau}_3 + \frac{\operatorname{th}(\varepsilon_p^{(0)}/2T)}{2\varepsilon_p^{(0)}}\frac{\Delta^0}{\varepsilon_p^{(0)}}\hat{\tau}_1\right\}\delta\tilde{\mu} +$$
$$+\left\{n_0'\left(\frac{\varepsilon_p^{(0)}}{T}\right)\frac{1}{T}\frac{\Delta^{(0)}}{\varepsilon_p^{(0)}}\hat{\tau}_3 + \frac{\operatorname{th}(\varepsilon_p^{(0)}/2T)}{2\varepsilon_p^{(0)}}\frac{\xi_p^{(0)}}{\varepsilon_p^{(0)}}\hat{\tau}_1\right\}\delta\Delta. \tag{2.44}$$

In (2.43), (2.44) $\hat{\varepsilon}_p$ is the true nonequilibrium spectrum unlike the equilibrium $\hat{\varepsilon}_p^{(0)}$.

Below we will drop the index (0); this will not cause confusion.

1) We will multiply both halves of (2.44) by $\hat{\sigma}_y = \hat{\tau}_2$ and will use the operation $\langle\ldots\rangle$; and we obtain

$$\langle\hat{\tau}_2\delta\tilde{\rho}\rangle = 0. \tag{2.45}$$

2) Operating on (2.44) with the operator $\langle\hat{\sigma}_x\{\ldots\}\rangle$, we discover that

$$n_\Delta\delta\tilde{\mu} - \left\{n_\mu - \left\langle\frac{1}{T}\left[-n_0'\left(\frac{\varepsilon_p}{T}\right)\right]\right\rangle\right\}\delta\Delta + \langle\hat{\sigma}_x\delta\tilde{\rho}\rangle = 0. \tag{2.46}$$

3) The result of applying the $\langle\hat{\sigma}_z\{\ldots\}\rangle$ operation to (2.44) together with conditions (2.39) takes the form

$$n_\mu\delta\tilde{\mu} + n_\Delta\delta\Delta + \langle\hat{\sigma}_z\delta\tilde{\rho}\rangle = 0. \tag{2.47}$$

We utilize the identities

$$n_\mu = \left\langle\frac{\Delta^2}{2\varepsilon_p^3}\operatorname{th}\frac{\varepsilon_p}{2T}\right\rangle + \left\langle\frac{1}{T}\frac{\xi_p^2}{\varepsilon_p^2}\left[-n_0'\left(\frac{\varepsilon_p}{T}\right)\right]\right\rangle,$$
$$n_\Delta = \left\langle\frac{\Delta\xi_p}{2\varepsilon_p^3}\operatorname{th}\frac{\varepsilon_p}{2T}\right\rangle - \left\langle\frac{1}{T}\frac{\Delta\xi_p}{\varepsilon_p^2}\left[-n_0'\left(\frac{\varepsilon_p}{T}\right)\right]\right\rangle.$$

4) By virtue of (2.40)

$$\left\langle\frac{\mathbf{p}}{m}\delta\tilde{\rho}\right\rangle = 0. \tag{2.48}$$

5) We obtain from (2.41) and (2.44)

$$\left\langle\frac{p^2}{2m}\hat{\sigma}_z\{\delta\hat{\rho}_0(\hat{\varepsilon}_p) + \delta\tilde{\rho}\}\right\rangle - \frac{2\Delta}{|g|}\delta\Delta = 0,$$

i.e.,

$$\tilde{\mu}\delta n + \langle\varepsilon_p\hat{\tau}_3\delta\hat{\rho}\rangle = 0.$$

We directly obtain from here and from (2.45), (2.46)

$$\left\langle\frac{\Delta}{T}\left[-n_0'\frac{\varepsilon_p}{T}\right)\right]\right\rangle\delta\Delta-\langle\varepsilon_p\hat{\tau}_3\delta\tilde{\rho}\rangle=0. \tag{2.49}$$

Equation system (2.45)-(2.49) makes it possible to completely solve the problem of determining the transport coefficients if the solution of the kinetic equation of the first approximation (2.42) is known.

Finally we will calculate the kinetic coefficients.

1. **The problem of thermal conductivity.** By definition the thermal conductivity coefficient is

$$\delta\mathbf{j}_\varepsilon=-\varkappa\nabla T, \tag{2.50}$$

where

$$\delta\mathbf{j}_\varepsilon=\left\langle\frac{\mathbf{p}}{m}\frac{p^2}{2m}\delta\tilde{\rho}\right\rangle.$$

With only the temperature gradient

$$\delta\tilde{\rho}=\nu_\alpha^{-1}\left[-n_0'\left(\frac{\varepsilon_p}{T}\right)\right]\left(\frac{n}{n_n}S-\frac{\xi_p}{T}\right)\frac{\mathbf{p}}{m}\frac{1}{T}\nabla T+\alpha_{\mathbf{u}}\frac{1}{T}\left[-n_0'\left(\frac{\varepsilon_p}{T}\right)\right]\frac{\mathbf{p}}{m}.$$

From (2.48) we have

$$\alpha_{\mathbf{u}}=\nu_\alpha^{-1}\left(\lambda-\frac{n}{n_n}S\right)\frac{1}{T}\nabla T,$$

so that

$$\delta\tilde{\rho}=\nu_\alpha^{-1}\left(\lambda-\frac{1}{T}\xi_p\right)\frac{1}{T}\nabla T\,\frac{\mathbf{p}}{m}\left[-n_0'\left(\frac{\varepsilon_p}{T}\right)\right], \tag{2.51}$$

where

$$\lambda=\frac{T}{\mu_F}\left\langle\frac{\xi_p^2}{T^2}\left[-n_0'\left(\frac{\varepsilon_p}{T}\right)\right]\right\rangle\left\langle\left[-n_0'\left(\frac{\varepsilon_p}{T}\right)\right]\right\rangle^{-1}.$$

We obtain the following expression from (2.50), (2.51) for $\varkappa$:

$$\varkappa=\frac{2}{3}\nu_\alpha^{-1}\frac{p_F^2}{m^2}\left\langle\frac{\xi_p^2}{T^2}\left[-n_0'\left(\frac{\varepsilon_p}{T}\right)\right]\right\rangle\left\{1-\frac{T}{\mu_F}\lambda\right\}. \tag{2.52}$$

Using estimates of the electron-electron collision frequencies found in Section 2 of Chapter 1 we find, finally, that when $T\to 0$

$$\varkappa\simeq\tilde{c}_\alpha\frac{n}{m}\frac{1}{(mgp_F)^2}\frac{\mu_F}{T}; \tag{2.53}$$

when $T\to T_c-0$

$$\varkappa\simeq c_\alpha'\frac{n}{m}\frac{T_c}{\mu_F}\left(1-\frac{\Delta}{T_c}\right), \tag{2.54}$$

where $\tilde{c}_\alpha$ and c_α' are dimensionless constants >0.

In the thermal conductivity problem the nonequilibrium addition to the density matrix is diagonal in the ν-representation. Study [28] does not account for the contribution to (2.51) from the "parasitic" solution; at the same time this addition reduces the result by precisely a factor of two.

2. The first viscosity problem. The first viscosity coefficient is determined by the tensor part of the general solution of equation (2.42). Consequently

$$\mathrm{Left}\,\{\hat{\rho}_l\} = -n_0'\left(\frac{\varepsilon_p}{T}\right)\frac{1}{T}\hat{\sigma}_z\frac{p_\alpha p_\beta}{2m}\left(\frac{\partial v_{n2}'}{\partial r_\beta}+\frac{\partial v_{n\beta}'}{\partial r_\alpha}-\frac{2}{3}\delta_{\alpha\beta}\,\mathrm{div}\,\mathbf{v}_n'\right). \tag{2.55}$$

By definition the correction to the tensor part of the momentum flux is

$$(\delta\overleftrightarrow{\pi}^{(0)})^{(t)}_{\alpha\beta} = \frac{1}{m}\left\langle\frac{p_\alpha p_\beta}{m}\left(1-\frac{1}{3}\delta_{\alpha\beta}\right)\hat{\sigma}_z\delta\tilde{\rho}\right\rangle =$$
$$= -\eta\left(\frac{\partial v_{n\alpha}'}{\partial r_\beta}+\frac{\partial v_{n\beta}'}{\partial r_\alpha}-\frac{2}{3}\delta_{\alpha\beta}\,\mathrm{div}\,\mathbf{v}_n'\right).$$

Solving equation (2.42) with the left half of (2.55) we find that the shear viscosity coefficient is determined by the contribution of the ν-diagonal components of the density matrix and is equal to

$$\eta = \frac{1}{15m}\nu_\alpha^{-1}\left\langle\frac{1}{T}\frac{\xi_p^2}{\varepsilon_p^2}\left(\frac{p^2}{m}\right)^2\left[-n_0'\left(\frac{\varepsilon_p}{T}\right)\right]\right\rangle. \tag{2.56}$$

When $T \to 0$ the leading term of the integral in (2.56) in angle brackets, $\langle(\ldots)\rangle \sim n\mu_F(\Delta_0/T)\; K_1(\Delta_0/T)$ [61]; when $T \to 0$ $\langle(\ldots)\rangle \sim n\mu_F$. Consequently,

$$\eta \simeq \begin{cases} C_\eta\dfrac{n}{m}\left(\dfrac{\mu_F}{\Delta_0}\right)^2\dfrac{1}{(mgp_F)^2}\dfrac{\Delta_0}{T}, & T\to 0,\\[2ex] C_\eta'\dfrac{n}{m}\dfrac{\mu_F}{\nu_e}\left(1-B_\alpha'\dfrac{\Delta}{T_c}\right). & T\to T_c-0,\end{cases} \tag{2.57}$$

where C_η, C_η', B_α' are positive dimensionless constants. The divergence $\sim 1/T$ of the shear viscosity coefficient η when $T \to 0$ was predicted in the study by Soda and Fudjiki [29] for the isotropic B-phase of superfluid helium-3 and is related to the rapid decay in frequency ν_α of the fermion-fermion collisions with a drop in temperature. This result is valid only in a range of perturbation frequencies much less than $\nu_\alpha \sim \frac{\Delta_0^2}{\mu_F}\left(\frac{T}{\Delta_0}\right)^{1/2}\exp\left(-\frac{\Delta_0}{T}\right)$, which significantly complicates its experimental test (see below). When $\omega \gtrsim \nu_\alpha$ the first viscosity manifests reactive behavior; we may show that in this case when $T \to 0$

$$\eta \sim \frac{n\mu_F}{m}\frac{\Delta_0}{T}K_1\left(\frac{\Delta_0}{T}\right)\frac{1}{i\omega+\nu_\alpha}$$

at a given perturbation frequency $\omega \gg \nu_\alpha$ the coefficient η behaves when $T \to 0$ as $\left(\frac{\Delta_0}{T}\right)^{1/2}\frac{\mu_F}{i\omega}\exp\left(-\frac{\Delta_0}{T}\right)$. When $T = T_c$ the coefficient η coincides with the viscosity coefficient of normal Fermi-fluid; the Cooper correlations result in a reduction in η when $T \lessdot T_c$. Shumeyko [28] came to the same conclusion by solving the kinetic equation for the excitation distribution function.

3. **The second viscosity problem.** The existence of volumetric viscosity of Fermi superfluid results from the existence of a scalar part of the general solution of equation (2.42):

$$\text{Left}\,\{\hat{\rho}_l\} = n_0'\left(\frac{\varepsilon_p}{T}\right)\left\{\frac{1}{T}\frac{\xi_p}{\varepsilon_p}\left[\left(\frac{\partial\mu}{\partial n}\right)_\Delta \operatorname{div}\mathbf{j}' - \frac{p^2}{3m}\operatorname{div}\mathbf{v}_n'\right] + \right.$$
$$\left. + \frac{\varepsilon_p}{T^2}\left(\frac{\partial T}{\partial n}\right)_\Delta \operatorname{div}\mathbf{j}'\right\}\hat{\tau}_3 + \left\{- n_0'\left(\frac{\varepsilon_p}{T}\right)\frac{1}{T}\frac{\Delta}{\varepsilon_p}\frac{p^2}{3m}\operatorname{div}\mathbf{v}_n' - \right.$$
$$\left. - \frac{\operatorname{th}\varepsilon_p/2T}{2\varepsilon_p}\frac{\Delta}{\varepsilon_p}\left(\frac{\partial\mu}{\partial n}\right)_\Delta \operatorname{div}\mathbf{j}'\right\}\hat{\tau}_1 \equiv A\hat{\tau}_3 + B\hat{\tau}_1. \tag{2.58}$$

The second viscosity coefficients are contained in the dissipative correction to the scalar part of the momentum flux tensor which we may write, accounting for the condition $\delta\mathscr{E}_0 = 0$, as

$$(\delta\overleftrightarrow{\pi}^{(0)})^{(sc)}_{\alpha\beta} = -\delta_{\alpha\beta}\frac{n}{m}\frac{2\Delta}{3n\,|g\,|}\delta\Delta, \tag{2.59}$$

and in the correction $\delta\tilde{\mu}$ in the equation for $\mathbf{v}_s$. According to Khalatnikov [62] in the coordinate system K'

$$-\left(\frac{m}{n}\right)^{-1}\frac{2\Delta}{3n\,|g\,|}\delta\Delta = -\zeta_1\operatorname{div}(\mathbf{j}' - n\mathbf{v}_n') - \zeta_2\operatorname{div}\mathbf{v}_n, \tag{2.60}$$

$$\delta\tilde{\mu} = -\zeta_3\operatorname{div}(\mathbf{j}' - n\mathbf{v}_n') - \zeta_4\operatorname{div}\mathbf{v}_n'. \tag{2.61}$$

According to the Onsager symmetry principle $\zeta_1 = \zeta_4$. The entropy does not diminish if the kinetic coefficients $\varkappa$, η, ζ_2 and ζ_3 are positive, while ζ_1 - ζ_3 are such that

$$\zeta_1^2 \leqslant \zeta_2\zeta_3. \tag{2.62}$$

We will write the general solution of equation (2.42) with the left half of (2.58):

$$\delta\rho_{11} = -\delta\rho_{22} = -\frac{1}{\nu_\alpha}A + (\delta\tilde{\rho}_\mu + \delta\tilde{\rho}_\Delta + \delta\tilde{\rho}_u)_{11}. \tag{2.63}$$

$$\delta\rho_{12} = \delta\rho_{21}^* = -(\nu_\gamma^- + i2\varepsilon_p)^{-1}\left\{B + \frac{\nu_\delta}{\nu_\alpha}A + \nu_\gamma[-\delta\tilde{\rho}_\mu - \delta\tilde{\rho}_\Delta - \delta\tilde{\rho}_u]_{12}\right\}. \tag{2.64}$$

Obviously (2.48) is satisfied identically when $\delta\tilde{\rho}_u = 0$. The remaining equations for solutions (2.63)-(2.64) take the following form; equation (2.45):

$$\alpha_\mu\left\langle\frac{\Delta^2}{2\varepsilon_p^3}\operatorname{th}\frac{\varepsilon_p}{2T}\right\rangle + \alpha_\Delta\left\langle\frac{\Delta\xi_p}{2\varepsilon_p^3}\operatorname{th}\frac{\varepsilon_p}{2T}\right\rangle = \frac{1}{\nu_\gamma}\left\langle\frac{\Delta}{\varepsilon_p}\left(B + \frac{\nu_\delta}{\nu_\alpha}A\right)\right\rangle; \tag{2.65}$$

equation (2.46):

$$n_\Delta\delta\tilde{\mu} - \left\{n_\mu - \left\langle\frac{1}{T}\left[-n_0'\left(\frac{\varepsilon_p}{T}\right)\right]\right\rangle\right\}\delta\Delta + \left\langle\frac{\Delta}{T^2}\left[-n_0'\left(\frac{\varepsilon_p}{T}\right)\right]\right\rangle \times$$

$$\times\left\{\left(\frac{\partial\Delta}{\partial\mu}\right)_T \alpha_\mu - \alpha'_\Delta\right\} = -\frac{1}{\nu_\alpha}\left\langle\frac{\Delta}{\varepsilon_p}A\right\rangle, \quad (2.66)$$

$$\alpha'_\Delta = \left(\frac{\partial\Delta}{\partial\mu}\right)_T \alpha_\Delta;$$

equation (2.47):

$$n_\mu\delta\tilde{\mu} + n_\Delta\delta\Delta + \left\langle\frac{1}{T}\frac{\xi_p^2}{\varepsilon_p^2}\left[-n'_0\left(\frac{\varepsilon_p}{T}\right)\right]\right\rangle\alpha_\mu = \frac{1}{\nu_\alpha}\left\langle\frac{\xi_p}{\varepsilon_p}A\right\rangle; \quad (2.67)$$

equation (2.49):

$$\left\langle\frac{\Delta}{\varepsilon_p}\frac{\varepsilon_p}{T}\left[-n'_0\left(\frac{\varepsilon_p}{T}\right)\right]\right\rangle\delta\Delta + \left\langle\frac{\varepsilon_p^2}{T^2}\left[-n'_0\left(\frac{\varepsilon_p}{T}\right)\right]\right\rangle\left(\frac{\partial T}{\partial\Delta}\right)_\mu \times$$
$$\times\left(\alpha_\mu\left(\frac{\partial\Delta}{\partial\mu}\right)_T - \alpha'_\Delta\right) = \frac{1}{\nu_\alpha}\langle\varepsilon_p A\rangle. \quad (2.68)$$

In solving equation system (2.65)-(2.68) we pass to the symmetric integration limits in integrals with respect to $(d\xi_p)$ that rapidly diverge when $|\xi_p| \to \infty$; the resulting error when $T \to 0$ is of the order $[\mu_F/|g|\ \exp(-\mu_F/|g|)]^2$, while when $T \to T_c - 0$ it is $\sim(\mu_F/|g|)\exp(-2\mu_F/|g|)$. We may state that the coefficients ζ_1, ζ_2, and ζ_4 in (2.60), (2.61) are equal to zero with identical accuracy. For the coefficient ζ_3 we obtain the following estimates:

when $T \to 0$

$$\zeta_3 \simeq C_\alpha \frac{1}{m}\frac{p_F^2}{3mn}\frac{1}{(mgp_F)^2}\frac{\mu_F}{\Delta_0^2}, \quad C_\alpha > 0 \quad (2.69)$$

(determined by the contribution of the components $\delta\tilde{\rho}$ that are diagonal in the ν-representation);

when $T \to T_c - 0$

$$\zeta_3 = \frac{4}{\pi}\frac{T_c}{\Delta}\frac{1}{\nu_\gamma}\frac{p_F^2}{3mn}, \quad (2.70)$$

where $\nu_\gamma = \nu_e$ is the electron-electron collision frequency in a normal metal when $T = T_c$. Obviously the second viscosity coefficient in this limiting case is related to the existence of $\delta\tilde{\rho}$ components that are nondiagonal in the ν-representation; the contribution of the "diagonal channel" is equal to zero with $(\mu_F/|g| \times \exp(-2\mu_F/|g|)$ accuracy. ζ_3 is singular in the vicinity of the critical point $(\sim T_c/\Delta)$; the dissipative correction in the equation for the $\mathbf{v}_s$-combination $\zeta_3\mathrm{div}(\mathbf{j} - n\mathbf{v}_n)$ when $T \to T_c - 0$ vanishes entirely as Δ/T_c, following the behavior of the correction to the dissipative term describing the shear viscosity.

Thus, the hydrodynamic equations of a nonideal superconducting fluid take the following form:

$$\frac{\partial n}{\partial t} + \mathrm{div}\,\mathbf{j} = 0, \qquad \mathbf{j} = n_n\mathbf{v}_n + n_s\mathbf{v}_s, \quad (2.71)$$

$$\frac{\partial}{\partial t}(\mathbf{j})_\alpha + \frac{\partial}{\partial r_\beta}\left\{\left(n_n v_{n\alpha} v_{n\beta} + n_s v_{s\alpha} v_{s\beta} + \frac{1}{m}P\delta_{\alpha\beta}\right) - \right.$$
$$\left. -\eta\left(\frac{\partial v_{n\alpha}}{\partial r_\beta} + \frac{\partial v_{n\beta}}{\partial r_\alpha} - \frac{2}{3}\delta_{\alpha\beta}\,\mathrm{div}\,\mathbf{v}_n\right)\right\} = 0, \quad (2.72)$$

$$\frac{\partial \mathscr{E}}{\partial t} + \operatorname{div}\{\mathbf{j}_\varepsilon^{(0)} - \varkappa \nabla T - \zeta_3 [\operatorname{div}(\mathbf{j} - n\mathbf{v}_n)](\mathbf{j} - n\mathbf{v}_n)\} -$$
$$- \frac{\partial}{\partial r_\alpha}\left\{\eta v_{n\beta}\left[\frac{\partial v_{n\alpha}}{\partial r_\beta} + \frac{\partial v_{n\beta}}{\partial r_\alpha} - \frac{2}{3}\delta_{\alpha\beta}\operatorname{div}\mathbf{v}_n\right]\right\} = 0, \qquad (2.73)$$

$$m\frac{\partial}{\partial t}\mathbf{v}_s + \operatorname{grad}\left\{\mu_0(P, T, u) + \frac{1}{2}mv_s^2 -\right.$$
$$\left. - \frac{1}{2}m(\mathbf{v}_s - \mathbf{v}_n)^2 - m\zeta_3 \operatorname{div}(\mathbf{j} - n\mathbf{v}_n)\right\} = 0, \qquad (2.74)$$

where the kinetic coefficients $\varkappa$, η and ζ_3 are found in sections 1-3, respectively, of Section 3 of this chapter.

4. First and second sound damping

The propagation of first sound is accompanied by perturbations in the total density of the Fermi-fluid; activation of Coulomb interaction transforms this mode into plasma oscillations [62, 64].

The possibility for the existence of second sound in superconductors appearing as temperature (and entropy) oscillations and independent of oscillations in fluid density was considered by Bardeen [65] and Ginzburg [66]. They demonstrated that the damping of second sound may be small only in extraordinarily rigid conditions: $\max\{\nu_i, \nu_E\} \ll \omega\nu_{e-e}v_2/v_F$, where ω is the perturbation frequency, v_2 is the velocity of second sound. Since $v_2 \ll v_F$, and, as a rule $\nu_{e-e} < \nu_E \ll \nu_i$, this condition on the frequency is virtually impossible to satisfy. Above we found the kinetic coefficients for a neutral Fermi superfluid assuming that the primary source of relaxation are fermion-fermion collisions. Following Khalatnikov [67] linearizing the system of hydrodynamic equations (2.71)-(2.74) and using estimates of the kinetic coefficients, we find the damping coefficients of the first ($\alpha_1(\omega, T)$) and second sounds in Fermi superfluid when $T \to 0$ in the lower temperature range of the phase transition as well:

$$\alpha_1(\omega, T) = \frac{2\omega^2\eta}{\sqrt{3}\, nv_F^2} \sim \frac{\omega^2}{mv_F^2}\begin{cases}\left(\frac{\mu_F}{\Delta_0}\right)^2 \frac{1}{(mgp_F)^2}\frac{\Delta_0}{T}, & T \to 0,\\ \frac{\mu_F}{\nu_e}\left(1 - B_\alpha'\frac{\Delta}{T_c}\right), & T \to T_c - 0;\end{cases} \qquad (2.75)$$

$$\alpha_2(\omega, T) = \frac{\omega^2}{2nv_2^2}\frac{\varkappa}{C_V} \sim \frac{\omega^2}{v_F}\begin{cases}\frac{\mu_F^4}{(T\Delta_0)^{5/2}}\frac{1}{(mgp_F)^2}\exp\left(\frac{\Delta_0}{T}\right), & T \to 0,\\ \frac{1}{\nu_e}\left(\frac{\mu_F}{T_c}\right)^3\left(1 - \frac{T}{T_c}\right)^{-3/2}, & T \to T_c - 0,\end{cases} \qquad (2.76)$$

where C_V is the specific thermal capacity.

Analysis reveals that the damping coefficient of first sound $\alpha_1(\omega, T)$ is determined by the shear viscosity, and the damping of second sound is determined by the thermal conductivity. Measurements of the first viscosity of superfluid helium-3 were carried out in study [68]. Its results qualitatively confirm our derived temperature dependence of the shear viscosity $\eta \sim \eta_n\left(1 - B_\alpha'\frac{\Delta}{T}\right)$ when $T \to T_c - 0$ for the A-phase of He^3 and growth of η when $T \to 0$ in the quasi-isotropic β-phase. Study [28] alternately predicts η approaching a constant in the latter case.

The behavior of the damping coefficient of second sound is consistent with the conclusion of studies [65, 66] regarding the impossibility of second sound propagation in superconductors, while the relation $\alpha_2(\omega, T) \sim (1 - T/T_C)^{-3/2}$ when $T \to T_c - 0$ was observed in study [69].

Thus, the kinetic equation for a density matrix with a collision integral that conserves particle number derived in Chapter 1 makes it possible to formulate a correct hydrodynamic scheme for describing superconducting Fermi-systems and allows us to make predictions that are confirmed experimentally. We note that in the hydrodynamic stage the ν-nondiagonal components of the density matrix make the primary contribution to the dissipative correction describing the second viscosity near T_C. As noted by Khalatnikov [70] the existence of bulk viscosity is a direct consequence of the fact that slow equilibrium processes occur in the excitation gas. Such processes in superconductors are well known and are related to the equalization of the populations of the electron and hole branches of the excitation spectrum. Part of the next chapter is devoted to their examination. We note that this range of issues, evidently, cannot be correctly described within the scope of a kinetic equation for the excitation distribution function.

Chapter 3

THE COLLISIONAL RELAXATION OF THE ORDER PARAMETER MODULUS, AND THE RELAXATION, DIFFUSION AND FLUCTUATIONS OF THE BRANCH IMBALANCE OF THE SPECTRUM OF EXCITATIONS AND NONEQUILIBRIUM SPIN DENSITY IN SUPERCONDUCTORS

In many technical applications of superconductivity a situation often occurs in which the superconductor is influenced by external perturbations that vary slowly in time and are weakly-inhomogeneous in space. The electron scattering processes from electron-electron, electron-phonon, and electron-impurity collisions play a significant role in the evolution of the nonequilibrium state of a superconductor in such conditions. A correct consideration of scattering processes is, therefore, decisive for a correct description of system behavior.

Weakly-inhomogeneous states of pure superconductors, as a rule, are investigated on the basis of the kinetic equation for the excitation distribution function derived in studies [4-6, 8-10]. In many such studies (for example, [71-74]) in formulating the solution of the linearized kinetic equation the need to observe the solvability conditions of the derived inhomogeneous integrodifferential equation are not accounted for; these conditions are related to the existence of the additive invariants of the collision integral. The situation is complicated by the fact that the quasi particle integral used in these studies does not have the additive invariant corresponding to conservation of real particle number.

In this chapter we will apply a kinetic scheme based on an equation for a single-electron density matrix with a collision integral that conserves particle number to a number of problems of nonequilibrium superconductivity. As noted previously the cross-terms of the kinetic equation derived in Chapter 1 of this study contain a complete set of additive invariants. This fact allows us to apply the Chapman-Enskog method to its solution and to obtain two-fluid hydrodynamic equations. In this chapter we propose a generalized Chapman-Enskog method which may be used to derive a system of hypergeometric distribution equations accounting for retardation that is valid in the frequency range $\omega \ll \Delta$ and in the wave number range $k \ll v_F^{-1}\backslash i\omega + \nu_p$, where ν_p is the momentum relaxation frequency. It is important to emphasize that the relation between the frequencies ω and ν_p here may take any form. This makes it possible to describe the spatial and temporal behavior of the superconductors in this range (ω, k) in complete detail using a simpler hydrodynamic scheme than the complex kinetic equation. We apply this scheme to investigate the evolution of the branch imbalance of the excitation spectrum of the superconductor with a temperature gradient in tunnel injection conditions and we investigate the features of the penetration of the longitudinal electrical field into the superconductor and the generation of a Carlson-Goldman mode for the case where the external action has a time dependence and we calculate the spectrum of fluctuations in the branch imbalance of the excitation spectrum of the superconductor. The generalized Chapman-Enskog method makes it possible to consider the problem of collisional relaxation of the order parameter modulus. Here we will analyze the behavior of small spatially-homogeneous perturbations of Δ. There is significant interest in the problem of interpreting the thermoelectric effect in multiply coupled inhomogeneous superconducting systems. This chapter provides a description of this phenomenon in a thin-walled cylindrical superconductor that is based on an investigation of the nonequilibrium state penetrating current-carrying superconductors under the action of thermal inhomogeneities. In this chapter we also investigate the behavior of the nonequilibrium spin density and we calculate the frequency-dependent spin diffusion coefficient in superconductors with nonequilibrium-oriented spins.

An analysis of the research preceding this study will be provided as the material is presented.

1. Generalized Chapman-Enskog method

For definiteness we will consider a pure superconductor ($\nu_i \ll T_c$) without paramagnetic impurities; as our source of inelastic collisions we will consider electron scattering by phonons as the most effective energy relaxation mechanism in metals. We will assume that the phonons are equilibrium phonons and the external actions do not excite a spin degree of freedom. Then the collision integral $\hat{R}^{(2)}\{\hat{\rho}\}$

$$\hat{R}^{(2)} = \hat{R}^{(2)}_{e-ph} + \hat{R}^{(2)}_{i} \tag{3.1}$$

describes the relaxation of the electron subsystem to the state

$$\hat{\rho}_0 = n_0\,(\hat{\varepsilon}_p/T). \tag{3.2}$$

We note that $\hat{R}^{(2)}$ vanishes by the equilibrium distribution $\hat{\rho}_0$ as a function of the real nonequilibrium energy. We will call the state of the superconductor corresponding to density matrix $\hat{\rho}_l$ of the type in (3.2) implicitly dependent on the coordinates and time a locally-equilibrium state. We will select the parameters of the locally-equilibrium state so that the bulk density of the electrons in $(\mathbf{r}t)$, the energy density of the electron-phonon system $\mathcal{E}\,(\mathbf{r}t)$ and the superfluid velocity $\mathbf{v}_s(\mathbf{r}t)$ are true nonequilibrium values of the corresponding macroscopic quantities.

We have the conservation laws of particle number and energy from kinetic equation (1.28) and identities (1.39) and (1.53) (we are assuming the phonons are in equilibrium and consistent with this assumption we ignore phonon contributions to the macroscopic quantities). Arguments analogous to those given in Chapter 2 produce an equation for the superfluid velocity $\mathbf{v}_s(\mathbf{r}t)$. Unlike (2.15) we will assign the chemical potential $\mu_0(P, T)$ to the coordinate system where $\mathbf{v}_s = 0$. Then we will have

$$m\frac{\partial}{\partial t}\mathbf{v}_s = e\mathbf{E} - \frac{\partial}{\partial \mathbf{r}}\left(\mu_0(P,T) + \frac{1}{2}mv_s^2 + \delta\tilde{\mu}\right), \tag{3.3}$$

where

$$\mathbf{E} = -\frac{\partial}{\partial \mathbf{r}}\varphi - \frac{\partial}{\partial t}\mathbf{A}.$$

Equation (3.3) is valid when the nonequilibrium addition to the chemical potential $\varepsilon\tilde{\mu}$ is small compared to its unperturbed value.

The conservation laws and equation (3.3) in which all bulk densities and fluxes are calculated by means of $\hat{\rho}_l$ will be called the local equilibrium equations. In superconductors containing impurities the momentum of the electron-phonon system is not conserved (even with reciprocal lattice vector accuracy); hence an equation analogous to (2.19) becomes the momentum source in the right half. In place of this equation we normally consider an expression for the particle current that is calculated microscopically. For example, in local equilibrium, as we know,

$$\mathbf{j}^{(l)} = n_s\mathbf{v}_s, \tag{3.4}$$

where

$$n_s = n_l\left(1 + 2\int_0^\infty d\xi_p \frac{\partial n_0(\varepsilon_p/T)}{\partial \varepsilon_p}\right) \tag{3.5}$$

is the density of the superfluid component; n_l is the locally-equilibrium electron density. Formulae (3.4)-(3.5) are valid in the linear approximation in $p_F v_s/\Delta$.

Let the gradients of the macroscopic quantities be small; the smallness criterion will be noted below. We will assume that the solution $\hat\rho$ of kinetic equation (1.28) with collision integral (3.1) takes the form

$$\hat\rho = \hat\rho_l + \delta\hat\rho = \hat\rho_l + \delta\hat\rho_0 + \delta\tilde\rho, \tag{3.6}$$

where $\hat\rho_0 = n_0(\varepsilon_p/T)$, ε_p is the nonequilibrium spectrum. For the first order gradient correction we obtain a linear inhomogeneous integrodifferential equation

$$\frac{\partial}{\partial t}\hat\rho_l + \frac{\partial}{\partial t}(\delta\hat\rho_0) + \frac{1}{2}\left\{\left[\frac{\partial\hat\varepsilon_p}{\partial \mathbf{p}}, \frac{\partial}{\partial \mathbf{r}}\hat\rho_l\right]_+ - \left[\frac{\partial\hat\varepsilon_p}{\partial \mathbf{r}}, \frac{\partial}{\partial \mathbf{p}}\hat\rho_l\right]_+\right\} =$$
$$= \hat L\{\delta\tilde\rho\} - \frac{\partial}{\partial t}(\delta\tilde\rho), \tag{3.7}$$

where we designate

$$\hat L = -i\,[\varepsilon_p, \;.\;.\;.\;]_- + \hat{\mathscr{L}}^{(2)} \tag{3.8}$$

($\hat{\mathscr{L}}^{(2)}$ is the linearized collision integral corresponding to (3.1)).

We will search the solution of equation (3.7) in a set of two-dimensional Hermitian matrices satisfying the conditions

$$\langle\sigma_y\delta\tilde\rho\rangle = 0, \tag{3.9}$$

$$\delta\Delta = {}^1/_2\,|g|\,\langle\hat\sigma_x\delta\hat\rho\rangle, \tag{3.10}$$

$$\delta n = 0, \tag{3.11}$$

$$\delta\mathscr{E} = 0, \tag{3.12}$$

where δn and $\delta\mathscr{E}$ are the corrections to the (nonequilibrium!) quantities n and $\mathscr{E}$ related to the existence of the nonequilibrium addition to the density matrix. The energy density $\mathscr{E}$ takes the form

$$\mathscr{E} = \mathscr{E}_0 + m\mathbf{v}_s\mathbf{j}_0 + \frac{1}{2}mv_s^2, \tag{3.13}$$

$$\mathscr{E}_0 = \left\langle\frac{p^2}{2m}\left(\hat\sigma_z\hat\rho + \frac{1}{2}\right)\right\rangle - \frac{1}{2}|g|\left\{\frac{n^2}{2} + \frac{2}{g^2}\Delta^2\right\}, \tag{3.14}$$

$$\mathbf{j}_0 = \left\langle\frac{\mathbf{p}}{m}\left(\hat\rho - \frac{1}{2}\right)\right\rangle, \tag{3.15}$$

we have written out only the electron part of the energy density. Identities (1.39) and (1.54) appearing with any density matrix are also valid for the linearized collision integral $\hat{\mathscr{L}}^{(2)}$. These identities mean that the matrices $\hat\sigma_z$ and $\hat\varepsilon_p$ form a subspace of the left eigenvectors of the $\hat{\mathscr{L}}^{(2)}$ operator belonging to the zero eigenvalue.

When condition (3.9) is imposed on $\delta\tilde{\rho}$ this spectral property applies to the operator $\hat{L}$ and determines the set of solvability conditions of equation (3.7). We may easily test that in set (3.9)-(3.12) these solvability conditions of equation (3.7) with respect to $\delta\tilde{\rho}$ are nothing other than local equilibrium equations for the macroscopic quantities. Hence in calculating the left half of equation (3.7) we should substitute all time derivatives of the macroscopic quantities in $(\partial/\partial t)(\hat{\rho}_l)$ with spatial gradients consistent with the local equilibrium equations. In the classical scheme we drop the term $(\partial/\partial t)(\delta\hat{\rho})$ in the first approximation equation requiring that the perturbation vary in time with frequency ω much less than the collision frequency. However normalization of the local-equilibrium solution to the real values of n and $\mathcal{E}$ (and in the self-consistent problem to the matching equations (3.9), (3.10)) together representing the range of determination of the Hermitian addition $\delta\tilde{\rho}$ by relations (3.9)-(3.12), makes it possible to conserve in the first approximation equation the term $(\partial/\partial t)(\delta\hat{\rho})$: by subjecting $\delta\tilde{\rho}$ to conditions (3.9)-(3.12) we automatically satisfy the solvability conditions. For example, by operating on (3.7) by the operator $\langle\sigma_z(\ldots)\rangle$ we obtain from $(\partial/\partial t)(\delta\hat{\rho})$ the addition $(\partial/\partial t)(\delta n)$, where $\delta n = 0$ in accordance with (3.11). Hence the condition $\omega \ll \nu$, where ν is the characteristic collision frequency is not necessary; the first order correction $\delta\tilde{\rho} \sim [kv_F/(i\omega + \nu)]\tilde{\rho}_l$, where k is the wave vector of the perturbation, v_F is the particle velocity on the Fermi surface. Thus, when $kv_F \ll |i\omega + \nu|$ the correction to the locally-equilibrium solution is small compared to the solution itself, while ρ_l is indeed the solution of the kinetic equation in the zeroth approximation in the parameter $kv_F|i\omega + \nu|$. Below we will demonstrate that in superconductors $\nu \sim \nu_p$ (ν_p is the momentum relaxation frequency). However when $\omega \geqslant \nu_p$ the function $\hat{\rho}_l$ is normalized to macroscopic quantities that vary rapidly in time and in this sense they differ from the locally-equilibrium state of classical hydrodynamics.

After solving first approximation equation (3.7) in condition (3.9)-(3.12), we may easily find the dissipative corrections to the fluxes of the macroscopic quantities; the kinetic coefficients here will have a dependence on the perturbation frequency. The derived hydrodynamic equations with retardation in the first order with respect to spatial gradients serve as the solvability conditions of the equation for the second order correction to the density matrix, etc.

It is important to emphasize that the solution of the zeroth approximation is nonnormalized to the particle number and energy fluxes; hence, in equation (3.7) we cannot conserve the spatial derivatives $\delta\hat{\rho}$ without then violating the solvability conditions.

The operator $\hat{\mathcal{L}}^{(2)}$ is nonHermitian; its left and right eigenvectors corresponding to the zeroth eigenvalue do not coincide. We may show that $\hat{\mathcal{L}}^{(2)}$ vanishes by the functions

$$\delta\tilde{\rho}_\mu = C_\mu \frac{\partial}{\partial\mu}(\hat{\rho}_l)_\Delta, \quad \delta\hat{\rho}_\Delta = C_\Delta \frac{\partial}{\partial\Delta}(\hat{\rho}_l)_\mu. \tag{3.16}$$

Introducing the convention

$$\text{Left}\,\{\hat{\rho}_l\} = \left(\frac{\partial}{\partial t} + \frac{1}{2}\left\{\left[\frac{\partial \hat{\varepsilon}_p}{\partial \mathbf{p}}, \frac{\partial}{\partial \mathbf{r}} \cdots\right]_+ - \left[\frac{\partial \hat{\varepsilon}_p}{\partial \mathbf{r}}, \frac{\partial}{\partial \mathbf{p}} \cdots\right]_+\right\}\right)\hat{\rho}_l \qquad (3.17)$$

and assuming that in (3.17) all time derivatives are dropped using the local equilibrium equations, we may write equation (3.7) subject to (3.16) in the form (3.18)

$$\text{Left}\,\{\hat{\rho}_l\} + \frac{\partial}{\partial t}(\delta\hat{\rho}_0) = -\frac{\partial}{\partial t}(\delta\tilde{\rho}) - i\,[\hat{\varepsilon}_p, \delta\tilde{\rho}]_- + \hat{\mathscr{L}}^{(2)}\{\delta\tilde{\rho} - \delta\tilde{\rho}_\mu - \delta\tilde{\rho}_\Delta\}. \qquad (3.18)$$

The general solution of equation (3.18) will be expressed through the spatial gradients of the macroscopic quantities, the nonequilibrium corrections to the excitation spectrum $\delta\tilde{\mu}$ and $\delta\Delta$ and two unknown functions C_Δ and C_μ. Determining these for unknown quantities from the four equations (3.9)-(3.12) we may satisfy the solvability conditions of equation (3.18).

2. Tunnel electron source

Bearing in mind that we will henceforth consider the tunnel current through a superconductor/normal metal junction as one possible source of the nonequilibrium state in the system, we will determine how the scheme examined above changes with a tunnel particle source. Tunneling will be described by a tunnel Hamiltonian which in the Eliashberg representation takes the form

$$H_T = \Sigma \mathscr{T}_{pp'} \psi_p^{\alpha} N_{\alpha\beta} \psi_{p'\beta} + (\text{h.c.}), \qquad (3.19)$$

where the primed momentum belongs to the electron field operators in the metal N. Accounting for interaction (3.19) by solving the Dyson equation for Green's functions (1.3), (1.4) causes the nonequilibrium source to appear in the right half of equation (1.19). Neglecting the possible spin polarization of the electrons in the normal metal we find that this source has a structure of the type (1.27). Designating its element (1, 1) in terms of the spin index by the symbol $I_{NS}(\mathbf{r}\mathbf{p}t)$, we obtain in the ν-representation

$$(\hat{I}_{NS})_{\nu\nu'} = \pi \sum |\mathscr{T}_{pp'}|^2 \{(\hat{1} - \hat{\rho}_p)_{\nu\sigma} (\hat{\sigma}_z)_{\substack{p\,p' \\ \sigma\,\sigma'}} f_{\substack{p' \\ \sigma'\alpha'}} (\hat{\sigma}_z)_{\substack{p'\,p \\ \alpha'\,\nu'}} -$$
$$- \hat{\rho}_{\substack{p \\ \nu\,\sigma}} (\hat{\sigma}_z)_{\substack{p\,p' \\ \sigma\,\sigma'}} (\hat{1} - \hat{f}_{p'})_{\sigma'\alpha'} (\hat{\sigma}_z)_{\substack{p'\,p \\ \alpha'\,\nu'}} \delta(\hat{E}_{\substack{p \\ \nu\,\sigma}} - \hat{E}_{\substack{p' \\ \sigma'\,\alpha'}}) + (\text{h. c.})\}, \qquad (3.20)$$

where $(\hat{f}_{p'})_{\sigma\sigma'} = f_{p'}\delta_{\sigma\sigma'}$; $(\hat{\sigma}_z)_{\substack{pp' \\ \sigma\sigma'}} = (\hat{U}_p \hat{\sigma}_z \hat{1})_{\sigma\sigma'}$, $f_{p'}$ is the electron distribution function in the metal N. It is convenient to assume that in the normal metal

$$(\hat{E}_{p'})_{\sigma\sigma'} = (-1)^{\sigma+1}\,[\xi_{p'} + V(t)]\,\delta_{\sigma\sigma'},$$

where $V(t)/e$ is the potential of the metal N with respect to S. In examining the first order effects with respect to the tunneling frequency $\nu_T \sim |\mathcal{T}_{pp'}|^2$, which, as a rule, is small compared to the energy relaxation frequency ν_E, it is sufficient to find $\hat{I}_{NS}$ in local equilibrium. In this approximation we will have in the ν-representation

$$(\hat{I}_{NS})_{\sigma\sigma} = \frac{\nu_T}{2}\left\{ n_0\left[\left(\frac{E_{p\,\sigma\sigma} - V(t)}{T}\right) + n_0\left(\frac{E_{p\,\sigma\sigma} + V(t)}{T}\right) - \right.\right.$$

$$\left.\left. - 2n_0\left(\frac{E_{p\,\sigma\sigma}}{T}\right)\right] + \frac{\xi_p}{\varepsilon_p}\left[n_0\left(\frac{E_{p\,\sigma\sigma} - V(t)}{T}\right) - n_0\left(\frac{E_{p\,\sigma\sigma} + V(t)}{T}\right)\right]\right\}, \tag{3.21}$$

$$(\hat{I}_{NS})_{\sigma\bar{\sigma}} = \frac{\nu_T}{2}\frac{\Delta}{2\varepsilon_p}\left\{ n_0\left[\frac{\varepsilon_p + \mathbf{p}\mathbf{v}_s - V(t)}{T}\right] + n_0\left[\frac{-\varepsilon_p + \mathbf{p}\mathbf{v}_s - V(t)}{T}\right] - \right.$$

$$\left. - n_0\left[\frac{\varepsilon_p + \mathbf{p}\mathbf{v}_s + V(t)}{T}\right] - n_0\left[\frac{-\varepsilon_p + \mathbf{p}\mathbf{v}_s + V(t)}{T}\right]\right\}. \tag{3.22}$$

Such a result was obtained by Volkov [25] who selected the representation in Keldysh space somewhat differently. In (3.21)-(3.22) $\nu_T = \frac{mp_F}{\pi^2}|\mathcal{T}|^2$, $\sigma = 1 \div 2$, $\bar{\sigma} = 3 - \sigma$. We obtain the continuity equation from equation (1.21) with source (3.20) for the electron density:

$$(\partial n/\partial t) + \operatorname{div}\mathbf{j} = I, \tag{3.23}$$

where the intensity of the tunnel particle source - the tunnel current - is determined by the formula

$$I = \langle \hat{\sigma}_z I_{NS} \rangle, \tag{3.24}$$

and the energy balance equation containing the intensity of the energy source. The equation for $\mathbf{v}_s$ conserves its form of (3.3) when a particle source exists as well.

3. Kinetic coefficients accounting for retardation effects

The application of the generalized Chapman-Enskog method (Section 2 of this chapter) to solving kinetic equation (1.28) produces a system of two-fluid hydrodynamic retardation equations; after calculation of the kinetic coefficients this system becomes closed and allows us in principle to completely describe weakly-nonequilibrium states of pure superconductors in external fields characterized by frequencies $\omega \ll \Delta$ and wave numbers $k \ll v_F^{-1}|i\omega + \nu_p|$. With tunnel injection there is the obvious smallness condition of the intensity of the tunnel particle source. This system of generalized hydrodynamic equations has a standard form and differs from the classical system (see, for example, [76]) by the existence of a frequency dependence of its kinetic coefficients of bulk viscosity, conductivity, thermal conductivity and the thermoelectric coefficient.

In calculating the inhomogeneous term of equation (3.7) in accordance with the discussion above, we obtain the following equation for the nonequilibrium addition $\delta\tilde{\rho}$ to the density matrix:

$$A\hat{\tau}_3 + B\hat{\tau}_1 + C\hat{1} = -\left(\frac{\partial}{\partial t} + i[\varepsilon_p, \ldots]_-\right)\delta\tilde{\rho} + \hat{\mathcal{L}}^{(2)}\{\delta\tilde{\rho} - \delta\tilde{\rho}_\mu - \delta\tilde{\rho}_\Delta\}, \qquad (3.25)$$

where

$$A = \frac{1}{T}\left[-n_0'\left(\frac{\varepsilon_p}{T}\right)\right]\left\{\frac{\xi_p}{\varepsilon_p}\frac{\partial}{\partial t}(\delta\tilde{\mu}) - \frac{\Delta}{\varepsilon_p}\frac{\partial}{\partial t}(\delta\Delta)\right\} - (\hat{I}_{NS})_{11} -$$

$$-\frac{1}{T}\left[-n_0'\left(\frac{\varepsilon_p}{T}\right)\right]\left\{\frac{\xi_p}{\varepsilon_p}\left(\frac{\partial\mu}{\partial n}\right)_\Delta + \frac{\varepsilon_p}{T^2}\left(\frac{\partial T}{\partial n}\right)_\Delta\right\}[\operatorname{div}(n_s\mathbf{v}_s) - I],$$

$$B = \frac{\operatorname{th}\varepsilon_p/2T}{2\varepsilon_p}\left\{\left[\frac{\Delta}{\varepsilon_p}\frac{\partial}{\partial t}(\delta\tilde{\mu}) + \frac{\xi_p}{\xi_p}\frac{\partial}{\partial t}(\delta\Delta)\right] - \frac{\Delta}{\varepsilon_p}\left(\frac{\partial\mu}{\partial n}\right)_\Delta(\operatorname{div}(n_s\mathbf{v}_s) - I)\right\} -$$

$$-\left\{\frac{1}{T}\left[-n_0'\left(\frac{\varepsilon_p}{T}\right)\right] - \frac{\operatorname{th}\varepsilon_p/2T}{2\varepsilon_p}\right\}\frac{\Delta}{\varepsilon_p}\frac{p_F^2}{3m}\operatorname{div}\mathbf{v}_s - (\hat{I}_{NS})_{12}, \qquad (3.27)$$

$$C = -\frac{1}{T}\left[-n_0'\left(\frac{\varepsilon_p}{T}\right)\right]\frac{\mathbf{p}}{m}\left\{e\mathbf{E}_N - \frac{\xi_p}{\varepsilon_p}\nabla T\right\}. \qquad (3.28)$$

We will assume that v_s is small and we provided in (3.26)-(3.28) the leading terms of the expansion of the left half of equation (3.7) in terms of the parameter $p_F v_s/\Delta$. Following study [77] we designate as $\mathbf{E}_N$ the "effective" electrical field in the superconductor whose existence produces the electromotive force:

$$\mathbf{E}_N = \mathbf{E} - \frac{1}{e}\frac{\partial}{\partial\mathbf{r}}\mu_0(P; T). \qquad (3.29)$$

We are interested in the stationary solution of the initial problem for equation (3.25) in conditions (3.9)-(3.12). In order to find such a solution it is convenient to carry out a Fourier time transform in this equation.

1. Conductivity and thermal conductivity coefficients. The thermoelectric coefficient. We will calculate the dissipative correction to the particle number flux, that by definition is equal to

$$\mathbf{j}_n(\mathbf{r}\omega) = \frac{1}{e}\sigma(\omega)\mathbf{E}_N - \frac{1}{e}\eta_T(\omega)\nabla T \qquad (3.30)$$

(where σ is the conductivity and η_T is the thermoelectric coefficient), by substituting the general solution of equation (3.25) into the equation of microscopic theory:

$$\mathbf{j}_n = \left\langle\frac{\mathbf{p}}{m}\delta\tilde{\rho}\right\rangle. \qquad (3.31)$$

In calculating the nonequilibrium addition $\delta\tilde{\rho}$ to the density matrix, as in Chapter 2, we will use for the linearized collision integral $\hat{\mathcal{L}}^{(2)}$ relaxation time approximation (1.57) and will account for the "parasitic" solutions of (3.16). It follows from (3.31) that the quantity $\mathbf{j}_n$ is determined by the ν-diagonal part of the general solution of $\delta\tilde{\rho}$ proportional to the unity matrix. This part of $\delta\tilde{\rho}$ takes the form

$$\delta\tilde{\rho} = \frac{1}{T}\left[-n_0'\left(\frac{\varepsilon_p}{T}\right)\right]\frac{1}{i\omega+\nu_p}\frac{\mathbf{p}}{m}\left(e\mathbf{E}_N - \frac{\xi_p}{T}\nabla T\right). \tag{3.32}$$

Consequently

$$\sigma(\omega) = \left\langle \frac{1}{T}\frac{[-n_0'(\varepsilon_p/T)]}{i\omega+\nu_p}\frac{p^2}{m}\right\rangle\frac{e^2}{3}, \tag{3.33}$$

$$\eta_T(\omega) = \left\langle \frac{1}{T}\frac{[-n_0'(\varepsilon_p/T)]}{i\omega+\nu_p}\frac{\xi_p}{m^2}\right\rangle\frac{e}{3}. \tag{3.34}$$

Here ν_p is the momentum relaxation frequency that, in superconductors, as a rule, is determined by elastic collisions (with the exception of very pure materials). In the case $T - T \ll T_c$ which will be of interest to us we obtain from (3.33), (3.34)

$$\sigma(\omega) = \frac{ne^2}{m}\frac{1}{i\omega+\nu_p}, \qquad \eta_T(\omega) = \frac{\pi^2}{3e}\frac{T_c}{\mu_F}\sigma(\omega). \tag{3.35}$$

Analogously calculating the dissipative correction to the thermal flux

$$\mathbf{q} = \left\langle \frac{\mathbf{p}}{m}\frac{p^2}{2m}\delta\tilde{\rho}\right\rangle, \tag{3.36}$$

we find that it takes the form

$$\mathbf{q}(\mathbf{r}\omega) = \eta_T(\omega)T\mathbf{E}_N - \varkappa(\omega)\nabla T, \tag{3.37}$$

where $\varkappa$ is the thermal conductivity coefficient equal to

$$\varkappa(\omega) = \frac{1}{3}\left\langle \frac{p^4}{2m^2}\frac{\xi_p}{T^2}\frac{[-n_0'(\varepsilon_p/T)]}{i\omega+\nu_p}\right\rangle. \tag{3.38}$$

When $T \to T_c - 0$

$$\varkappa(\omega) = \frac{\pi^2}{3e^2}T_c\sigma(\omega). \tag{3.39}$$

The evolution of ν-diagonal density matrix components determining charge and energy transport is the result of momentum relaxation and is characterized by the transport relaxation time ν_p^{-1}. Hence formulae (3.33), (3.34), and (3.38) for the kinetic coefficients are applicable when $kv_F \ll |i\omega + \nu_p|$.

2. The bulk viscosity coefficient. Analysis reveals that when $p_F v_s \ll \Delta$ accurate to corrections of the order (T_c/μ_F) the behavior of the nonequilibrium correction to the chemical potential $\delta\tilde{\mu}$ is determined by the part of the ν-diagonal components of the general solution of equation (3.25) that is odd with respect to the variable ξ_p and by the part of its ν-nondiagonal components that is even with respect to ξ_p. We may show that when $2p_F v_s < \Delta$ the density matrix with components of this parity does not relax through the nonmagnetic impurities and the action of the electron-phonon collision integral conserves the ξ_p-parity of the density matrix components if its ν-diagonal and ν-nondiagonal components have equal parity with respect to the variable ξ_p.

Hence $\hat{\mathscr{L}}^{(2)}$ in equation (3.25) in the bulk viscosity problem infers the linearized electron-phonon collision integral.

We will limit ourselves to the temperature range $T_c - T \ll T_c$ and will drop the cumbersome calculations, immediately proceeding to the result in the $(\mathbf{r}\omega)$-representation:

$$\delta\tilde{\rho}_{kk}(\mathbf{r}\omega) = (-1)^k (i\omega + \nu_\alpha^{ph})^{-1} D_{kk}, \tag{3.40}$$

$$\delta\tilde{\rho}_{k\bar{k}}(\mathbf{r}\omega) = -(i\omega + \nu_\gamma^{ph} - (-1)^k i2\varepsilon_p)^{-1} D_{k\bar{k}}, \tag{3.41}$$

where

$$D_{kk} = \frac{1}{T}\frac{\xi_p}{\varepsilon_p}\left[-n_0'\left(\frac{\varepsilon_p}{T}\right)\right]\left\{\left(\frac{\partial\mu}{\partial n}\right)_\Delta [\operatorname{div}(n_s\mathbf{v}_s) - I] + i\omega\delta\tilde{\mu} - \nu_\alpha^{ph} C_\mu(\omega)\right\} - \\ - \nu_T \frac{\xi_p}{\varepsilon_p}\left[n_0\left(\frac{\varepsilon_p - V}{T}\right) - n_0\left(\frac{\varepsilon_p + V}{T}\right)\right](\mathbf{r}\omega), \tag{3.42}$$

$$D_{k\bar{k}} = \frac{\operatorname{th}\varepsilon_p/2T}{2\varepsilon_p}\frac{\Delta}{\varepsilon_p}\left\{\left(\frac{\partial\mu}{\partial n}\right)_\Delta [\operatorname{div}(n_s\mathbf{v}_s) - I] + i\omega\delta\tilde{\mu} - \nu_\gamma^{ph} C_\mu(\omega)\right\} - \\ - \left\{\frac{1}{T}\left[-n_0'\left(\frac{\varepsilon_p}{T}\right)\right] - \frac{\operatorname{th}\varepsilon_p/2T}{2\varepsilon_p}\right\}\frac{\Delta}{\varepsilon_p}\frac{p_F^2}{3m}\operatorname{div}(\mathbf{v}_s) - \\ - \nu_\delta^{ph}\left\{\frac{1}{2}\operatorname{Sp}(\hat{\tau}_3\delta\tilde{\rho}) + \frac{1}{T}\left[-n_0'\left(\frac{\varepsilon_p}{T}\right)\right]C_\mu(\omega)\right\} - (\hat{I}_{NS})_{12} \tag{3.43}$$

(k = 1-2, $\bar{k}$ = 3 - k). Equations (3.10) and (3.12) yield $\delta\Delta = C_\Delta = 0$; dropping C_μ from (3.9) and (3.11) we obtain

$$\delta\tilde{\mu} = -m\zeta_3(\omega)\operatorname{div}(n_s\mathbf{v}_s), \tag{3.44}$$

where $\zeta_3(\omega)$ is the frequency-dependent second viscosity coefficient:

$$\zeta_3(\omega) = \frac{4T_c}{\pi\Delta}\left(\frac{\partial\mu}{\partial n}\right)_\Delta \frac{1}{m}\frac{1}{i\omega + \nu_\gamma^{ph}}. \tag{3.45}$$

A similar expression was proposed phenomenologically by Putterman [78]; although in this study the "imbalance relaxation frequency" $\nu_Q \sim$ figures in this study in place of ν_γ^{ph}, where ν_E is the energy relaxation frequency and there is no common multiplier $(4T_c/\pi\Delta)$ in formula (4.35). The result of (3.44)-(3.45) results from the contribution of the ν-nondiagonal components of the general solution of equation (3.25) and cannot be found within the scope of the kinetic equation normally used for the perturbation distribution function which is obtained from (1.28) by ignoring precisely the ν-nondiagonal components of the density matrix and the collision integral.

The existence of the commutator $[\hat{\varepsilon}_p, \hat{\rho}]_-$ making a contribution to the equation for the ν-nondiagonal components of $\delta\hat{\rho}$ causes the quantity ε_p' to function as the characteristic frequency in this equation; across the entire temperature range $0 < T < T_c$ we have $\varepsilon_p \sim T_c$. Consequently result (3.45) is valid when $kv_F \ll \max\{\omega, T_c\}$. Since in pure superconductors $\nu_p \ll T_c$ the range of frequencies and wave perturbation vectors in which the generalized hydrodynamic equations with kinetic coefficients (3.35), (3.45) are valid is determined by the relations $\omega \ll \Delta$, $kv_F \ll |i\omega + \nu_p|$. The kinetic coefficients in this section that are frequency-dependent represent the variety of the memory functions (see study [79] where, specifically, the issue of satisfaction of the Onsager relations is considered). Making the transformation to the t-representation we obtain, for example, for the thermal flux with only a temperature gradient

$$\mathbf{q}(\mathbf{r}t) = -\int_0^t dt' \varkappa(t - t') \nabla T(\mathbf{r}t'), \tag{3.46}$$

where $T_c - T \ll T_c$

$$\varkappa(\tau) = \frac{\pi^2}{3m} n T_c \exp(-\tau\nu_p), \qquad \tau \geqslant 0. \tag{3.47}$$

Consistent with the formulation of the problem we assume that the initial perturbation is adiabatically activated with negative values of t.

4. Collisional relaxation of the order parameter modulus

In order to complete the formulation of the generalized hydrodynamic equations of superconductors, we must consider the issue of the behavior of the order parameter modulus in the collisional frequency range $\omega \leqslant \nu_E$. This phenomena was investigated by many authors. Schmid and Schön[19] and Gorkov and Eliashberg (see study [50]) used the energy-integrated Green's function technique in the τ-approximation for the collision integral to show that near T_c the small spatially-homogeneous perturbations in Δ decay over times $\tau_\Delta^{(0)} = \frac{\pi^3}{7\xi(3)} \frac{T_c}{\Delta} \frac{1}{\nu_E}$. These studies considered the situation when the nonequilibrium phonons rapidly exit the superconductor (the corresponding frequency $\nu_{es} \gg \nu_E$) and its phonon subsystem may be considered to be in equilibrium; the superconductor temperature was assumed to be fixed. Eckern and Schön [80] developed a more exact approximation for $\hat{\mathscr{L}}^{(2)}$ and obtained a solution of the kinetic equation that is orthogonal to the zero subspace of the exact linearized collision integral. Analysis [80] has revealed that the relaxation of Δ in superconductors, generally speaking, is accompanied by a change in its temperature.

The approach developed in Section 1 of this chapter allows us to investigate the joint behavior of $\delta\Delta$ and the nonequilibrium addition to the temperature of the electron system δT where an energy balance equation may be used to find the latter quantity. We will limit our examination to the case of poor heat exchange ($\nu_{es} \ll \nu_E$) and here over relaxation times of Δ the superconductor energy is constant. For simplicity we will consider the situation when $C_{ph}/C_{el} \ll 1$, where C_{ph} and C_{el} are the thermal capacities of the phonons and the electrons.

We may show that the contribution from the nonequilibrium nature of the phonons to the energy balance equation is small to a like degree.

Assume that at initial time $t_0 = 0$ the system experiences a small spatially-homogeneous perturbation described by the small addition to the equilibrium density matrix $\delta\hat{\rho}(t_0)$. In the zeroth order in T_c/μ_F when $\nu_s = 0$ the behavior of Δ is determined by the ξ_p-even additions to the ν-diagonal components of the density matrix and the ξ_p-odd additions to the ν-nondiagonal components. In the right half of equation (3.15) when $\delta T \neq 0$ we have

$$\delta\hat{\rho}_0 = \frac{1}{T}\left[-n_0'\left(\frac{\varepsilon_p}{T}\right)\right]\left\{-\frac{\Delta}{\varepsilon_p}\delta\Delta + \frac{\xi_p}{\varepsilon_p}\delta\tilde{\mu} + \frac{\varepsilon_p}{T}\delta T\right\}\hat{\tau}_3 + \\ + \frac{\operatorname{th}\varepsilon_p/2T}{2\varepsilon_p}\left\{\frac{\xi_p}{\varepsilon_p}\delta\Delta + \frac{\Delta}{\varepsilon_p}\delta\tilde{\mu}\right\}\hat{\tau}_1.$$

Accounting for the fact that the isotropic perturbations to the density matrix of this parity do not relax by the impurities, we find the solution for $\delta\tilde{\rho}(\omega)$ in the form of (4.1), (4.2) where

$$D_{kk} = (-1)^{k+1}\left\{\left[-n_0'\left(\frac{\varepsilon_p}{T}\right)\right]\left[\frac{i\omega}{T}\left(-\frac{\Delta}{\varepsilon_p}\delta\Delta + \frac{\varepsilon_p}{T}\delta T\right) - \right.\right. \\ \left.\left. - \frac{\nu_\alpha}{T}\frac{\Delta}{\varepsilon_p}C_\Delta\right]\right\} + [\delta\hat{\rho}(0)]_{kk}, \tag{3.48}$$

$$D_{k\bar{k}} = \frac{\operatorname{th}\varepsilon_p/2T}{2\varepsilon_p}\frac{\xi_p}{\varepsilon_p}[i\omega\delta\Delta + \nu_\gamma C_\Delta] + [\delta\hat{\rho}(0)]_{k\bar{k}}. \tag{3.49}$$

Equations (3.7) and (3.9) are identically satisfied when $\delta\tilde{\mu} = 0$, $C_\mu = 0$. We obtain from (3.8) and (3.10) the following, respectively:

$$-\frac{n_s}{n}N(0)\,\delta\Delta + \frac{1}{T}\mathscr{E}_\Delta\delta T + \langle\hat{\sigma}_x\delta\tilde{\rho}\rangle = 0, \tag{3.50}$$

$$\mathscr{E}_\Delta\delta\Delta + \mathscr{E}_T\delta T + \langle\varepsilon_p\hat{\tau}_3\delta\tilde{\rho}\rangle = 0, \tag{3.51}$$

where

$$\mathscr{E}_\Delta = -\left\langle\frac{\Delta}{\varepsilon_p}\left[-n_0'\left(\frac{\varepsilon_p}{T}\right)\right]\right\rangle; \quad \mathscr{E}_T = \left\langle\frac{\varepsilon_p^2}{T^2}\left[-n_0'\left(\frac{\varepsilon_p}{T}\right)\right]\right\rangle;$$
$$N(0) = mp_F/\pi^2.$$

The energy balance equation of the superconductor in a thermostat takes the following form in the homogeneous case

$$\frac{\partial}{\partial t}(\delta\tilde{\mathscr{E}}_e + \delta\tilde{\mathscr{E}}_{ph}) + \nu_{es}\delta\tilde{\mathscr{E}}_{ph} = 0, \tag{3.52}$$

where $\delta\tilde{\mathscr{E}}_e$ and $\delta\tilde{\mathscr{E}}_{ph}$ are the nonequilibrium additions to the electron and phonon energy densities. In the case $\nu_{es} \ll \nu_E$ over times $\tau_\Delta \ll \nu_{es}^{-1}$ we may neglect phonon drift from the superconductor; when $C_{ph} \ll C_{el}$ (such as in Al) $\varepsilon\tilde{\mathscr{E}}_{ph} \ll \delta\tilde{\mathscr{E}}_e$ [80]. Dropping the quantities C_Δ and δT from (3.50)-(3.52) we obtain for $\delta\Delta(\omega)$:

$$\left\{\frac{n_s}{n} + \frac{\pi}{4}\frac{\Delta}{T_c}\frac{i\omega}{i\omega + \nu_E} + \frac{\mathscr{E}_\Delta^2}{\mathscr{E}_T}\frac{1}{TN(0)}\frac{\nu_E}{i\omega + \nu_E} + \right.$$

$$+\frac{1}{N(0)}\left\langle\frac{i\omega(i\omega+\nu_\gamma)}{(i\omega+\nu_\gamma)^2+(2\varepsilon_p)^2}\frac{\xi_p^2}{\varepsilon_p^2}\frac{\operatorname{th}\varepsilon_p/2T}{2\varepsilon_p}\right\rangle\Bigg\}\delta\Delta(\omega)=\Phi'(\omega), \tag{3.53}$$

where $\Phi(\omega)$ is the familiar function of the initial conditions varying at frequencies $\sim\nu_E$ and 2Δ. The third component in the multiplier for $\delta\Delta(\omega)$ in (3.53) results from the change in superconductor temperature while the fourth component represents the contribution of the ν-non-diagonal components of the matrix $\delta\tilde{\rho}$; there is no such term in studies [15, 19, 80]. When $\omega \lessgtr \nu_E$ the contribution of $\delta\tilde{\rho}_{k\bar{k}}$ to (3.53) is not significant; relaxation of $\delta\Delta$ is determined by the singularity of the Fourier-transform of $\delta\Delta(\omega)$ closest to the real axis from (3.53) and is characterized by time

$$\tau_\Delta=\frac{\pi}{4}\frac{\Delta}{T_c}\left\{\frac{n_s}{n}+\frac{1}{TN(0)}\frac{\mathscr{E}_\Delta^2}{\mathscr{E}_T}\right\}^{-1}\tau_E. \tag{3.54}$$

When $T\to T_C-0$ we obtain $\tau_\Delta=\left[1+\frac{12}{7\xi(3)}\right]^{-1}\tau_\Delta^{(0)}$. Here the factor with $\tau_\Delta^{(0)}$ represents the ratio of the electron thermal capacities of the normal metal and the superconductor when $T = T_C$ in accordance with the result from study [80]. Unlike study [80] formula (3.54) is valid across the entire temperature range $T < T_C$.

In the inverse limit $\omega \gg \nu_E$ the contribution of $\delta\tilde{\rho}_{k\bar{k}}$ to (3.53) becomes significant; accounting for this contribution produces the result of studies [12, 46] carried out in a collisionless approximation.

5. The relaxation of the population imbalance of the excitation spectrum of a superconductor and the possibilities for its experimental investigation

In the preceding section we determined that with low values of the superfluid velocity the order parameter modulus evolves independent of the macroscopic quantities; the behavior of the latter when $\omega \ll \Delta$, $kv_F \ll (i\omega + \nu_E)$ may be described completely by solving a system of generalized two-fluid hydrodynamic equations. Here this problem will be implemented with respect to the problem of the relaxation of the branch imbalance of the excitation spectrum of the superconductor, which is closely related to the issue of the penetration of a longitudinal electrical field into the superconductor.

Current flow through a superconductor junction with another metal, as we know, causes a nonequilibrium state to arise in the system. It has been demonstrated in experiments [81, 82] that such a state is characterized by a shift in the chemical potential of the superconductor. Tinkham [83] related this phenomenon to the relaxation of the population imbalance of the electron and hole branches of the excitation spectrum. The value of the nonequilibrium addition $\delta\tilde{\mu}$ to the chemical potential of the superconductor when $T \ll T_C$ was calculated in study [84], and was investigated for the case near T_C in studies [83, 71, 19, 80]. A comparison of the results of these studies reveals that the predictions based on the energy-integrated Green's function technique [19, 80] differ somewhat from those produced by applying to this problem the kinetic equation for the excitation distribution function. As noted in study [85] the quantity $\delta\Delta$ obtained in studies [19, 80] is proportional to the total intensity of the injection particle source I, at the same time that in studies [83, 71] it is proportional to its part related to quasi particle injection. This difference which is insignificant when $T_c - T \ll T_c$ may become significant in the intermediate temperature range, as indicated by results from a comparison of numerical solutions of the appropriate equations (see, for example, [26]). Here we will identify one possible cause of this discrepancy.

1. **Relaxation frequency, diffusion coefficients and thermal diffusion of the branch population imbalance of the excitation spectrum of a superconductor.** We will derive an equation for the nonequilibrium addition $\delta\tilde{\mu}$ characterizing the branch population imbalance of the excitation spectrum of a superconductor which arises at the superconductor border of a tunnel N-S-junction with thermal inhomogeneities in the nonstationary case. Let the voltage $V_{NS}(t)$ be activated across the junction at time $t = 0$ and be equal to $V_{NS}(t) = V(t)\theta(t)$. The linearized generalized hydrodynamic equations take the form

$$m\frac{\partial}{\partial t}\mathbf{v}_s = e\mathbf{E}_N - \frac{\partial}{\partial \mathbf{r}}(\delta\tilde{\mu}), \tag{3.55}$$

$$\frac{\partial}{\partial t}(\delta\tilde{n}) + \operatorname{div}\mathbf{j} = I, \tag{3.56}$$

$$\operatorname{div}(\mathbf{E}) = 4\pi e(\delta\tilde{n}), \tag{3.57}$$

where

$$e\mathbf{E}_N = e\mathbf{E} - \nabla(\mu_0(P, T)), \tag{3.58}$$

$$\mathbf{j} = n_s\mathbf{v}_s + \mathbf{j}_n, \tag{3.59}$$

$$\mathbf{j}_n = \frac{\sigma}{e}\mathbf{E}_N - \frac{\eta_T}{e}\nabla(\delta T), \tag{3.60}$$

$$\delta\tilde{\mu} = -m\hat{\zeta}_3 \operatorname{div}(n_s\mathbf{v}_s) \tag{3.61}$$

(we are limiting our examination to the case $T_c - T \ll T_c$). We will assume that v_s is a quantity of the first order of smallness, together with the nonequilibrium additions $\delta\tilde{\nu}$, δT, and $\delta\tilde{\mu}$; hence in (3.59), (3.61) it is the equilibrium value of the density of the superconducting component. In the frequency range $\omega \ll \omega_p$ where ω_p is the electron plasma frequency, the neutrality condition $\varepsilon\tilde{n} = 0$ is satisfied. We will carry out a Fourier time transform in (3.55): when $\mathbf{v}_s(\mathbf{r}, t)|_{t=0} = 0$ we obtain the following equation for $\delta\tilde{\mu}$:

$$(i\omega + \nu_\mu(\omega) - D_\mu(\omega)\nabla^2)\delta\tilde{\mu}(\mathbf{r}\omega) = D_{\mu T}(\omega)\nabla^2\delta T(\mathbf{r}\omega) - \frac{e^2 n_s}{\sigma(\omega)}\zeta_3(\omega) I(\mathbf{r}\omega), \tag{3.62}$$

where the quantity

$$\nu_\mu(\omega) = e^2 n_s / m\sigma(\omega) \tag{3.63}$$

is the population imbalance relaxation frequency; the corresponding diffusion and thermal diffusion coefficients d_μ and $D_{\mu T}$ take the form

$$D_\mu(\omega) = \zeta_3(\omega) n_s, \tag{3.64}$$

$$D_{\mu T}(\omega) = -\frac{e\eta_T(\omega)}{\sigma(\omega)} D_\mu(\omega). \tag{3.65}$$

Near the critical temperature

$$\nu_\mu(\omega) = 2(1 - T/T_c)(i\omega + \nu_p), \tag{3.66}$$

$$D_\mu(\omega) = \frac{8}{3\pi^2}\left(\frac{7}{2}\zeta(3)\right)^{1/2} \frac{\mu_F}{i\omega + \nu_\gamma}(1 - T/T_c)^{1/2}, \tag{3.67}$$

$$D_{\mu T}(\omega) = -\frac{\pi^2}{3}\frac{T_c}{\mu_F} D_\mu(\omega). \tag{3.68}$$

The diffusion length $l_D = (D_\mu/\nu_\mu)^{1/2}$ determined by equation (3.62) is

$$l_D^{-2}(\omega) = \frac{3\pi}{4v_F^2}\frac{\Delta}{T_c}(i\omega + \nu_p)(i\omega + \nu_\gamma). \tag{3.69}$$

In the stationary case ($\omega = 0$) $l_D \equiv l_E$ determines the depth of penetration of a longitudinal electrical field into the superconductor.

An equation similar to (3.62) was derived in studies [85] (in the stationary case) and [4] in the limit $\omega \ll \max\{\nu_p, \nu_E\}$ by means of the kinetic equation for the excitation distribution function derived in studies [4-6, 8, 10] (see also the series of studies by Pethick and Smith [71, 72, 87, 88] and the subsequent studies [73, 74]). In the latter limiting case results (3.63)-(3.65) for frequency ν_μ and diffusion and thermal diffusion coefficients D_μ and $D_{\mu T}$ differ from those found in study [4] although they produce identical expressions for the depth of penetration of the longitudinal field in the stationary case. When $\omega = 0$ result (3.69) for l_D coincides with that found in study [21] using the Green's function technique; the temperature dependence $l_D \sim (\Delta/T_c)^{1/2}$ is confirmed by experiment [89]. In this regard it would be useful to carry out independent measurements of the quantities ν_μ and D_μ. In order to measure the nonequilibrium addition $\delta\tilde{\Delta}$ we may use, for example, a double N-S-N_p tunnel structure.

2. The imbalance in thin superconducting films. The relaxation of spatially-homogeneous perturbations. If the thickness of a superconducting film is small compared to the depth of penetration of the field l_E, a spatially-homogeneous distribution $\delta\tilde{\mu}(T)$ is established in the superconductor and a difference of potentials equal to $\delta\tilde{\mu}(t)$ arises between S and N_p [90]. In the stationary case

$$\delta\tilde{\mu}_{st} = -\frac{4}{\pi}\frac{T_c}{\Delta}\frac{1}{\nu_\gamma}\frac{p_F^2}{3mn} I(v); \tag{3.70}$$

consistent with the formula for $\delta\tilde{\mu}_{st}$ obtained by the Green's function technique. Result (3.70) contains the total intensity I of the tunnel source. The quantity $\left(\frac{4T_c}{\pi\Delta}\frac{1}{\nu_\gamma}\right)$, generally speaking, differs from the "branch mixing time" given in study [83].

We may measure the branch imbalance relaxation frequency ν_μ in experiments on an N-S-N_p-structure by investigating the spatially-homogeneous relaxation of $\delta\tilde{\mu}$(pt) with instantaneous switching on of an external voltage. If $V_{NS}(t) = V\theta(t_0 - t)$, then

$$\delta\tilde{\mu}(t) = \delta\tilde{\mu}_{st} \exp\left\{-\left(1 + \frac{n}{n_s}\right)^{-1} \nu_p (t - t_0)\right\}, \qquad t \geqslant t_0; \tag{3.71}$$

the potential difference between S and N_p will drop off near T_c with frequency $\nu_\mu(0) = \frac{n_s}{n}\nu_p \sim (1 - T/T_c)\,\nu_p$, which is determined by the density of the superfluid component and the inverse momentum relaxation time. Studies [4, 71, 87, 88, 73] predict a different relaxation frequency ν_Q of spatially-homogeneous perturbations $\delta/\tilde{\mu}$ in this experiment: $\nu_Q \sim \frac{\Delta}{T_c}\nu_E \sim (1 - T/T_c)^{1/2}\,\nu_E$. Another method of measuring the frequency ν_μ involves determining the fluctuations in the population imbalance $\delta\tilde{\mu}(\mathbf{r}, t)$ using, for example, the magnitude of the broadening of the Josephson radiation line (see Section 7 of this chapter). It is possible to experimentally determine which of these predictions is realized by means of the differential in the temperature relations $\nu_\mu(T)$ and $\nu_Q(T)$.

3. Collective modes. Penetration of a longitudinal electrical field into bulk superconductors. We will now assume that the superconductor temperature is maintained at a constant level: $\delta T = 0$, and that the tunnel source is removed; then from (3.62) we obtain a dispersion equation that in the frequency range

$$\max\left\{\nu_\gamma, \frac{n_s}{n}\nu_p\right\} \ll \omega \ll \nu_p \tag{3.72}$$

has a solution describing weakly-attenuated modes with a spectrum

$$\omega = ku_s + i\left\{\nu_\gamma + \frac{7\zeta(3)}{4\pi^2}\left(\frac{\Delta}{T_c}\right)^2 (\nu_\gamma + \nu_p)\right\}, \tag{3.73}$$

where

$$u_s = \left(\frac{7\zeta(3)}{3\pi^3}\frac{\Delta}{T_c}\right)^{1/2} v_F.$$

The upper limit in (3.72) derives from the applicability condition of generalized hydrodynamics.

A dispersion law of the type shown in (3.73) was first obtained for pure superconductors by Artemenko and Volkov [91] using a kinetic equation [4-6, 8, 10] ignoring energy relaxation. Ovchinnikov [92] considered the case of superconductors with an arbitrary impurity concentration and accounting for corrections to the excitation spectrum related to electron-impurity interaction obtained the more exact damping law of this mode observed experimentally by Carlson and Goldman [93, 94]. In the limit $\nu_p \ll T_c$ the result from study [92] becomes (3.73).

Thus, equation (3.62) describes, specifically, generation of a Carlson-Goldman mode. Its solution for the bulk S-region penetrating the superconductor and satisfying the condition υ_{sn} $(x = 0, t) = 0$ on the junction surface may easily be found for the case where the tunnel source is localized on the surface of the junction and has a harmonic time dependence. We will not provide the solution here due to its cumbersome size, and we note only that in this case the nature of the distribution of the longitudinal component $\mathbf{E}_N^{(l)}$ of the field $\mathbf{E}_N$ depends on the sign of the parameter

$$r = 1 - \Omega_0^2\tau_\gamma(\tau_0 + \tau_\gamma + \tau_p),$$

where ω_0 is the frequency of the external action; $\tau_0 = (n/n_s)\tau_p$; $\tau_p = \nu_p^{-1}$; $\tau_\gamma = \nu_\gamma^{-1}$:

a) if $r > 0$, the longitudinal field $E_N^{(l)}(xt)$ oscillating in time with frequency ω_0 is attenuated in penetrating the S-region over a characteristic length

$$l_E^{(-)}(\Omega_0, T) = l_E A_{(-)}^{-1}(\Omega_0, T), \tag{3.74}$$

where

$$A_{(\pm)}(\Omega_0, T) = \frac{1}{\sqrt{2}}[(r^2 + p^2)^{1/2} \pm r^2]^{1/2},$$
$$p = \Omega_0(\tau_0 + \tau_\gamma + \tau_p); \tag{3.75}$$

b) if $r < 0$, the field in the S-region becomes a superposition of waves with frequency Ω_0 and wavelength

$$\lambda_E(\Omega_0, T) = l_E A_{(+)}^{-1}(\Omega_0, T),$$

experiencing attenuation in penetrating the S-region over a distance

$$l_E^{(+)}(\Omega_0, T) = l_E A_{(+)}^{-1}(\Omega_0, T). \tag{3.76}$$

It is clear that if Ω_0 satisfies conditions (3.72), then $\lambda_E \ll l_E^{(+)}$ and such a mode is well-defined.

6. Thermoelectric effects in superconductors

The possibility of observing thermoelectric effects in inhomogeneous or homogeneous, yet anisotropic superconductors was predicted by Ginzburg [32] in spite of the widely-held view that such phenomena disappear in the transition to the superconducting state (see, for example, [76, 95]). Ginzburg's study laid the groundwork for a new direction in the investigation of superconductivity, which has produced new interesting discoveries of thermoelectric effects demonstrating the specific features of the superconducting state [96-98]. An understanding of the microscopic mechanisms of these phenomena may be useful for investigating nonequilibrium states in superconductors.

With a temperature and electrical field gradient in the superconductor, as in a normal metal, an excitation current arises

$$e\mathbf{j}_n = \sigma\mathbf{E}_N - \eta_T \nabla T.$$

Until recently it was believed that in the steady state a longitudinal electrical field in a superconductor is equal to zero. It directly follows from here and from the ideal diamagnetism property that with a time-constant temperature gradient in the superconductor, neither an electrical field nor total current arise.

Ginzburg [32] has noted that total compensation of normal current $\mathbf{j}_n$ by a superfluid current $\mathbf{j}_s = n_s\mathbf{v}_s$ is possible only in the simplest case of a homogeneous and isotropic superconductor; in all similar situations a nonzero total current and a corresponding magnetic field arise.

Gal'perin et al. [58] and Garland and van Harlingen [97] proposed using a topologic limit imposed by the ring geometry on the phase of the order parameter to measure the thermoelectric addition to the magnetic flux in a bimetallic ring consisting of two different superconductors S_1 and S_2.

Artemenko and Volkov [86] based on a kinetic equation for excitations [4-6, 8, 10] have demonstrated that the existence of thermal inhomogeneities causes a thermoelectric field to arise in superconductors where $\Delta \equiv 0$ (for gapless superconductors this was carried out before in study [100]) and, specifically, they accounted for the correction to the magnetic field related to the appearance of a nonequilibrium addition to the chemical potential caused by the spatial inhomogeneity of the time constant of the temperature gradient. It turns out that this correction has smallness (l_E/L) with respect to the unquantized part of the magnetic flux calculated in studies [58, 97].

In the first half of this section we used equation (3.62) and examined the nonstationary thermoelectric effect in a system generated by the time-dependence of the temperature of the junctions T_1 and T_2. Here it is discovered that in the appropriate range of variation in the temperature differential $(T_1 - T_2)(t)$ the Carlson-Goldman modes excited in the ring make a significant contribution to the nonquantized addition ΔW_T to the magnetic flux.

Then kinetic equation (1.28) is used to find the nonequilibrium addition to the chemical potential in the current state of the superconductor in the presence of a temperature gradient, and a new formation mechanism of the quantity δW_T in a thin-walled superconducting cylinder is proposed.

1. The nonstationary thermoelectric effect in a closed network of different superconductors. We will consider a bimetallic ring consisting of different bulk superconductors S_1 and S_2. We will limit ourself to the single-dimensional case and for convenience will decompose the ring into segments $-L \leqslant x \leqslant L$, and will place junction 1 at the coordinate origin.

We will assume that a temperature distribution such that

$$dT/dx = -a(t)\,\mathrm{sign}\,x,$$

is generated in the ring in a special manner, where $a(t) = \delta T(T)/L$. Let the transverse dimensions of the superconductor be large compared to the depth of penetration of the magnetic field; then there will be no total current **j** in the measurement circuit. When $T_c - T \ll T_c$ by solving equation (3.62) with a given temperature distribution we find

$$\delta\tilde{\mu}(x\omega) = \frac{\pi^2}{3e}\frac{T_c}{\mu_F}\frac{iA_{(-)} - A_{(+)}}{r^2 + p^2}\{\exp[-(A_{(+)} + iA_{(-)})|x|] - \exp[-(A_{(+)} + iA_{(-)})(L - |x|)]\}\frac{\delta T(\omega)}{L}, \quad (3.77)$$

where $A_{(\pm)}(\omega)$, $r(\omega)$, $p(\omega)$ are the parameters determined by relations (3.75) and the factor T_c/μ_F when $x > 0$ refers to metal 1 and when $x < 0$ it refers to metal 2. In the stationary case this result becomes the $\delta\tilde{\mu}(x)$ distribution obtained in study [96]. Let $\delta T(t) = (T_2 - T_1)\exp(i\Omega_0 t)$. Calculating the circulation of the total current **j** through the contour passing into the measurement circuit, we find the stationary solution for the time-dependent magnetic flux $W(t)$ in the contour:

$$W(t) = KW_0 + \delta W_T(t),$$

where

$$\delta W_T = [(i\Omega_0 + \nu_0)^{-1}\delta\tilde{\mu}(L, \Omega_0)]\Big|_1^2 - \frac{\pi^2}{3e}\left[\frac{T_c}{\mu_F}(i\Omega_0 + \nu_0)^{-1}\right]\Big|_1^2 \times (T_2 - T_1)\exp(i\Omega_0 t). \quad (3.78)$$

here $W_0 = \pi/|e|$ is the flux quantum, k is an integer ($c = \hbar = 1$), $\nu_0 = n_s\nu_p/n$; the permutation indices refer to the metals S_1 and S_2. We may see from (3.77)-(3.78) that when frequency Ω_0 satisfies equations (3.72) and Carlson-Goldman modes are excited in the ring, the depth of penetration l_E of the longitudinal field into the superconductors becomes comparable to its length L and the relative smallness $\sim(l_E/L)$ of the first term in formula (3.78) resulting from the thermoelectric field as noted in study [86] when $\Omega_0 = 0$ vanishes. Hence measurements in this frequency range are of interest for investigating the features of collective mode propagation in the system. As will be demonstrated in part 3 of this section in order to observe such features it is, evidently, necessary to provide a sufficiently low magnetic flux KW_0

contributed to the contour. When $\Omega_0 = 0$ formula (3.78) is the same as the result obtained by Artemenko and Volkov [86].

2. The thermal effect in a bimetallic ring: analysis of experimental data. Experiments [101-103] to verify relation (3.78) in the steady-state case confirm that the nonquantized addition to the magnetic flux is proportional to the temperature difference of the junctions and increases in the vicinity of the critical temperature $T \leqslant \min\{T_{c1}, T_{c2}\}$. Formula (3.78) in this case predicts the divergence $\delta W_T \sim (1 - T/T_c)^{-1}$; however experimental results indicate a sharper increase in δW_T in the vicinity of T_c and a magnitude of the effect that exceeds the prediction of (3.78) by a factor of several times. For example, the measurements from study [103] performed in a toroidal geometry that is topologically equivalent to a ring have revealed that $\delta W_T \sim (1 - T/T_c)^{-3/2}$; the magnitude of the observed magnetic flux δW_T was approximately five orders of magnitude (!) greater than that predicted by the stationary limit $\Omega_0 = 0$ (in (3.78)). The issue of the origin of such a serious discrepancy between theory and experiment has not yet become clear and is still the subject of intense discussion. Study [104] has investigated the possible role of "parasitic" effects associated with the quantized magnetic flux captured in the circuit; study [105] has proposed a kinetic model of the phenomena based on a kinetic equation for the excitation distribution function and according to this model the significant auxiliary currents appear on the internal surface of the ring in nonequilibrium conditions; study [106] solves the problem of the thermoelectric effect in the ring in a model formulation when the normal excitation current is given in place of the temperature gradient and it is demonstrated within the scope of Ginzburg-Landau theory that even without the "trapped" flux with an increase in j_n transitions to a state with an increasing number of flux quanta become more advantageous from the energy viewpoint in the system. The issue of whether or not such transitions occur is still an open question.

We note that in the first experiments to investigate the thermoelectric effect in rings carried out by Zavaritskiy [101], where small values of the thermoelectric flux $\delta W_T \sim 10^{-2} W_0$ ($W_0 = \pi/|e| = 2\cdot 10^{-7}$ G$\cdot$cm^2) were observed consistent with the prediction of (3.78), samples were specially selected so that the effect could be reversed by changing the sign of the temperature gradient, thereby guaranteeing no "trapped" flux in them.

Here we will show that if the ring contains a residual quantized flux KW_0, where K is an integer, there exists a kinetic mechanism that has not been examined before and whose action results in a large thermoelectric effect in a thin-walled bimetallic superconducting cylinder. This mechanism is related to the penetration of a longitudinal electrical field into the superconductor; this exists in a nonequilibrium state arising due to thermal inhomogeneities and superfluid motion.

3. The equilibrium shift in the chemical potential in the current state of a superconductor under temperature gradient action. This effect was predicted in study [107] based on a kinetic equation for the excitation distribution function derived in studies [4-6, 8, 10]. It was assumed in this study that when current was present in a thermally inhomogeneous system the primary relaxation mechanism of the branch imbalance are inelastic collisions, and the electron-impurity collision integral in the kinetic equation was dropped. However, results published simultaneously with [107] by Clarke et al. [108] using a double tunnel N-S-N_p-structure, revealed that a value of $\delta\tilde{\mu}$ that is two to three orders of magnitude smaller than that predicted in study [107] was recorded in the experiment. This fact indicates that the dominant role in the formation of the nonequilibrium addition to the chemical potential of the superconductor in such a situation is played by the collisions that result in "mixing" of the electron and hole branches of the excitation spectrum with superfluid velocity present as predicted by the Anderson theorem [109]. The existence of such a relaxation mechanism of the branch imbalance was discovered in study [110]. Calculations carried out in studies [111, 112] produced the following expression for the nonequilibrium addition $\delta\tilde{\mu}$ when $T_c - T \ll T_c$:

$$\delta\tilde{\mu} = \alpha \frac{\mu_F}{v_p} \mathbf{v}_s \frac{1}{T_c} \nabla T, \tag{3.79}$$

where $\sigma \sim 1$ is a dimensionless constant. The linear dependence on $\mathbf{v}_s$ and ∇T and the sign in formula (3.79) have been confirmed experimentally in study [108]. Measurements of $\delta\tilde{\mu}$ were carried out in conditions where the value of the superfluid velocity v_s is close to its critical value $v_c = \Delta/p_F$.

We will show that when $v_s < \Delta/2p_F$ the shift in the chemical potential in the current state of the superconductor in the presence of a temperature gradient results solely from inelastic collisions and its temperature dependence is different from (3.79). For simplicity we will consider the stationary case. As before let $p_F v_s \ll T_c$; however unlike the calculations in the preceding sections we will assume that the superfluid velocity $\mathbf{v}_s$ together with the macroscopic quantities n and $\mathcal{E}$ are quantities of zero order of smallness with respect to the spatial gradients. Again we will write the nonequilibrium density matrix $\tilde{\rho}$ as

$$\hat{\rho} = \hat{\rho}_l + \delta\hat{\rho}_0 (\hat{\varepsilon}_p/T) + \delta\tilde{\rho},$$

where now we must account for the fact that

$$\hat{\rho}_l = n_0 \left(\frac{\varepsilon_p + \mathbf{p}\mathbf{v}_s}{T} \right) \hat{\tau}_+ + n_0 \left(\frac{-\varepsilon_p + \mathbf{p}\mathbf{v}_s}{T} \right) \hat{\tau}_-,$$

$$\tau_\pm = \frac{1}{2} (\hat{1} \pm \hat{\tau}_3),$$

we obtain the following expression for the inhomogeneous term in the equation for the first order correction to the density matrix:

$$\text{Left}\,\{\hat{\rho}_l\} = n_0'\left(\frac{\varepsilon_p + \mathbf{p}\mathbf{v}_s}{T}\right)\left\{\left[\frac{1}{T}\frac{\xi_p}{\varepsilon_p}\left(\frac{\partial\mu}{\partial n}\right)_\Delta + \frac{\varepsilon_p + \mathbf{p}\mathbf{v}_s}{T^2}\left(\frac{\partial T}{\partial n}\right)_\Delta\right]\operatorname{div}(n_s\mathbf{v}_s) + \right.$$

$$+ \frac{1}{T}\left(\frac{\mathbf{p}}{m} + \frac{\xi_p}{\varepsilon_p}\mathbf{v}_s\right)e\mathbf{E}_N + \frac{1}{2}\frac{\partial}{\partial \mathbf{p}}[(\varepsilon_p + \mathbf{p}\mathbf{v}_s)^2]\frac{\partial}{\partial \mathbf{r}}\left(\frac{1}{T}\right)\Big\}\hat{\tau}_+ +$$

$$+ n_0'\left(\frac{-\varepsilon_p + \mathbf{p}\mathbf{v}_s}{T}\right)\left\{\left[-\frac{1}{T}\frac{\xi_p}{\varepsilon_p}\left(\frac{\partial\mu}{\partial n}\right)_\Delta + \frac{-\varepsilon_p + \mathbf{p}\mathbf{v}_s}{T^2}\left(\frac{\partial T}{\partial n}\right)_\Delta\right]\operatorname{div}(n_s\mathbf{v}_s) + \right.$$

$$+ \frac{1}{T}\left(\frac{\mathbf{p}}{m} - \frac{\xi_p}{\varepsilon_p}\mathbf{v}_s\right)e\mathbf{E}_N + \frac{1}{2}\frac{\partial}{\partial \mathbf{p}}[(-\varepsilon_p + \mathbf{p}\mathbf{v}_s)^2]\frac{\partial}{\partial \mathbf{r}}\left(\frac{1}{T}\right)\Big\}\hat{\tau}_- +$$

$$+ \left\{\frac{1}{2}\left[n_0\left(\frac{\varepsilon_p + \mathbf{p}\mathbf{v}_s}{T}\right) - n_0\left(\frac{-\varepsilon_p + \mathbf{p}\mathbf{v}_s}{T}\right)\right]\left[\left(\frac{\Delta}{\varepsilon_p^2}\left(\frac{\partial\mu}{\partial n}\right)_\Delta \operatorname{div}(n_s\mathbf{v}_s) + \right.\right.\right.$$

$$\left.\left.\left. + \mathbf{v}_s e\mathbf{E}_N\right) - \mathbf{v}_s\left(\frac{\Delta}{\varepsilon_p^2}\nabla\Delta + \frac{\xi_p}{\varepsilon_p^2}\nabla\Delta\right) - \frac{\Delta}{\varepsilon_p^2}\frac{\mathbf{p}}{m}\nabla(\mathbf{p}\mathbf{v}_s)\right]\right\} +$$

$$+ \frac{\Delta}{2\varepsilon_p}\frac{\mathbf{p}}{m}\left[n_0'\left(\frac{\varepsilon_p + \mathbf{p}\mathbf{v}_s}{T}\right)\frac{\partial}{\partial \mathbf{r}}\left(\frac{\varepsilon_p + \mathbf{p}\mathbf{v}_s}{T}\right) + n_0'\left(\frac{-\varepsilon_p + \mathbf{p}\mathbf{v}_s}{T}\right)\times\right.$$

$$\times \frac{1}{T}\left(\frac{\xi_p}{\varepsilon_p}\frac{\mathbf{p}}{m} + \mathbf{v}_s\right) + n_0'\left(\frac{-\varepsilon_p + \mathbf{p}\mathbf{v}_s}{T}\right)\frac{1}{T}\left(-\frac{\xi_p}{\varepsilon_p}\frac{\mathbf{p}}{m} + \mathbf{v}_s\right)\Big]\times$$

$$\times\left(\frac{\Delta}{\varepsilon_p^2}\frac{\partial}{\partial \mathbf{r}}(\mu) + \frac{\xi_p}{\varepsilon_p^2}\frac{\partial}{\partial \mathbf{r}}(\Delta)\right)\hat{\tau}_1. \tag{3.80}$$

In this formula $n_0'(x) = (d/dx)\, n_0(x)$. Retaining the ξ_p-odd part of the ν-diagonal elements and the ξ_p-even part of the ν-nondiagonal elements of inhomogeneous term (3.80) of equation (3.7) and focusing solely on the contribution of $\simeq \mathbf{v}_s \nabla T$, we obtain in the lowest nonvanishing order in $p_F v_s/T$ an equation of the type shown in (3.25) for the nonequilibrium addition $\delta\tilde{\rho}$, where

$$A = \left\{n_0''\left(\frac{\varepsilon_p}{T}\right)\frac{(\mathbf{p}\mathbf{v}_s)}{T}\xi_p - \left[-n_0'\left(\frac{\varepsilon_p}{T}\right)\frac{\xi_p}{\varepsilon_p}(\mathbf{p}\mathbf{v}_s)\right]\right\}\frac{\mathbf{p}}{m}\nabla\left(\frac{1}{T}\right), \tag{3.81}$$

$$B = \frac{1}{T}n_0''\left(\frac{\varepsilon_p}{T}\right)\frac{(\mathbf{p}\mathbf{v}_s)}{T}\left(\frac{\partial\Delta}{\partial T}\right)_n\left(\frac{\mathbf{p}}{m}\nabla T\right) +$$

$$+ \frac{\Delta}{\varepsilon_p}\left\{n_0''\left(\frac{\varepsilon_p}{T}\right)\frac{\varepsilon_p}{T} - \left[-n_0'\left(\frac{\varepsilon_p}{T}\right)\right]\right\}(\mathbf{p}\mathbf{v}_s)\left(\frac{\mathbf{p}}{m}\nabla T\right). \tag{3.82}$$

We may show that when $v_s < \Delta/2p_F$ the elastic collisions do not make a contribution to the frequency ν_γ calculated in the vicinity of the Fermi surface, and in the zeroth order in the parameter $p_F v_s/\Delta$ near T_c we have $\nu_\gamma = 1/2\nu_E$. Solving equation (3.25) with the left half of (3.81)-(3.82) and subjecting the derived solution to conditions (3.9)-(3.12), we find that in these conditions ($v_s < \Delta/2p_F$, $T_c - T \ll T_c$) the nonequilibrium addition to the chemical potential is determined by the contribution of the first component in (3.82) and is equal to

$$\delta\tilde{\mu} = -\frac{4}{\pi}\frac{T_c}{\Delta}\frac{p_F^2}{3mN(0)}\frac{1}{3\nu_\gamma}\left\langle\frac{1}{T}\frac{\Delta}{\varepsilon_p}n_0''\left(\frac{\varepsilon_p}{T}\right)\right\rangle\left(\frac{\partial\Delta}{\partial T}\right)_n(\mathbf{v}_s\nabla T), \tag{3.83}$$

where $N(0) = 3mn/p_F^2$.

We will calculate the leading term of the expansion in terms of the parameter Δ/T_c of the integral that appears in (3.84). We have

$$\left\langle \frac{1}{T}\frac{\Delta}{\varepsilon_p}\left(n_0''\frac{\varepsilon_p}{T}\right)\right\rangle = -\frac{1}{T}\frac{\partial}{\partial(1/T)}\left\{\left\langle \frac{1}{T}\frac{\Delta^2}{\varepsilon_p^2}\left[-n_0'\left(\frac{\varepsilon_p}{T}\right)\right]\right\rangle\right\}_\Delta \simeq$$

$$\simeq \frac{3mp_F}{3\pi^2}\frac{7\zeta(3)}{2\pi^2}\left(\frac{\Delta}{T}\right). \tag{3.84}$$

Subject to (3.84) we finally obtain for $\delta\tilde{\mu}$:

$$\delta\tilde{\mu} = \frac{8}{\pi}\frac{T_c}{\Delta_i^*}\frac{1}{\nu_\gamma}\frac{p_F^2}{3m}\left(\mathbf{v}_s\frac{1}{T_c}\nabla T\right), \tag{3.85}$$

where in accordance with the discussion above $\nu = 1/2\nu_E$. Thus, with sufficiently small values of υ_s the chemical potential shift is singularly dependent on temperature ($\sim(1 - T/T_c)^{-1/2}\upsilon_s$) and unlike (3.79) is not proportional to the elastic collision time, but rather to the energy relaxation time which is significantly greater in real superconductors. The smallness condition of υ_s necessary for observation of relation (3.85) may therefore be partially compensated by the significant differential of ν_E and ν_p. We obtained result (3.85) only for pure superconductors $\nu_p \ll T_c$; with large impurity concentrations pair breaking effects become significant; these were not accounted for in deriving formula (3.85).

With growth of υ_s by virtue of the smallness of (ν_E/ν_p) the elastic collisions begin to play a determinant role, and $\delta\tilde{\mu}$ is described by relation (3.79); the transition near the point $\upsilon_s^{(0)} = \Delta/2p_F$ will therefore result in a "change of state": a transition from relation (3.85) to relation (3.79).

We note that result (3.85) produced by the contribution of the ν-nondiagonal components of the density matrix, although it does contain the inelastic collision time, nonetheless has a sharper temperature dependence when $T \to T_c - 0$ than that obtained in the initial study by Pethick and Smith [107]. Formula (3.85) cannot be derived by means of a kinetic equation for the excitation distribution function obtained from the equation for $\hat{\rho}$ (1.28) by ignoring the components of the density matrix and the collision integral nondiagonal in the ν-representation.

4. The strong thermoelectric effect in an inhomogeneous doubly-connected superconducting network. As was demonstrated above a non-equilibrium shift in the chemical potential (3.85) occurs everywhere in a superconductor where superfluid motion with $\upsilon_s < \Delta/2p_F$ exists together with a temperature gradient. In view of the smallness of the ratio (ν_E/ν_p) this phenomenon may be registered by using a technique traditional for such problems based on the use of a double tunnel N-S-N_p junction [113]. However the phenomenon will, evidently, be most clearly manifest, as we will see, in experiments to measure the thermoelectric addition to the magnetic flux in a doubly-connected network of different superconductors. As our example we will again consider a bimetallic cylinder of radius R; we will assume that the transverse dimensions d of the superconductors S_1 and S_2 comprising the cylinder are comparable to the depth of penetration of the magnetic field $\lambda(T)$. We will integrate the expression $((2m/en_s)\mathbf{j})$ with respect to the closed contour C passing through the cylinder and encompassing the cavity; we remember that the total charge current is

$$e\mathbf{j} = en_s\mathbf{v}_s - \eta_T\nabla T + \sigma\mathbf{E}_N. \tag{3.86}$$

We will limit our examination to the stationary case; it follows then from the equation for superfluid velocity that

$$e\mathbf{E}_N = \nabla(\sigma\tilde{\mu}). \tag{3.87}$$

It is convenient to rewrite expression (3.86) as

$$e\mathbf{j} = en_s\left(1 + \alpha\frac{1}{T}\nabla^2 T\right)\mathbf{v}_s + \mathbf{j}_n,$$

$$\mathbf{j}_n = -\eta_T\nabla T, \quad \alpha = 4\frac{n}{n_s}l_E^2, \quad l_E^2 = \frac{4v_F^2}{3\pi\nu_E\nu_p}\frac{T_c}{\Delta}. \tag{3.88}$$

It is assumed that the correction term $\sim\alpha$ is small compared to unity. In (3.88) we write only the contribution of (3.85) to the nonequilibrium shift of the chemical potential and we drop the components $(\sim\nabla T \times \mathrm{rot}\,\mathbf{A})$. We will have

$$W = KW_0 + \frac{4\pi}{c}\oint_C dl\lambda^2\mathbf{j}_n - \frac{4\pi}{c}\oint_C dl\lambda^2\mathbf{j}, \tag{3.89}$$

where $W = \oint_C d\mathbf{l}\mathbf{A}$ is the magnetic flux in the cavity, K is an integer, $W_0 = hc/2e$ is the flux quantum, $\lambda^2 = \frac{mc^2}{4\pi e^2}\frac{1}{n_s}\left[1 + \alpha\frac{1}{T}\nabla^2 T\right]^{-1}$. The current density in a thin-walled $(d \leqslant \lambda)$ cylinder is constant in the cross-section of the conductor, and $\gamma = cW/4\pi^2 R^2 d$ (there is no external magnetic field). Finally from (3.89) we find for W the expression

$$W = \frac{KW_0 + W_n}{1 + P}, \tag{3.90}$$

$$W_n = \frac{4\pi}{c}\oint_C dl\lambda^2\mathbf{j}_n, \quad P = \oint_C \frac{\lambda^2}{Rd}\frac{dl}{L} \ll 1 \tag{3.91}$$

($L = \pi R$ is the semiperimeter of the ring). The contribution of a regular thermoelectric field to W is small and amounts to $\sim(l_E/L)W_n$ [86], where $W_n \sim 10^{-2}\,W_0$ [96]. Let $K \gg 1$ in the cylinder; then we may ignore the W_n term in (3.91). In view of the smallness of the parameter P (in a temperature range that is not too close to T_c), we have

$$W = KW_0[1 - P_1(T) + P_2(T)], \tag{3.92}$$

$$P_1 = \oint_C \frac{\lambda_0^2}{Rd}\frac{dl}{L}, \quad P_2 = \oint_C \frac{\lambda_0^2}{Rd}\alpha\frac{\nabla^2 T}{T_c}\frac{dl}{L}, \quad \lambda_0^2 = \frac{mc^2}{4\pi e^2 n_s}. \tag{3.93}$$

The term $P_1(T)$ in (3.92) corresponds to a reduction of the flux trapped in the cylinder due to its "leakage" in the wall; the component $P_2(T)$ corresponds to an increase in the flux of the system due to additional electron acceleration by the longitudinal electrical field $\mathbf{E}_N \sim \nabla\delta\tilde{\mu}$.

Assume the cylinder is now in a thermostat and heat exchange is described by Newton's law with characteristic frequency ν_{es}; then the temperature distribution in the cylinder, for example, when $0 \leqslant x \leqslant L$ takes the form

$$T(x) = T_1 + (T_2 - T_1)\frac{\mathrm{sh}\,(x/l_T)}{\mathrm{sh}\,(L/l_T)}, \qquad l_T = \left(\frac{v_F^2}{\nu_p \nu_{es}}\right)^{1/2}, \tag{3.94}$$

where T_2 is the temperature of a hot junction, $x = R\varphi$, φ is the polar angle in the "natural" cylindrical coordinate system; l_T is the characteristic temperature decay length; T_1 is the thermostat temperature. The correction $\sim \alpha \nabla^2 T$ that is insignificant in the case of weak heat exchange ($l_T \gg L$) must be accounted for in the opposite case of good heat exchange ($l_T \ll L$). Using (3.88), (3.92)-(3.94) we find the relations $P_1(T)$ and $P_2(T)$ (the index < refers to the metal with the lower critical temperature):

$$P_1(T) = \frac{\lambda_0^2(0)}{Rd}\frac{1}{1-t_1}\left[1 + \frac{l_T}{L}\ln\frac{1-t_1}{1-t_2}\right],$$

$$P_2(T) = \frac{2}{3}\frac{\lambda_0^2(0)}{Rd}\alpha_0\frac{l_T}{L}\left[\frac{1}{(1-t_2)^{3/2}} - \frac{1}{(1-t_1)^{3/2}}\right], \tag{3.95}$$

$$\alpha_0 = 4\frac{l_E^2(0)}{l_T^2}, \qquad t_i = \frac{T_i}{T_{c<}}, \qquad i = 1 \div 2.$$

In order to estimate the magnitude of the effect we will use lead (T_{c2} = 7.2 K) and indium (T_{c1} = 3.4 K) as our superconductors S_1 and S_2, respectively; let $d \sim 10^{-4}$ cm, $L \sim 1$ cm, $1 - t_1 \sim 10^{-2}$, $\lambda_0(0) \sim 5 \cdot 10^{-6}$ cm; $v_F \sim 10^8$ cm/s, $\nu_{es} \sim 5 \cdot 10^{-6}$ cm, $l_T \sim 10^{-1}$ cm, $l_E(0) \sim 10^{-2}$ cm. Here the quantity P_1 with a weak temperature dependence T_2 will be of the order $P_1 \sim 10^{-4}$. The quantity $P_2 = 0$ when $T_2 = T_1$ although P_2 increases sharply with temperature and when $1 - t_2 = 10^{-3}$ amounts to $P_2 \sim 0.5 \cdot 10^{-4}$. Thus, in the temperature range $10^{-2} = 1 - t_1 \leqslant 1 - t_2 \leqslant 10^{-3}$ s with an increase in temperature of the hot junction there is an increase in the magnetic flux in the system as $(1 - t_2)^{-3/2}$. If, for example, the initially trapped flux was $W = 2 \cdot 10^6 \times W_0$ (i.e., $H \sim 2 \cdot 10^{-1}$ erg), the increase in the flux $\delta W = KW_0 P_2$ amounts to $\delta W \sim 100 W_0$, which may easily be observed in experiments. The terrestrial magnetic field flux through the cylinder is $\sim 10^7 W_0$.

We note that in the case of a bulk cylinder ($d \gg \lambda$) this effect is very small and hence the increase in the magnetic flux as $(1-t_2)^{-3/2}$ in toroidal samples observed in thermoelectric experiments [103] may be due to other causes (see, for example, [106]).

5. The thermomagnetic effect in superconductors. The existence of a nonequilibrium addition to the chemical potential proportional to superfluid velocity makes possible thermomagnetic effects in superconductors. We will write $\delta\tilde{\mu}$ as

$$\delta\tilde{\mu} = \beta\,(\mathbf{v}_s \nabla T),$$

where $\beta(T)$ is a function of temperature. We will drop the contribution of the regular thermoelectric field to $\delta\tilde{\mu}$; this will not be needed below. For the thermoelectric field E_N we will have in the stationary case

$$eE_N = (\mathbf{v}_s\nabla)\beta\nabla T + \beta[(\nabla T\cdot\nabla)\mathbf{v}_s + \beta\frac{e}{m}\mathbf{B}\times\nabla T,$$

where $\mathbf{B}$ = rot $\mathbf{A}$. For definiteness we will consider a superconducting region between two parallel planes, one of which is Oxy; let the external magnetic field on the axis Ox be nonzero only in one of the free half-spaces. We will represent the temperature gradient $\partial t/\partial y$ in the superconductor, then when $q_z = 0$ a temperature gradient appears in the superconductor and on the O_z axis:

$$\frac{\partial T}{\partial z} = \left(\frac{\eta_T T}{m\varkappa}\beta\right)\frac{\partial T}{\partial y}B_x(z).$$

This is the Righi-Leduc effect; the quantity $K_{LR} = \eta_T T\beta/m\varkappa$ serves as the corresponding coefficient . This situation is easily implemented by "trapping" the magnetic flux in a hollow superconducting cylinder; if the external field differs from the trapped field and the temperature distribution in the cylinder is dependent on the polar angle in the "natural" cylindrical coordinate system, its external and internal surfaces will be heated differently. If the induction value of v_s in the cylinder is sufficiently small ($v_s < \Delta/2p_F$), then near T_c by virtue of (3.85)

$$K_{LR} = \frac{32}{3\pi}\frac{e}{m}\frac{T_c}{\Delta}\tau_E \simeq 1{,}8\cdot 10^{11}\tau_E(c)(1 - T/T_c)^{-1/2} \ (\mathrm{a\cdot s^2/kg}).$$

When $T/T_c \sim 0{,}99$; $\tau_E \sim 10^{-9}$ s, $K_{LR} \sim 10^3$ a·s^2/kg.

7. Fluctuations in the branch imbalance of the excitation spectrum of a superconductor. Josephson radiation linewidth

The spectrum of fluctuations in the branch imbalance of the excitation spectrum in pure superconductors was found in the study by Aronov and Katilyus [114]; this study developed the Langevin method based on a kinetic equation for the distribution function of normal excitations [4-6, 8, 10]. In this section we find the spectral function of fluctuations in the interbranch imbalance using a system of hydrodynamic retardation equations accounting for the temporal and spatial correlation of fluctuations in the frequency range $\omega \ll \Delta$ and wave vectors $k \ll v_F^{-1}|i\omega + \nu_p|$. We will limit our examination to the case $T_c - T \ll T_c$.

In Section 5 of this chapter we derived the following "equation of motion" for the quantity $\delta\tilde{\mu}(rt)$ characterizing the branch imbalance of the excitation spectrum of a superconductor:

$$\left(\frac{\partial}{\partial t} + \hat{\nu}_{\mu}(t) - \hat{D}_{\mu}(t)\nabla_{\mathbf{r}}^{2}\right)\delta\tilde{\mu}(\mathbf{r}t) - \hat{D}_{\mu T}\nabla_{\mathbf{r}}^{2}T(\mathbf{r}t) = 0, \tag{3.96}$$

where, for example, for any reasonable function $f(t)$

$$\hat{D}_{\mu}(t)f(t) = \int_{0}^{t} D_{\mu}(t-t')f(t')\,dt'; \tag{3.97}$$

the kernel of the integral operator in (3.97) when $T_c - T \ll T_c$ takes the form

$$D_{\mu}(t) = \frac{4T_c}{\pi\Delta}\left(\frac{\partial\mu}{\partial n}\right)_{\Delta}\frac{n_s}{n}\exp(-\nu_{\gamma}t), \qquad t \geqslant 0. \tag{3.98}$$

We will consider a superconductor in the absence of tunnel injection. According to the general theory of quasi-stationary fluctuations (see, for example, [115, Sections 118, 119]), the correlation function $\langle\delta\tilde{\mu}(\mathbf{r}_1t_1)\,\delta\tilde{\mu}(\mathbf{r}_2t_2)\rangle$ satisfies the equation

$$\left(\frac{\partial}{\partial t_1} + \hat{\nu}_{\mu}(t_1) - D_{\mu}(t_1)\nabla_{\mathbf{r}_1}^{2}\right)\langle\delta\tilde{\mu}(\mathbf{r}_1t_1)\,\delta\tilde{\mu}(\mathbf{r}_2t_2)\rangle -$$
$$- \hat{D}_{\mu T}(t_1)\nabla_{\mathbf{r}_1}^{2}\langle\delta T(\mathbf{r}_1t_1)\,\delta\tilde{\mu}(\mathbf{r}_2t_2)\rangle = 0, \qquad t_1 - t_2 > 0. \tag{3.99}$$

As is normally the case the angle brackets in (3.99) designate averaging over t_2 with a given difference $\tau = t_1 - t_2$. The initial condition for equation (3.99) is the value of the simultaneous correlator $\langle\delta\tilde{\mu}(\mathbf{r}_1t_1)\delta\tilde{\mu}(\mathbf{r}_2t_1)\rangle$ and the problem of calculating this correlator in the general nonequilibrium case is an independent problem. Here we will consider fluctuations in the near-equilibrium state (see, for example, [41, Section 19]). Then all correlation functions are dependent only on τ and $\mathbf{r}_1 - \mathbf{r}_2 = \mathbf{R}$. We are interested in the spectral fluctuation function

$$\langle\delta\tilde{\mu}^2\rangle_{\omega\mathbf{k}} = \int_{-\infty}^{\infty} d\tau \int d\mathbf{R}\,\langle\delta\tilde{\mu}(R,\tau)\,\delta\tilde{\mu}(0,0)\rangle\exp(i\mathbf{k}\mathbf{R} - i\omega\tau), \tag{3.100}$$

which is related to the average value of the product of the Fourier components of the quantity $\delta\tilde{\mu}(\mathbf{r}t)$ by the formula

$$\langle\delta\tilde{\mu}(\mathbf{k}\omega)\,\delta\tilde{\mu}(\mathbf{k}'\omega')\rangle = (2\pi)^4\,\delta(\omega+\omega')\,\delta(\mathbf{k}+\mathbf{k}')\,\langle\delta\tilde{\mu}^2\rangle_{\omega\mathbf{k}}. \tag{3.101}$$

It follows from (3.100) that

$$\langle\delta\tilde{\mu}^2\rangle_{\omega\mathbf{k}} = \langle\delta\tilde{\mu}^2\rangle_{\omega\mathbf{k}}^{(+)} + \langle\delta\tilde{\mu}^2\rangle_{-\omega,-\mathbf{k}}^{(+)}, \tag{3.102}$$

where the symbol $(\ldots)^{(+)}$ designates the Fourier transform

$$\langle\delta\tilde{\mu}^2\rangle_{\omega\mathbf{k}}^{(+)} = \int_{0}^{\infty} d\tau \int d\mathbf{R}\,\langle\delta\tilde{\mu}(\mathbf{R}\tau)\,\delta\tilde{\mu}(0,0)\rangle\exp(i\mathbf{k}\mathbf{R} - i\omega\tau). \tag{3.103}$$

We will carry out Fourier transform (3.103) on equation (3.99); as a result we obtain

$$\langle\delta\tilde{\mu}^2\rangle^{(+)}_{\omega k} = \frac{\langle\delta\tilde{\mu}^2\rangle_{0k} - D_{\mu T} k^2 \langle\delta T \delta\tilde{\mu}\rangle_{\omega k}}{i\omega + \nu_\mu(\omega) + D_\mu(\omega) k^2}. \tag{3.104}$$

We may easily see that in this approximation the generalized hydrodynamic equation for $t(\mathbf{r}t)$ does not contain the quantity $\delta\mu(\mathbf{r}t)$. Calculation of the equilibrium simultaneous correlators [114, 116] when $T \to T_c - 0$ yields

$$\langle\delta\tilde{\mu}^2\rangle_{0k} = \frac{4T^2}{\pi\Delta}\left(\frac{\partial\mu}{\partial n}\right)_\Delta, \tag{3.105}$$

$$\langle\delta T \delta\tilde{\mu}\rangle^{-}_{0k} = 0. \tag{3.106}$$

Hence the second component of the numerator of formula (3.104) vanishes; finally we will have for the spectral fluctuation function

$$\langle\delta\tilde{\mu}^2\rangle_{\omega k} = \frac{8T_c^2}{\pi^2\Delta}\left(\frac{\partial\mu}{\partial n}\right)_\Delta \frac{\nu_0 + \frac{4T_c}{3\pi\Delta}\frac{n_s}{n}\nu_\gamma \frac{k^2 v_F^2}{\omega^2 + \nu_\gamma^2}}{\left[\nu_0 + \frac{4T_c}{3\pi\Delta}\frac{n_s}{n}\nu_\gamma\frac{k^2 v_F^2}{\omega^2+\nu_\gamma^2}\right]^2 + \omega^2\left[1 + \frac{n_s}{n} - \frac{4T_c}{3\pi\Delta}\frac{k^2 v_F^2}{\omega^2+\nu_\gamma^2}\right]^2}, \tag{3.107}$$

where $\nu_0 = (n_s/n)\,\nu_p$, $\nu_\gamma = {}^1/_2\nu_E$. In the plane (ω, k) the spectral correlator (3.107) has a sharp maximum near the region of the Carlson-Goldman mode (3.72)-(3.73). We obtain from (3.107) for the spectrum of spatially-homogeneous fluctuations $(k = 0)$

$$\langle\delta\tilde{\mu}^2\rangle_\omega = \frac{8T_c^2}{\pi^2\Delta}\left(\frac{\partial\mu}{\partial n}\right)_\Delta \frac{\nu_0}{\omega^2 + \nu_0^2}. \tag{3.108}$$

when $T \to T_c - 0$ the width of the fluctuation spectrum drops in proportion to $(1 - T/T_c)$. The expression for $\langle\delta\xi\rangle_\omega$ obtained in study [114] takes the form of (3.108) with the difference that the frequency ν_0 is replaced with $\nu_\Phi \sim \frac{T_c^2}{s_0^2 p_F^2}\nu_E$. The reason for this differential, obviously, may be traced to ignoring the nondiagonal elements of the density matrix and collision integral underlying the approach [4-6, 8, 10].

Fluctuations in the quantity $\delta\tilde{\mu}$ may, as noted in study [114] determine the Josephson radiation linewidth. The problem of the influence of fluctuations on radiation linewidth was first examined by Kulik [117] for superconductors with paramagnetic impurities near T_c. Following approach [114] we may assume that in two superconductors forming a weak link the quantities $\delta\tilde{\mu}$ fluctuate independently. Further calculations are completely analogous to those carried out in study [114] (see also [118]). We immediately obtain the result for the function $J(\omega)$ given in study [118] and describing the shape of the radiation line:

$$J(\omega) = \frac{1}{\pi} \frac{\Gamma}{(\omega - \omega_0)^2 + \Gamma^2}, \tag{3.109}$$

where the Josephson frequency is $\omega_0 = 2|e|V_0$, V_0: the potential difference applied to the junction;

$$\Gamma = 4T_c \left(e^2 R_T + \frac{1}{\Omega} \frac{4T_c}{\pi\Delta} \left(\frac{\partial\mu}{\partial n} \right)_\Delta \frac{1}{\nu_0} \right). \tag{3.110}$$

In deriving relations (3.109)-(3.110) we assumed a junction consisting of identical superconductors and that the linewidth Γ is much less than the frequency ν_0 determining the dispersion of the fluctuation spectrum of the branch imbalance of the excitation spectrum everywhere within the applicability of formulae (3.109)-(3.110). In (3.110) R_T is the resistance of the junction $\Omega = (\Omega_1^{-1} + \Omega_2^{-1})^{-1}$, where Ω_i is the volume of the i-superconductor; the first component in this formula represents the contribution of fluctuations in the potential difference across the junction related to the finite resistance of the tunnel junction. The second component in (3.110) results from fluctuations in the interbranch imbalance; this term behaves as $(1-T/T_c)^{-3/2}$ when $T \to T_c - 0$ and is proportional to the momentum relaxation time unlike the result from study [14] in which the corresponding component $\sim(1 - T/T_c)^{-1/2}\tau_E$. For aluminum $\Omega_1 = \Omega_2 = 10^{-2}$ cm^3, $\nu_i \sim 10^{11}$ s^{-1}, $(1 - T/T_c) \sim 10^{-2}$ the second component in (3.110) is within an order of magnitude of 10_3 Hz.

8. Diffusion and convective transport of spin density in the current state of a superconductor with nonequilibrium-oriented spins

We will consider a pure superconductor in which some means is used to generate a nonequilibrium distribution of the spin density; for simplicity we will limit our examination to the case where only the z-component of this distribution $s_z(\mathbf{r}t)$ is nonzero. Let the electron-electron interaction be sufficiently weak $\left(|g| < \frac{4}{3}\mu_F\right)$ and antiferromagnetic ordering of electrons be impossible, with s_z appearing as a slow function of the coordinates on the $\xi(T)$ scale, nowhere exceeding the critical value of $s_c = 2\Delta \left\langle \frac{1}{T} [-n_0'(\varepsilon_p/T)] \right\rangle$ (here $\Delta = \Delta(T)$ is the order parameter when $s_z = 0$), where Δ vanishes. One possible method of generating a nonequilibrium distribution of the spin density may be injection of electrons to the superconductor from a magnetized ferromagnetic, as considered phenomenologically by Aronov [119]. The spin-orbital interaction in metals may be quite weak due to the small difference between the g-factor of the electron in a metal from the g_0 factor of a free electron. Hence, as our spin relaxation mechanism we will consider electron scattering by paramagnetic impurities randomly distributed in a superconductor with a low concentration N_m. Applying equation (1.28) to this problem allows us to show that over times $t \gg 1/\Delta$ the behavior of the nonequilibrium spin density in the superconductor may be described in the language of generalized hydrodynamics.

With paramagnetic impurities and a tunnel source $\hat{I}_{NS}$ the kinetic equation for a single-electron density matrix of the superconductor takes the form

$$\frac{\partial}{\partial t}\hat{\rho}+\frac{1}{2}\left\{\left[\frac{\partial\hat{\varepsilon}_p}{\partial\mathbf{p}},\frac{\partial}{\partial\mathbf{r}}\hat{\rho}\right]_+-\left[\frac{\partial\hat{\varepsilon}_p}{\partial\mathbf{r}}\frac{\partial}{\partial\mathbf{p}}\hat{\rho}\right]_+\right\}+i\,[\hat{\varepsilon}_p,\hat{\rho}]_-=$$
$$=\hat{R}^{(2)}_{e-ph}\{\hat{\rho}\}+\hat{R}^{(2)}_{\underset{i}{(n)}}\{\hat{\rho}\}+\hat{R}^{(2)}_{\underset{i}{(m)}}\{\hat{\rho}\}+\hat{I}_{NS}\{\hat{\rho}\}. \tag{3.111}$$

Electron-electron collisions represent a less effective energy relaxation mechanism in superconductors compared to electron-phonon collisions; hence, in (3.111) we drop the contribution of electron-electron scattering to the cross-term. Let the concentration N_m of paramagnetic impurities be small, so energy relaxation occurs more rapidly than relaxation of the spin degree of freedom (this assumption limits us to a temperature range determined by the inequality $\nu_E(T) \gg \tau_s^{-1}$ is the collision frequency with spin rotation which, with small values of $T \sim T_c$ is weakly dependent on temperature); we may also assume the frequency of the tunnel junctions ν_T to be small compared to ν_E. The electron-phonon collision integral describes relaxation of the electron subsystem to a state described by a density matrix of the type

$$\hat{\rho}_0=n_0\left(\frac{\hat{\varepsilon}_p-\mathrm{v}_s}{T}\right), \tag{3.112}$$

where when $\mathrm{v}_s \ll \Delta$, $p_F\mathrm{v}_s \ll \Delta$:

$$\mathrm{v}_s=\left\langle\frac{1}{T}\left[-n_0'\left(\frac{\varepsilon_p}{T}\right)\right\rangle^{-1} s_z.$$

As is normally the case we will identify as the local-equilibrium state the system state corresponding to the density matrix $\hat{\rho}_l$ of the type (3.112) all of whose parameters are dependent on the coordinates and time. We will assume that $\hat{\rho}_l$ is normalized to the true nonequilibrium values of $n(\mathbf{r}t)$, $\mathcal{E}(\mathbf{r}t)$ and $s_z(\mathbf{r}t)$. We obtain from (3.111) and identity (1.51) satisfied for $\hat{R}^{(2)}_{e-ph}$ and $\hat{R}^{(2)}_i(n)$ a continuity equation for the spin density:

$$\frac{\partial}{\partial t}s_z+\operatorname{div}\,(\mathbf{j}_s^{(0)}+s_z\mathbf{v}_s)=-\frac{s_z}{\tau_s}+I^s, \tag{3.113}$$

where $\mathbf{j}_s^{(0)}=\left\langle\frac{\mathbf{p}}{m}\left(\sigma_z\hat{\rho}+\frac{1}{2}\right)\right\rangle$; from (1.87)-(1.93)

$$\tau_s^{-1}=\frac{4}{3}\pi N_m\sum|J(\mathbf{q})|^2 s(s+1)C_{\underset{1\sigma'}{p'p'}}C_{\underset{\sigma'1}{p'p}}\,\delta(\hat{E}_{\underset{\sigma'\sigma'}{-\mathbf{p}'}}+\hat{E}_{\underset{11}{p}})\,\delta(\mathbf{p}'-\mathbf{p}-\mathbf{q});$$

in the zeroth order of $p_F\nu_s/\Delta$ in a model with point electron-impurity interaction

$$\tau_s^{-1}\simeq\left(1+\frac{\Delta^2}{\varepsilon_p^2}\right)\frac{\varepsilon_p}{|\xi_p|}(\tau_s^{(n)})^{-1},$$

where $\tau_s^{(n)}$ is the spin density relaxation time in the N-metal when $T = T_c$. In order to find the intensity of the tunnel source of the electron spin it is sufficient to calculate the tunnel N-S-source (3.20) with local-equilibrium density matrices of the electrons in the normal metal and the superconductor of the type shown in (3.112). We will have

$$I_{NS}^{s} = \langle \hat{I}_{NS} \rangle, \tag{3.114}$$

where in the ν-representation

$$\begin{aligned}(\hat{I}_{NS})_{\sigma\sigma} = \nu_T \Bigg\{ \Bigg[\Bigg(n_0 \frac{E_{p\,\sigma\sigma} - V(t) - \nu_s^N(t)}{T} \Bigg) + \\ + n_0 \Bigg(\frac{E_{p\,\sigma\sigma} + V(t) - \nu_s^N(t)}{T} \Bigg) - 2n_0 \Bigg(\frac{E_{p\,\sigma\sigma} - \nu_s}{T} \Bigg) \Bigg] + \\ + \frac{\xi_p}{\varepsilon_p} \Bigg[n_0 \Bigg(\frac{E_{p\,\sigma\sigma} - V(t) - \nu_s^N(t)}{T} \Bigg) - n_0 \Bigg(\frac{E_{p\,\sigma\sigma} + V(t) - \nu_s(t)}{T} \Bigg) \Bigg] \Bigg\}.\end{aligned} \tag{3.115}$$

In the linear approximation in ν_s, ν_s^N and $V(t)$:

$$I^s = 2\nu_T (s_z^N - s_z), \tag{3.116}$$

where the index N corresponds to the normal metal. It is natural to assume that this function is concentrated on the surface of the junctions. Calculating $\mathbf{j}_{s(l)}^{(0)}$ with any locally-equilibrium density matrix, we obtain

$$\mathbf{j}_{s(l)}^{(0)} = -\frac{1}{3} \left\langle \frac{1}{T^2} \frac{p^2}{m^2} \frac{\xi_p}{\varepsilon_p} n_0'' \left(\frac{\varepsilon_p}{T} \right) \right\rangle \nu_s \mathbf{v}_s = -\frac{2}{3} s_z \mathbf{v}_s. \tag{3.117}$$

Thus, the local equilibrium equation for $s_z(\mathbf{r}, t)$ takes the form

$$\frac{\partial}{\partial t} s_z + \operatorname{div}\left(\frac{1}{3} s_z \mathbf{v}_s \right) = -\frac{s_z}{\tau_s} + I^s, \tag{3.118}$$

where

$$\mathbf{j}_{s(l)} = \mathbf{j}_{s(l)}^{(0)} + s_z \mathbf{v}_s = \frac{1}{3} s_z \mathbf{v}_s. \tag{3.119}$$

We will search the solution of equation (3.111) as

$$\hat{\rho} = \hat{\rho}_l + \delta\hat{\rho} = \hat{\rho}_l + \delta\hat{\rho}_0 + \delta\rho,$$

where $\hat{\rho}_0 = n_0(\hat{\varepsilon}_p/T)$, $\hat{\varepsilon}_p$ is the nonequilibrium spectrum, $\delta\tilde{\rho}$ is the small addition. As in Section 1 of this chapter we may show that the equation for the first order correction $\delta\tilde{\rho}$ with respect to the gradients of the macroscopic quantities takes the form

$$\begin{aligned}\frac{\partial}{\partial t} \hat{\rho}_l + \frac{\partial}{\partial t} (\delta\hat{\rho}_0) + \frac{1}{2} \left\{ \left[\frac{\partial \hat{\varepsilon}_p}{\partial \mathbf{p}}, \frac{\partial}{\partial \mathbf{r}} \hat{\rho}_l \right]_+ - \left[\frac{\partial \hat{\varepsilon}_p}{\partial \mathbf{r}}, \frac{\partial}{\partial \mathbf{p}} \hat{\rho}_l \right]_+ \right\} - \\ - \hat{R}_{i(m)}^{(2)}\{\hat{\rho}_l\} - \hat{I}_{NS}\{\hat{\rho}_l\} = \hat{L}\{\delta\tilde{\rho}\} - \frac{\partial}{\partial t}(\delta\tilde{\rho}),\end{aligned} \tag{3.120}$$

where

$$\hat{L} = \hat{\mathcal{L}}^{(2)}_{e-ph} + \mathcal{L}^{(2)}_{imp(n)} - i\,[\varepsilon_p, \ldots]_-, \qquad (3.121)$$

where $\hat{\mathcal{L}}^{(2)}$ are the linearized collision integrals solved for $\delta\rho$ if we search its solution in a set of two-dimensional Hermitian matrices satisfying conditions (3.9)-(3.12) and the condition

$$\delta s_z = \langle \delta\rho \rangle = 0, \qquad (3.122)$$

while in the left half we replace the time-derivatives in $\partial/\partial t(\hat{\rho}_l)$ with the spatial gradients consistent with the local equilibrium equations (we must now add equation (3.118) for s_z to such equations). We will consider the phonons to be in equilibrium. In $\hat{\mathcal{L}}^{(2)}_{imp(n)}$, aside from the contribution of the nonmagnetic impurities, we also include the part of the paramagnetic impurity collision integral proportional to the matrices $B_{pp'}$ (see (1.87)) and resulting in momentum relaxation that is not accompanied by spin rotations. The normalization conditions of the solution of the zeroth approximation to the true values of the particle density, spin, and energy make it possible to conserve in the equation for the correction to the density matrix the time derivative $\partial/\partial(t)(\delta\hat{\rho})$ without violating the solvability conditions of this equation. The relation $kv_F \times |i\omega + \nu_p|^{-1}$, where (k, ω) is the wave number and perturbation frequency functions as the small parameter of the expansion of the solution of the kinetic equation in terms of the spatial gradients. The locally-equilibrium solution $\hat{\rho}_l$ is not normalized to the particle number, spin, or energy fluxes; hence we cannot conserve the convective terms with the correction $\delta\hat{\rho}$ in equation (3.120).

The operator $\mathcal{L}^{(2)} = \hat{\mathcal{L}}^{(2)}_{e-ph} + \mathcal{L}^{2}_{imp}$ is not Hermitian; its left and right eigenvectors corresponding to identical eigenvalues generally do not coincide. Identities (1.39), (1.51) and (1.53) corresponding to conservation of particle number, spin and energy in collisions infer that the left eigenvalues of the operator $\hat{\mathcal{L}}^{(2)}$ corresponding to a zero eigenvalue are the matrices $\hat{\sigma}_z$, 1, and $\hat{\varepsilon}_p$. We may show that the corresponding right eigenvectors take the form

$$\delta\tilde{\rho}_\mu = C_\mu \frac{\partial}{\partial\mu}(\hat{\rho}_l)_{\Delta, s_z}, \qquad \delta\tilde{\rho}_\Delta = C_\Delta \frac{\partial}{\partial\Delta}(\hat{\rho}_l)_{\mu, s_z}, \qquad \delta\tilde{\rho}_s = C_s \frac{\partial}{\partial s_z}(\hat{\rho}_l)_{\mu\Delta}.$$

In order to determine the five independent quantities $\delta\tilde{\mu}$, $\delta\Delta$, C_μ, C_Δ and C_S we use the matching equation for the phase and modulus of the order parameter and the three normalization conditions.

We may easily determine that the quantities $\delta\tilde{\mu}$, $\delta\Delta$, C_μ and C_Δ are determined independent of s_z, while the parasitic solutions make no contribution to the spin density flux. This, specifically, means that in a linear approximation the existence of s_z does not cause a branch imbalance in the excitation spectrum nor corrections to the energy gap in the superconductor.

Here we will provide only the components of the left half of the first approximation equation (3.120) which determine the evolution of the spin degree of freedom:

$$\left[-n_0'\left(\frac{\varepsilon_p}{T}\right)\right]\frac{1}{T}\left\{\frac{\partial}{\partial t_l}(\mathbf{v}_s)_l+\left[\frac{\xi_p}{\varepsilon_p}\tau_3-\frac{\Delta}{\varepsilon_p}\hat{\tau}_1\right]\frac{\mathbf{p}}{m}\nabla\mathbf{v}_s\right\}=$$
$$=-i\,[\hat{\varepsilon}_p,\delta\tilde{\rho}]_- -\frac{i\partial}{\partial t}\delta\tilde{\rho}+\hat{\mathscr{L}}^{(2)}\{\delta\tilde{\rho}-\delta\tilde{\rho}_s\}. \tag{3.123}$$

We have ignored the contribution of exchange interaction that is small in the parameter $mgp_F \ll 1$ in the excitation spectrum. Again we will use an approximation of the relaxation time for the collision integral. We may easily see that the primary contribution to the diffusion spin flux results from components of the density matrix that are diagonal in the ν-representation; the correction related to the nondiagonal components $\hat{\rho}$ in this representation has relative smallness $\sim \nu_\alpha\nu_\gamma/\Delta^2$. Solving equation (3.123) and comparing the result from the calculation of the diffusion spin flux to the definition of the spin diffusion coefficient D_s:

$$\delta\mathbf{j}_s{}^z = {}^z - D_s\nabla s_z, \tag{3.124}$$

we obtain the following expression for the frequency-dependent coefficient D_s:

$$D_s(\omega)=\frac{1}{3\varkappa_s}\left\langle\frac{1}{T}\frac{p^2}{m^2}\frac{\xi_p^2}{\varepsilon_p^2}\left[-n_0'\left(\frac{\varepsilon_p}{T}\right)\right]\frac{1}{i\omega+\nu_\alpha}\right\rangle, \tag{3.125}$$

where

$$\nu_\alpha=\frac{|\xi_p|}{\varepsilon_p}\nu_i+\frac{1}{2}\nu_m+\nu_\alpha^{ph}$$

is the energy-dependent momentum relaxation frequency which, as a rule, is determined by electron scattering by impurities;

$$\varkappa_s=\left\langle\frac{1}{T}\left[-n_0'\left(\frac{\varepsilon_p}{T}\right)\right]\right\rangle.$$

When $T \to T_c - 0$

$$D_s=\frac{v_F^2}{3(i\omega+\nu_p)}\left(1-\frac{\pi}{4}\frac{\Delta}{T_c}\right), \tag{3.126}$$

where ν_p^{-1} is the transport momentum relaxation time in a normal metal when $T = T_c$;

when $T \to 0$ when the concentration of the paramagnetic impurities is sufficiently small and elastic scattering is determined by electron-nonmagnetic impurity collisions, we obtain for D_s:

$$D_s=\frac{v_F^2}{3(i\omega+\nu_p)}\frac{T}{\Delta_0}. \tag{3.127}$$

It is demonstrated in study [29] that in the pure isotropic B-phase of He^{-3} in this limiting case when $\omega \ll \nu_{e-e}D_s \sim T^{-1}$ where ν_{e-e} is the fermion-fermion collision frequency.

Estimate (3.127) is valid in a temperature range that is not too close to absolute zero, so that energy relaxation is a faster process compared to spin relaxation: $\nu_E(T) \ll \tau_s^{-1}$. Otherwise it is necessary to account for the linearized collision integral with spin rotation in the right half of equation (3.121).

Results (3.126)-(3.127) reveal that the Cooper correlations retard the diffusion of spin density in the superconductors.

Thus, the evolution of the nonequilibrium distribution of the spin density in superconductors with a low concentration of paramagnetic impurities is described by the equation

$$\frac{\partial}{\partial t} s_z + \operatorname{div}\left\{\frac{1}{3} s_z \mathbf{v}_s - \hat{D}_s \nabla s_z\right\} = \frac{s_z}{\tau_s} + I^s; \tag{3.128}$$

the diffusion spin flux in the key representation takes the form of a time convolution (see, for example, formula (3.46)) of the memory function $D_s(t)$ accounting for the relatively rapid momentum relaxation process and the spin density gradient. We may obtain from the matching equation near T_c for a nonequilibrium state characterized by a shift in the chemical potential $\delta\tilde{\mu}$ with nonequilibrium spin density s_z in the presence of superfluid motion at velocity $\mathbf{v}_s$ in the conditions $\delta\tilde{\mu} \ll \Delta$, $s_z \ll \varkappa_s \Delta$, $p_F v_s \ll \Delta$ the following relation for the order parameter

$$\tilde{\Delta}^2(T) = \Delta^2(T) - 2(\delta\tilde{\mu})^2 - 2\left(\frac{s_z}{\varkappa_s}\right)^2 - \frac{2}{3} p_F v_s^2, \tag{3.129}$$

where $\tilde{\Delta}(T)$ is the order parameter in the nonequilibrium state, and Δ is its value in the absence of nonequilibrium and superfluid motion. Specifically, the applicability condition of equation (3.128) follows from (3.129):

$$s_z \ll \frac{\varkappa_s}{\sqrt{2}} \Delta(T). \tag{3.130}$$

This condition, which is very rigid in the immediate proximity of zero temperature is easily satisfied when $0 \ll T \leqslant T_c$ when $s_c = \frac{3}{\sqrt{2}} \frac{\Delta}{2\mu_F} n$. For example, when $\Delta/T_c \sim 10^{-1}$, $n \sim 10^{23}$ cm^{-3}

We are grateful to V.L. Ginzburg and the staff of the I.E. Tamm Division of Theoretical Physics of the Physics Institute of the Academy of Sciences for their interest in this study and their stimulating conversation and to I.O. Kulik and A.N. Omel'yanchuk for their useful discussion of the results of this study.

BIBLIOGRAPHY

1. Geylikman, B.T. "The thermal conductivity of superconductors " ZhETF,. 1958, V. 34, No. 4, p. 1042-1044.

2. Bardeen, J., Rickayzen, G., Tewordt, L. "Theory of thermal conductivity of superconductors" PHYS. REV., 1959, V. 113, No. 4, p. 982-994.

3. Geylikman, B.T., Kresin, V.Z. "Kineticheskie i nestatsionarnye yavleniya v sverkhprovodnikakh" [Kinetic and nonstationary phenomena in superconductors] Moscow: Nauka, 1972, 176 p.

4. Aronov, A.G., Galperin, Yu.M., Gurevich, V.L., Kozub, V.I. "The Boltzmann-equation description of transport in superconductors" ADV. PHYS., 1981, V. 30, No. 4, p. 539-592.

5. Betbeder-Matibet, O., Nozieres, P. "Transport equations in clean superconductors" ANN. PHYS. (US), 1969, V. 51, No. 3, p. 392-417.

6. Galiyko, V.P. "Kinetic equations for relaxation processes in superconductors" ZhETF, 1971, V. 60, No. 1(7), p. 382-397.

7. Eliashberg, G.M. "Inelastic electron collisions and nonequilibrium steady states in superconductors" ZhETF, 1971, V. 61, No. 3(9), p. 1254-1271.

8. Aronov, A.G., Gurevich, V.L. "The theory of the response of pure superconductors to slowly-varying perturbations" FTT, 1974, V. 16, No. 9, p. 2656-2665.

9. Keldysh, L.V. "The diagrammatic technique for nonequilibrium processes" ZhETF, 1964, V. 47, No. 4(10), p. 1515-1527.

10. Bar'yakhtar, V.G., Buchkova, N.N., Seminozhenko, V.P. "Kinetic equations for electron excitations and phonons in a superconductor" TMF, 1979, V. 38, No. 2, p. 251-262.

11. Bogolyubov, N.N. "A new method in superconductivity theory" In: Izbrannye trudy po statisticheskoy fizike [Selected works on statistical physics] Moscow: Izd-vo MGU, 1979, p. 132-142.

12. Kulik, I.O. "Nonlinear dynamical properties of superconductors", FNT, 1976, V. 2, No. 8, p. 962-978.

13. Tilenberger, G. "Transformation of Gor'kov's equations for type II superconductors into transport-like equations" ZTSCHR. PHYS., 1968, V. 214, p. 199-213.

14. Lapkin, A.I., Ovchinnikov, Yu.N. "The nonlinear conductivity of superconductors in a mixed state" ZhETF, 1975, V. 68, No. 5, p. 1915-1923.

15. Eliashberg, G.M. "Teoriya neravnovesnykh sostoyaniy i nelineynaya elektrodinamika sverkhprovodnikov: [The theory of nonequilibrium states and the nonlinear electrodynamics of superconductors] Dissertation for the doctorate of physics and mathematics, Chernogolovka, 1971, 180 p.

16. Gor'kov, L.P., Kopnin, N.B. "The problem of nonlinear ther-

modynamics of thin superconducting films" ZhETF, 1970, V. 59, No. 1, p. 234-245.

17. Gor'kov, L.P., Kopnin, N.B. "The features of viscous vortex flow in superconducting alloys near T_C" ZhETF, 1973, V. 64, No. 1, p. 356-370.

17a. Gor'kov, L.P., Kopnin, N.B. "Viscous vortex flow in superconducting alloys" ZhETF, 1973, V. 65, No. 1(7), p. 396-410.

18. Larkin, A.I., Ovchinnikov, Yu.N. "Fluctuation conductivity in the vicinity of the superconducting transition" J. LOW TEMP. PHYS., 1973, V. 10, p. 407-421.

19. Schmid, A., Schön, G. "Linearized kinetic equations and relaxation processes of superconductors near T_C" J. LOW TEMP. PHYS., 1975, V. 20, No. 1/2, p. 207-227.

20. Lapkin, A.I., Ovchinnikov, Yu.N. "Nonlinear effects with vortex motion in superconductors" ZhETF, 1977, V. 73, No. 1(7), p. 299-312.

21. Ovchinnikov, Yu.N. "Penetration of an electric field into a superconductor" II. J. LOW TEMP. PHYS., 1978, V. 31, p. 785-802.

22. Schmid, A. "Superconductors out of thermal equilibrium" J. PHYS., 1978, colloq. C-6, No. 3, p. 1360-1367.

23. Schmid, A. "Kinetic equation for dirty superconductors" NATO ADV. STUDY INST. SER., 1981, V. B65, p. 423-480.

24. Shelankov, A.L. "Increasing the normal component by a condensate in nonequilibrium superconductors" ZhETF, 1980, V. 78, V. 6, p. 2359-2373.

25. Entin-Wohlman, O., Orbach, R. "On the microscopic and Boltzmann equation approaches to non-equilibrium superconductors" ANN. PHYS. (US), 1979, V. 122, p. 64-73.

26. Pals, J.A., Weiss, K., Attekum, P.T.M.T. van, et al. "Nonequilibrium superconductivity in homogeneous thin films" PHYS. REP., 1982, V. 80, No. 4, p. 323-390.

27. Eliashberg, G.M. "Electron/lattice vibration interaction in a superconductor" ZhETF, 1960, V. 38, No. 3, p. 966-972.

28. Shumeyko, V.S. "The kinetic coefficients of superfluid" ZhETF, 1972, V. 63, No. 2(8), p. 621-633.

29. Soda, T., Fudjiki, K. "Transport properties of superfluid ^{3}He" PROGR. THEOR. PHYS., 1974, V. 52, No. 5, p. 1405-1430.

30. Chepmen, S., Kauling, T. "Matematicheskaya teoriya neodnorodnykh gazov" [Mathematical theory of inhomogeneous gases] Moscow: Izd-vo inostr. lit., 1960, 512 p.

31. Ivlev, B.I., Lisitsin, S.G., Eliashberg, G.M. "Nonequilibrium excitations in superconductors in high frequency field" J. LOW TEMP. PHYS., 1973, V. 10, No. 3/4, p. 449-468.

32. Ginzburg, V.L. "The thermoelectric effects in superconductors" ZhETF, 1944, V. 14, No. 6, p. 177-183.

33. Balescu, R. "Irreversible processes in ionized gases" PHYS. FLUIDS, 1960, V. 3, No. 1, p. 52-63.

34. Lenard, A. "On Bogolubov's kinetic equation for a spatially homogeneous plasma" ANN. PHYS. (US), 1960, V. 3, p. 390-400.

35. Kadanov, L., Beym, G. "Kvantovaya statisticheskaya mekhanika" [Quantum statistical mechanics] Moscow: Mir, 1964, 256 p.

36. Migdal, A.B. "Electron/lattice vibration interaction in a normal metal" ZhETF, 1958, V. 34, No. 6, p. 1439-1446.

37. Eliashberg, G.M. "Temperature Green's functions of electrons in a superconductor" ZhETF, 1960, V. 39, No. 5(11), p. 1437-1441.

38. Abrikosov, A.A., Gor'kov, L.P. "A theory of superconducting alloys with paramagnetic impurities" ZhETF, 1960, V. 39, No. 6(12), p. 1781-1796.

39. Abrikosov, A.A., Gor'kov, L.P., Dzyaloshinskiy, I.Ye. "Metody kvantovoy teorii polya v statisticheskoy fizike" [Quantum field theory techniques in statistical physics] Moscow: Fizmatgiz, 1962, 443 p.

40. Fröhlich, M. "Superconductivity in lattice vibrations" PHYSICS, 1953, V. 19, No. 9, p. 755-764.

41. Lifshits, Ye.M., Pitaevskiy, L.P. "Fizicheskaya kinetika" [Physical kinetics] Moscow: Nauka, 1969, 527 p.

42. Kaplan, S.B., Chi, C.C., Langenberg, D.N., et al. "Quasi particle and phonon lifetimes in superconductors" PHYS. REV. B, 1976, V. 14, No. 11, p. 4854-4873.

43. Gradshtein, I.S., Ryzhik, I.M. "Tablitsy integralov, Summ, Ryadov i priozvedeniy" [Tables of integrals, sums, series, and products] Moscow: Fizmatgiz, 1963, 1100 p.

44. Bar'yakhtar, V.G., Klepikov, V.D., Setinozhenko, V.P. "A theory of relaxational processes in superconductors" FTT, 1973, V. 15, No. 4, p. 1213-1222.

45. Uayt, R.M. "Kvantovaya teoriya magnetizma" [Quantum theory of magnetism] Moscow: Mir, 1972, 306 p.

46. Volkov, A.F., Kogan, Sh.M. "Collisionless relaxation of the energy gap in superconductors" ZhETF, 1973, V. 65, No. 5(11), p. 2038-2045.

47. Landau, L.D. "Superfluid theory of helium-II" ZhETF, 1941, V. 11, No. 6, p. 592-614.

48. Bogolyubov, N.N. "The issue of the hydrodynamics of superfluid" Preprint of the Joint Institute of Nuclear Research, Dubna R-1395. 1963, 32 p.

49. Khalatnikov, I.M. "Teoriya sverkhtekuchesti" [Superfluidity theory] Moscow: Nauka, 1971, 320 p.

50. Galyasevich, Z. "Hydrodynamic equations of a Fermi superfluid and two-particle functions" Preprint of the Joint Institute of Nuclear Research, R-1953. Dubna, 1965, 26 p.

51. Hohenberg, P.C., Martin, P.C. "Microscopic theory of superfluid helium" ANN. PHYS. (US), 1965, V. 34, No. 1, p. 291-359.

52. Kadanoff, L.P., Martin, P.C. "Hydrodynamic equations and correlation functions" ANN. PHYS. (US), 1963, V. 24, No. 3, p. 419-469.

53. Svidzinskiy, A.V., Slyusarev, V.A. "Hydrodynamic equations in superconductivity theory" DAN SSSR, 1967, V. 172, No. 2, p. 322-325.

54. Stephen, M.J. "Transport equations for superconductors" PHYS. REV., 1965, V. 139, No. 1A, p. 197-205.

55. Morozov, V.G. "Hydrodynamic equations and kinetic coefficients for a Bose superfluid" TMF, 1976, V. 28, No. 2, p. 267-280.

56. Zubarev, D.N. "Neravnovesnaya statisticheskaya termodinamika" [Nonequilibrium statistical thermodynamics] Moscow: Nauka, 1971, 415 p.

57. Seminozhenko, V.P., Yatsenko, A.A. "Self-consistent interaction of electrons and elastic waves in superconductors" Preprint of the Institute of Theoretical Physics of the Academy of Sciences of the Ukranian SSR, No. 155, Kiev, 1979, 21 p.

58. Gal'perin, Yu.M., Gurevich, V.L., Kozub, V.I. "Thermoelectric effects in superconductors" ZhETF, 1976, V. 66, No. 4, p. 1387-1397.

59. Galiyko, V.P., Glushchuk, N.I. "Kinetic equations for superconductors with paramagnetic impurities" FNT, 1976, V. 2, No. 10, p. 1269-1276.

60. Libov, R. "Vvedenie v teoriyu kineticheskikh uravneniy" [An introduction to the theory of kinetic equations] Moscow: Mir, 1974, 371 p.

61. Prudnikov, A.P., Brychkov, Yu.A., Marichev, O.I. "Integraly i ryudy" [Integrals and series] Moscow: Nauka, 1981, 800 p.

62. Khalatnikov, I.M. "The hydrodynamics of extraneous particle solutions in helium-II" ZhETF, 1952, V. 23, No. 2(8), p. 169-181.

63. Bogolyubov, N.N. "The phonon-electron of compensation and the self-consistent field technique" In: Izbrannye trudy po statisticheskoy fizike" [Selected works on statistical physics] Moscow: Izd-vo MGU, 1979, p. 143-189.

64. Anderson, P.W. "Random phase approximation in the theory of superconductivity" PHYS. REV., 1958, V. 112, No. 6, p. 1900-1916.

65. Bardeen, J. "Two-fluid model of superconductivity" PHYS. REV. LETT., 1958, V. 1, No. 11, p. 399-400.

66. Ginzburg, V.L. "The second sound and the convective mechanism of thermal conductivity and exciton excitations in superconductors" ZhETF, 1961, V. 41, No. 3(9) p. 828-833.

67. Khalatnikov, I.M. "Thermal conductivity and sound absorption in helium-II" ZhETF, 1952, V. 26, No. 1(7) p. 21-34.

68. Alvesalo, T.A., Collan, H.K., Loponin, M.T., et al. "The viscosity and some properties of liquid ^{3}He at the melting curve between 1 and 100 mK" J. LOW TEMP. PHYS., 1975, V. 19, No. 1/2, p. 1-37.

69. Corruccini, L.R. "Second sound in superfluid ^{3}He" PHYSICA B, 1982, V. 109/110, p. 1590-1599.

70. Khalatnikov, I.M., Chernikova, D. "Relaxational phenomena in helium superfluid" ZhETF, 1965, V. 49, No. 6(12), p. 1957-1972.

71. Pethick, C.J., Smith, H. "Relaxation and collective motion in superconductors: A two-fluid description" ANN. PHYS. (US), 1979, V. 119, No. 1, p. 133-169.

72. Pethick, C.J., Smith, H. "Charge imbalance: its relaxation, diffusion and oscillation" NATO ADV. STUDY INST. SET., 1981, V. B65, p. 481-520.

73. Mattoo, B.A., Singh, Y. "Effect of electromagnetic perturbation on charge imbalance in a superconductor: A two-fluid description" PRAMANA, 1982, V. 19, No. 5, p. 483-496.

74. Mattoo, B.A., Singh, Y. "Charge imbalance due to thermoelectric effects in superconductors: a two-fluid description" PRAMANA, 1983, V. 20, No. 5, p. 393-403.

75. Volkov, A.F. "Nonequilibrium phenomena in superconducting tunnel structures" ZhETF, 1975, V. 68, No. 2, p. 756-765.

76. Patterman, S. "Gidrodinamika sverkhtekuchey zhidkosti" [The hydrodynamics of superfluid] Moscow: Mir, 1978, 520 p.

77. Schmid, A. "Phenomenological theory of dissipation in superconductor" J. LOW TEMP. PHYS., 1980, V. 41, No. 1/2, p. 37-44.

78. Putterman, S. "Phenomenological theory of collective modes and relaxation effects in type II superconductors" J. LOW TEMP. PHYS., 1977, V. 28, No. 3/4, p. 339-347.

79. Forster, D. "Gidrodinamicheskie fluktuatsii, narushennaya simmetriya i korrelyatsionnye funktsii" [Hydrodynamic fluctuations, broken symmetry, and correlation functions] Moscow: Atomizdat, 1980, 288 p.

80. Eckern, U., Schön, G. "Relaxation processes in superconductors" J. LOW TEMP. PHYS., 1978, V. 32, No. 5/6, p. 821-838.

81. Clarke, J. "Experimental observation of pair-quasi particle potential difference in nonequilibrium superconductors" PHYS. REV. LETT., 1972, V. 28, p. 1363-1366.

82. Clarke, J., Paterson, J.L. "Measurements of the relaxation of quasi particle branch imbalance in superconductors" J. LOW TEMP. PHYS., 1974, V. 15, No. 5/6, p. 491-522.

83. Tinkham, M. "Tunneling generation and tunneling detection of hole electron imbalance in superconductors" PHYS. REV. B, 1972, V. 6, No. 5, p. 1747-1756.

84. Bulyzhenkov, I.E., Ielev, B.I. "nonequilibrium phenomenon in superconductor junctions" ZhETF, 1978, V. 74, No. 1, p. 224-235.

85. Clarke, J., Eckern, U., Schmid, A., et al. "Branch-imbalance relaxation times in superconductors" PHYS. REV. B, 1979, V. 20, No. 9, p. 3933-3937.

86. Artemenko, S.N., Bolkov, A.F. "A thermoelectric field in superconductors" ZhETF, 1976, V. 70, No. 3, p. 1051-1060.

87. Pethick, G.J., Smith, H. "Charge imbalance in nonequilibrium superconductors" J. PHYS. C: SOLID STATE PHYS., 1980, V. 13, p. 6313-6347.

88. Nielsen, J.B., Pethick, C.J., Rammer, J., Smith, M. "Pair breaking and charge relaxation in superconductors" J. LOW TEMP. PHYS., 1982, V. 46, No. 516, p. 565-597.

89. Dmitriev, V.M., Khristenko, YeV. "The temperature dependence of the depth of penetration of a nonequilibrium longitudinal electrical field into a superconductor" VNT, 1977, V. 3, No. 9, p. 1210-1213.

90. Volkov, A.F., Zaytsev, A.V. "The asymmetry in quasi particle distribution in superconductors and normal metals" ZhETF, 1975, V. 69, No. 6(12), p. 2222-2230.

91. Artemenko, S.N., Volkov, A.F. "Collective modes with an acoustic spectrum in superconductors" ZhETF, 1975, V. 69, No. 5(11), p. 1764-1767.

92. Ovchinnikov, Yu.N. "Longitudinal modes in superconducting alloys" ZhETF, 1977, V. 72, No. 2, p. 773-782.

93. Carlson, R.V., Goldman, A.M. "Propagating order-parameter collective modes in superconducting films" PHYS. REV. LETT., 1975, V. 34, No. 1, p. 11-15.

94. Goldman, A.M. "Collective modes of the superconducting order parameter" NATO ADV. STUDY INST. SER., 1981, V. B65, p. 541-558.

95. Vinen, W.F. "A comparison of the properties of superconductors and superfluid helium" In: Superconductivity. NY, 1969, V. 2, Chapter 20.

96. V.L., Zharkov, G.F. "Thermoelectric effects in superconductors" UFN, 1978, V. 125, No. 1, p. 19-56.

97. Harlingen, D.J., van. "Thermoelectric effect in the superconducting state" PHYSICA B, 1982, V. 109/100, p. 1710-1721.

98. Falco, C.M., Garland, J.C. "Thermoelectric effects in superconductors" NATO ADV. STUDY INST. SER., 1981, V. B65, p. 521-540.

99. Garland, J.C., Harlingen, D.J. van. "Thermoelectric generation of flux in a bimetallic superconductor ring" PHYS. LETT. A, 1974, V. 47, No. 5, p. 423-424.

100. Gor'kov, L.N., Eliashberg, G.M. "A generalization of equations from Ginzburg-Landau theory for nonstationary problems in the case of alloys containing paramagnetic impurities" ZhETF, 1968, V. 54, No. 2, p. 612-626.

101. Zavaritskiy, N.V. "Observation of superconducting current excited in a superconductor by thermal flux" PIS'MA V ZhETF, 1974, V. 18, No. 4, p. 205-208.

102. Falco, C.M. "Thermally induced magnetic flux in a superconducting ring" SOLID STATE COMMUN., 1976, V. 19, p. 623-625.

103. Harlingen, D.J. van, Heidel, D.F., Garland, J.C. "Experimental study of thermoelectricity in superconducting indium" PHYS. REV. B - SOLID STATE, 1980, V. 21, No. 5, p. 1842-1857.

104. Pegrum, C.M., Guenault, A.M. "Thermoelectric flux effects in superconducting bimetalic loops" PHYS. LETT. A, 1976, V. 59, p. 393-395.

105. Kozub, V.I. "Surface thermoelectric effects in superconductors" ZhETF, 1978, V. 74, No. 1, p. 344-363.

106. Arutyunyan, R.M., Zharkov, G.F. "The thermoelectric effect in superconductors" ZhETF, 1982, V. 83, No. 3(9), p. 1115-1133.

107. Pethick, C.J., Smith, H. "Generation of charge imbalance in a superconductor by a temperature gradient" PHYS. REV. LETT., 1979, V. 43, No. 9, p. 640-642.

108. Clarke, J., Fjordbo ge B.R., Lindelof, P.E. "Supercurrent-induced charge imbalance measured in a superconductor in the presence of a thermal gradient" PHYS. REV. LETT., 1979, V. 43, No. 9, p. 642-645.

109. Anderson, P.W. "Theory of dirty superconductors" J. PHYS. AND CHEM. SOLIDS, 1959, V. 11, No. 1/2, p. 26-30.

110. Galayko, V.P., Shumeyko, V.S. "Supercritical current states in pure superconductors" ZhETF, 1976, V. 71, No. 2(8), p. 671-678.

111. Schmid, A., Schön, G. "Generation of branch imbalance by the interaction between supercurrent and thermal gradient" PHYS. REV. LETT., 1980, V. 44, No. 2, p. 106-109.

112. Clarke, J., Tinkham, M. "Theory of quasi particle charge imbalanced in a superconductor by a supercurrent in the presence of a thermal gradient" PHYS. REV. LeTT., 1980, V. 44, No. 2, p. 106-109.

113. Clarke, J. "Charge imbalance" NATO ADV. STUDY INST. SER., 1981, V. B65, p. 353-442.

114. Aronov, A.G., Katilyus, R. "The kinetics of fluctuations in pure superconductors" ZhETF, 1975, V. 68, No. 6, p. 2208-2223.

115. Landau, L.D., Lifshits, Ye.M. "Statisticheskaya fizika" [Statistical physics] Moscow: Nauka, 1976, Ch. 1, 583 p.

116. Lemberger, T.R. "Charge imbalance fluctuations in superconductors" PHYS. REV. B, 1981, V. 24, No. 7, p. 4105-4108.

117. Kulik, I.O. "The broadening mechanism of Josephson radiation related to pair relaxation" FIZIKA KONDENSIROVANNOGO SOSTOYANIYA, 1970, V. 8, p. 183-190.

118. Kulik, I.O., Yanson, I.K. "Effekt Dzhozefsona v sverkhprovodyashchikh tunnel'nykh strukturakh" [The Josephson effect in superconducting tunnel structures] Moscow: Nauka, 1970, 272 p.

119. Aronov, A.G. "Spin injection and excitation and nucleus polarization in superconductors" ZhETF, 1976, V. 71, No. 1(7), p. 370-376.

SUBJECT INDEX